Image
Super-Resolution
and Applications

Image
Super-Resolution
and Applications

Fathi E. Abd El-Samie
Mohiy M. Hadhoud
Said E. El-Khamy

CRC Press
Taylor & Francis Group
Boca Raton London New York

CRC Press is an imprint of the
Taylor & Francis Group, an **informa** business

CRC Press
Taylor & Francis Group
6000 Broken Sound Parkway NW, Suite 300
Boca Raton, FL 33487-2742

First issued in paperback 2019

© 2013 by Taylor & Francis Group, LLC
CRC Press is an imprint of Taylor & Francis Group, an Informa business

No claim to original U.S. Government works

ISBN-13: 978-1-4665-5796-3 (hbk)
ISBN-13: 978-0-367-38049-6 (pbk)

Visit the Taylor & Francis Web site at
http://www.taylorandfrancis.com

and the CRC Press Web site at
http://www.crcpress.com

Contents

Preface

We know what you are asking yourself. There are a lot of books available on image processing so what is novel in this book? We can summarize the answer with the following points:

1. This book is devoted to the issue of obtaining high-resolution images from single or multiple low-resolution images.
2. You always see different algorithms for image interpolation and super-resolution without a common thread between the two processes. This book presents interpolation as a building block in the super-resolution reconstruction process.
3. You see research papers on image interpolation either as a polynomial-based problem or an inverse problem without a comparison of the two trends. This book presents this comparison.
4. Two chapters are devoted to two complementary steps that are used to obtain high-resolution images. These steps are image registration and image fusion.
5. This book presents two directions for image super-resolution; super-resolution with *a priori* information and blind super-resolution.
6. This book presents applications for image interpolation and super-resolution in medical and satellite image processing.
7. MATLAB® codes for most of the simulation experiments discussed in this book are included in Appendix D at the end of the book.

Finally, we hope that this book will be helpful for the image processing community.

MATLAB® is a registered trademark of The MathWorks, Inc. For product information, please contact:

The MathWorks, Inc.
3 Apple Hill Drive
Natick, MA 01760-2098 USA
Tel: 508-647-7000
Fax: 508-647-7001
E-mail: info@mathworks.com
Web: www.mathworks.com

Acknowledgments

We would like to thank Amr Ragheb, Mohamad Metwalli, Maha Awad, Fatma Ali, Amira Shafik, Fatma Hashad, Heba Osman, and Huda Ashiba for their contributions of programs and simulation experiments during the preparation of this book.

Acknowledgments

Authors

Fathi E. Abd El-Samie, earned his BSc (Hons) in 1998, MSc in 2001, and PhD in 2005 all from Menoufia University, Menouf, Egypt.

Since 2005, he has been a teaching staff member with the Department of Electronics and Electrical Communications, Faculty of Electronic Engineering, Menoufia University. He is a coauthor of 160 papers published in international conference proceedings and journals.

His current research areas of interest include image enhancement, image restoration, image interpolation, super-resolution reconstruction of images, data hiding, multimedia communications, medical image processing, optical signal processing, and digital communications.

Dr. Abd El-Samie was a recipient of the Most Cited Paper Award from the *Digital Signal Processing* journal in 2008.

Mohiy M. Hadhoud PhD, received his BSc (Hons) in 1976 and MSc in 1981 from the Faculty of Electronic Engineering, Menoufia University, Menouf, Egypt, and his PhD from Southampton University in 1987. He joined the teaching staff of the Department of Electronics and Electrical Communications, Faculty of Electronic Engineering, Menoufia University, Menouf, Egypt from 1981 to 2001. He is currently a professor in the Department of Information Technology, Faculty of Computers and Information, Menoufia University, Shiben El-Kom.

Dr. Hadhoud has published more than 100 scientific papers in national and international conferences and journals. His current research areas of interest include adaptive signal and image processing techniques, image enhancement, image restoration, super-resolution reconstruction of images, data hiding and image coloring.

Said El-Khamy PhD, received his PhD from the University of Massachusetts, Amherst, in 1971. He is currently a professor emeritus, Department of Electrical Engineering, Faculty of Engineering, Alexandria University, Alexandria, Egypt. He served as the chairman of the Electrical Engineering Department from September 2000 to September 2003. While on academic leaves, he taught in Saudi Arabia, Iraq, Lebanon, and the Arab Academy for Science and Technology (AAST).

His current research areas of interest include mobile and personal communications, wave propagation in different media, smart antenna arrays, image processing and watermarking, modern signal processing techniques including neural networks, wavelets, genetic algorithms, fractals, HOS and fuzzy algorithms, and their applications in image processing, communication systems, antenna design and wave propagation problems. He has published more than 300 scientific papers in national and international conferences and journals.

Dr. El-Khamy participated in the organization of many local and international conferences including the yearly series of NRSC (URSI), ISCC 1995, ISCC 1997, ISSPIT 2000, and MELECON 2002. He also chaired technical sessions in many local and international conferences including ISSSTA 1996, Mainz, Germany, Sept. 1966; IGARSS'98, Seattle, Washington, USA, July 1998; and AP-S'99, Orlando, Florida, USA, July 1999. He is the chairman of National Radio Science conferences (NRSC2011 and NRSC2012).

He has earned many national and international research awards among which are the Alexandria University Research Award, 1979; the IEEE, R.W.P. King best paper award of the Antennas and Propagation Society of IEEE, in 1980; the Egypt's National Engineering Research award (twice) in 1980 and 1989; Egypt's State Science & Art Decoration of the first class, 1981; the A. Schoman's—Jordan's award for Engineering Research in 1982; and Egypt's State Excellence Decoration of the first class in 1995. He was selected as the National Communication Personality for 2002. Recently, he received three major prestigious national prizes, namely, the State Scientific Excellence award in Engineering Sciences for 2002, the Alexandria University Appreciation of Engineering Sciences for 2004 and finally, the State Appreciation Award of Engineering Sciences for 2004 and the IEEE Region 8 Volunteer Award for 2011.

Prof. El-Khamy is a Fellow Member of the IEEE since 1999 and he obtained the IEEE Life Fellow certificate in 2010. He is a Fellow of the Electromagnetic Academy and a member of Tau Beta Pi, Eta Kappa Nu and Sigma Xi. He is the founder and former chairman of the Alexandria/Egypt IEEE Subsection. Currently, he is the president of Egypt's National URSI Committee (NRSC) and Egypt's National URSI Correspondent for Commission C.

Chapter 1

Introduction

High-resolution (HR) images are required in most electronic imaging applications. The high resolution means that the pixel density within the image is high, and therefore an HR image can offer more details than those obtained from a low-resolution (LR) image. HR images are of great importance in applications such as medical imaging, satellite imaging, military imaging, underwater imaging, remote sensing, and high-definition television (HDTV).

In the past, traditional image vidicon and orthicon cameras have been the only available image acquisition devices. These cameras are analog cameras. Since the 1970s, charge-coupled devices (CCDs) and complementary metal oxide semiconductor (CMOS) image sensors have been widely used to capture digital images. Although these sensors are suitable for most imaging applications, the current resolution levels and their associated prices are not suitable for future demands. It is desirable to have very HR levels with prices as low as possible. The demands for HR levels have been the motivations to find methodologies for increasing the LR levels obtained using the current image acquisition devices.

The direct solution to increase the resolution level is to reduce the pixel size in sensor manufacturing. Therefore, the number of pixels per unit area is increased. The drawback of this solution is that the amount of light available from each pixel is decreased. The decrement of light amount leads to the generation of shot noise that seriously degrades the image quality. Unfortunately, the pixel size cannot be reduced beyond a certain level (40 μm^2 for 0.35 μm CMOS processes) to avoid shot noise. This level has already been reached in the manufacturing process.

Another solution to the problem of resolution level increment is to increase the chip size with the pixel size fixed. This solution leads to an increase in chip capacitance. It is well known that the large capacitance limits the speeding up of

1

the charge transfer rate. The slow rate of charge transfer leads to a great problem in the image formation process. Generally, all hardware solutions to this problem are limited by the costs of high-precision optics and required image sensors.

The most feasible solution to this problem is to integrate both the hardware and software capabilities to obtain the required HR level. Making use of as high an HR level as possible from the hardware can carry part of this task. The rest of the task is performed using software. This is the new trend in most up-to-date image capturing devices. Image processing algorithms can be used effectively to obtain HR images. Using a single LR image to obtain an HR image is known as image interpolation. On the other hand, when multiple degraded observations of the same scene are used to generate a single HR image, the process is known as image super-resolution.

1.1 Image Interpolation

Image interpolation is the process by which a single HR image is obtained from a single LR image. Interpolation can be classified as polynomial interpolation and interpolation as an inverse problem. Polynomial interpolation depends on the concepts of the sampling theory. In polynomial interpolation, estimated pixels are inserted between existing pixels using polynomial expansions. Different algorithms have been presented for polynomial image interpolation. The most famous of these algorithms is spline interpolation. Polynomial image interpolation depends on a finite neighborhood around the pixel to be estimated. Traditional polynomial image interpolation algorithms do not consider the LR image degradation model, and hence their performance is limited. Chapter 2 gives a discussion of polynomial image interpolation.

Adaptive variants of polynomial image interpolation have been presented in the literature. Some of these variants depend on distance adaptation without consideration of the LR image degradation model, while the others consider that this model yields better interpolation results. Chapter 3 is devoted to adaptive polynomial image interpolation, and Chapter 4 gives a neural modeling method for polynomial image interpolation.

Color image interpolation is a newly considered issue that makes use of simple polynomial image interpolation, but with color images. In a digital imaging process, not all color components are available in the acquired image, and hence there is a need for the interpolation of missing color components from the existing components of neighboring pixels. Chapter 5 is devoted to color image interpolation.

Polynomial image interpolation has found an application in pattern recognition. Instead of saving database images with their original sizes, decimation can be used to reduce their sizes. At the recognition step, image interpolation can be carried out to return images to their original sizes. The features extracted from the interpolated images must be robust to interpolation estimation errors. Chapter 6 is devoted to image interpolation for pattern recognition.

The limited performance of polynomial image interpolation led to the evolution of a new trend for image interpolation as an inverse problem. In this trend, the LR image degradation model is considered in the interpolation process. Four solutions have been developed for image interpolation as an inverse problem. The results obtained have shown a great success compared to polynomial image interpolation results. Chapter 7 gives an explanation of image interpolation as an inverse problem.

1.2 Image Super-Resolution

Image super-resolution is the process by which a single HR image is obtained from multiple degraded LR images. Image super-resolution can be carried out with or without *a priori* information. The problem of super-resolution reconstruction of images can be solved in successive steps: image registration, multi-channel image restoration, image fusion, and finally image interpolation, as shown in Figure 1.1.

Image registration aims at overlaying the LR degraded images prior to the super-resolution reconstruction process. This step is very important as it is responsible for the correct integration of the information in the multiple observations. Chapter 8 is devoted to image registration methodologies.

Image fusion is the process of integrating the information in multiple images into a single image. It can be used as a step in the super-resolution reconstruction process. Different algorithms based on transforms such as the wavelet and curvelet transforms can be used for image fusion and image fusion can be used to obtain HR images directly, which is the case in satellite image fusion. Image fusion can also be used with images of different modalities. Chapter 9 is devoted to image fusion and its applications in image super-resolution.

Image super-resolution can be carried out using some *a priori* information about the degradations in the available LR images, such as information about blurring, registration shifts, and noise. With this information available, the solution to the super-resolution reconstruction problem can be carried out easily. Chapter 10 is devoted to image super-resolution with *a priori* information. If no information is available, the problem is more difficult and it is known as blind image super-resolution. This is the topic of Chapter 11.

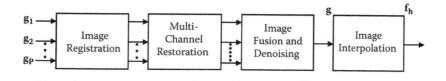

Figure 1.1   Successive steps of image super-resolution reconstruction.

Chapter 2

Polynomial Image Interpolation

2.1 Introduction

Image interpolation is a well known topic to most researchers who are interested in image processing. It pervades several applications. It is almost never a goal in itself, yet it affects both the desired results and the ways to obtain them. Interpolation may appear as a simple step in many image processing applications. Therefore, some authors give it less importance than it deserves. We try in this chapter to highlight the importance of image interpolation and cover the traditional work in this field.

Interpolation has been previously treated in the literature using different approaches [1–28]. Several authors have presented definitions for interpolation. One of the simplest definitions is that interpolation is an informed estimate of the unknown. This definition is simple and does not reveal what steps are performed in interpolation. We adopt another definition for interpolation as a model-based recovery of continuous data from discrete data within a known range of abscissa. The continuous data can then be sampled at a higher rate. This definition gives a distinction between interpolation and extrapolation. The former postulates the existence of a known data range where the model applies, and asserts that the deterministically recovered continuous data is entirely described by the discrete data. On the other hand, the latter authorizes the use of the model outside the known range, with the explicit assumption that the model is good near data samples and possibly worse elsewhere.

Image interpolation has a wide range of applications in image processing systems. It allows a user to vary the size of images, interactively to concentrate on certain details. Hence, interpolation can be used to obtain a high-resolution (HR) image from a low-resolution (LR) one. HR images are required in many fields such as medical imaging, remote sensing, satellite imaging, image compression and decompression, and high-definition television (HDTV). Image interpolation is also required in most geometric transformations such as translation, rotation, scaling, and registration.

In this chapter, we deal with interpolation in the context of exact separable interpolation of regularly sampled data. This is the traditional view of interpolation, which represents an arbitrary continuous function as a discrete sum of weighted and shifted synthesis functions. In other words, interpolation is treated as a digital convolution operation. Strictly speaking, digital image interpolation is treated as a digital convolution based on the choice of a suitable synthesis or basis function. This digital convolution operation is implemented in the world of digital signal processing using the digital filtering approach, row-by-row and then column-by-column separately. Another implementation of image interpolation can be performed using polynomials of a spacing parameter; therefore the polynomial image interpolation term is used.

2.2 Classical Image Interpolation

The process of image interpolation aims at estimating intermediate pixels between known pixels as shown in Figure 2.1. To estimate the intermediate pixel value at position x, the neighboring pixels and the distance s are incorporated into the estimation process. For an equally-spaced one-dimensional (1-D) sampled data sequence $f(x_k)$, many interpolation functions can be used. The $\hat{f}(x)$ value to be estimated can generally be written in the form [15–21]:

$$\hat{f}(x) = \sum_{k=-\infty}^{\infty} c(x_k)\beta(x - x_k) \qquad (2.1)$$

where $\beta(x)$ is the interpolation basis function and x and x_k represent the continuous and discrete spatial distances, respectively. The values of $c(x_k)$ are called the

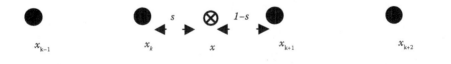

Figure 2.1 One-dimensional signal interpolation. The pixel at position x is estimated using its neighborhood pixels and the distance s.

interpolation coefficients, and they need to be estimated prior to the interpolation process.

The basis functions are classified into two main categories: interpolating and non-interpolating basis functions. Interpolating basis functions do not require the estimation of the coefficients $c(x_k)$. The sample values $f(x_k)$ are used instead of $c(x_k)$. On the other hand, non-interpolating basis functions require the estimation of the coefficients $c(x_k)$.

From the classical sampling theory, if $f(x)$ is band-limited to $[-\pi, \pi]$, then [21,29]:

$$\hat{f}(x) = \sum_k f(x_k)\operatorname{sinc}(x - x_k) \tag{2.2}$$

This is known as ideal interpolation. From a numerical computations perspective, the ideal interpolation formula is not practical because it relies on the use of ideal filters that are not commonly in use. Also, the band-limited hypothesis is in contradiction with the idea of finite duration signals. The band-limiting operation tends to generate Gibbs oscillations that are visually disturbing. A great problem is that the rate of decay of the interpolation kernel $\operatorname{sinc}(x)$ shown in Figure 2.2 is slow, which makes computations in time domains inefficient [20]. Approximations such as basic splines (B-splines), and Keys' basis functions are used as alternatives.

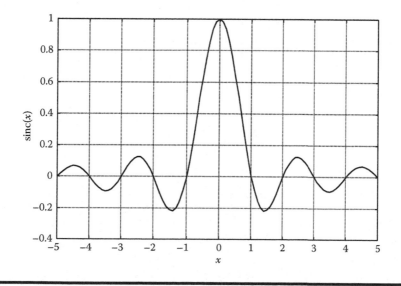

Figure 2.2   Interpolation kernel sinc(x).

2.3 B-Spline Image Interpolation

2.3.1 Polynomial Splines

Splines are piecewise polynomials with pieces that are smoothly connected. The joining points of these polynomials are called knots. For a spline of degree n, each segment is represented by a polynomial of degree n. As a result, $n + 1$ coefficients are required to represent each piece.

The continuity of the spline and its derivatives up to order $n - 1$ must be preserved at the knots [20]. Only splines with uniform knots and unit spacing will be considered. These splines are uniquely characterized in terms of the B-spline expansion given by the following equation [20]:

$$\hat{f}(x) = \sum_{k \in Z} c(x_k) \beta^n (x - x_k) \tag{2.3}$$

where Z is a finite neighborhood around x.

The B-spline basis function $\beta^n(x)$ is a symmetrical bell-shaped function obtained by $n + 1$ convolutions of a rectangle pulse β^0 given by [20]:

$$\beta^0(x) = \begin{cases} 1 & -\dfrac{1}{2} < x < \dfrac{1}{2} \\[2mm] \dfrac{1}{2} & |x| = \dfrac{1}{2} \\[2mm] 0 & \text{otherwise} \end{cases} \tag{2.4}$$

Thus, we have [20]:

$$\beta^n(x) = \underbrace{\beta^0 * \beta^0 * \ldots\ldots * \beta^0(x)}_{(n+1) \text{ times}} \tag{2.5}$$

Thanks to this convolution property used to derive a higher order spline in terms of the zero order spline, the Fourier transform of a B-spline basis function of order n is given by:

$$\hat{\beta}^n(\omega) = \left(\frac{\sin(\omega/2)}{(\omega/2)} \right)^{n+1} = \frac{\left(e^{j\omega/2} - e^{-j\omega/2} \right)^{n+1}}{(j\omega)^{n+1}} \tag{2.6}$$

To go to a closed-form expression for each B-spline basis function from its frequency domain representation, we define the one-sided power function given by:

$$(x)_+^n = \begin{cases} x^n & x \geq 0 \\ 0 & x < 0 \end{cases} \tag{2.7}$$

The Fourier transform of this function is given by [20]:

$$X_+^n(\omega) = \frac{n!}{(j\omega)^{n+1}} \tag{2.8}$$

Substituting from Equation (2.8) into Equation (2.6) gives:

$$\hat{\beta}^n(\omega) = \frac{\left(e^{j\omega/2} - e^{-j\omega/2}\right)^{n+1}}{(j\omega)^{n+1}} \frac{(j\omega)^{n+1} X_+^n(\omega)}{n!} = \frac{1}{n!}\left(e^{j\omega/2} - e^{-j\omega/2}\right)^{n+1} X_+^n(\omega) \tag{2.9}$$

Expanding the parenthesized term using the binomial theory gives:

$$\hat{\beta}^n(\omega) = \frac{1}{n!} \sum_{k=0}^{n+1} \binom{n+1}{k} (-1)^k e^{-j\omega[k-(n+1)/2]} X_+^n(\omega) \tag{2.10}$$

Applying the inverse Fourier transform to the above equation gives the closed form for the B-spline basis function of order n as follows [20]:

$$\beta^n(x) = \frac{1}{n!} \sum_{k=0}^{n+1} \binom{n+1}{k} (-1)^k \left(x - k + \frac{n+1}{2}\right)_+^n \tag{2.11}$$

This equation can be written in the following form [20]:

$$\beta^n(x) = \sum_{k=0}^{n+1} \frac{(-1)^k}{n!} \binom{n+1}{k} \left(x - k + \frac{n+1}{2}\right)^n \mu\left(x + \frac{n+1}{2} - j\right) \quad (x \in R) \tag{2.12}$$

where R is the set of real continuous indices and $\mu(x)$ is defined as follows:

$$\mu(x) = \begin{cases} 0, & x < 0 \\ 1, & x \geq 0 \end{cases} \tag{2.13}$$

The simplest interpolating B-spline basis function is the zero order function. In this case, the interpolation process is well known as nearest neighbor interpolation or just pixel repetition. Another candidate from the B-spline family is the linear basis function giving the so-called bilinear interpolation.

2.3.2 B-Spline Variants

Several B-spline basis functions can be used in image interpolation. These basis functions are shown in Figure 2.3. The simplest one is the zero order or nearest neighbor function. Another basis function is the linear function. The most famous B-spline basis function is the cubic spline function.

2.3.2.1 Nearest Neighbor Interpolation

Nearest neighbor interpolation is the simplest interpolation scheme. The basis function associated with nearest neighbor interpolation is given in Equation (2.4).

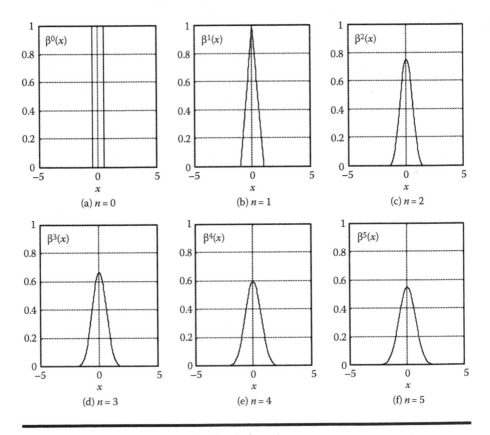

Figure 2.3 B-spline interpolation basis functions.

This scheme of interpolation is merely pixel repetition and the basis function used is interpolating.

2.3.2.2 Linear Interpolation

Linear interpolation enjoys large popularity due to its simplicity of implementation. It is commonly known as bilinear interpolation. The basis function used in bilinear interpolation is interpolating, given by [20]:

$$\beta^1(x) = \begin{cases} 1-|x| & |x| < 1 \\ 0 & 1 \le |x| \end{cases} \tag{2.14}$$

As shown in Figure 2.1, we define the distance between x and x_k, and between x_{k+1} and x as:

$$s = x - x_k \quad \text{and} \quad 1-s - x_{k+1} \quad x \tag{2.15}$$

Substituting from Equations (2.14) and (2.15) into Equation (2.3), we get the bilinear interpolation formula as follows [21,28]:

$$\hat{f}(x) = (1-s)f(x_k) + sf(x_{k+1}) \tag{2.16}$$

2.3.2.3 Cubic Spline Interpolation

In the family of polynomial splines, cubic spline tends to be the most popular. Its closed form basis function is given by [20]:

$$\beta^3(x) = \begin{cases} \dfrac{2}{3} - |x|^2 + \dfrac{|x|^3}{2} & 0 \le |x| < 1 \\[2mm] \dfrac{(2-|x|)^3}{6} & 1 \le |x| < 2 \\[2mm] 0 & 2 \le |x| \end{cases} \tag{2.17}$$

The cubic spline basis function is non-interpolating, and thus the coefficients in Equation (2.3) must be estimated prior to the interpolation process. This estimation can be implemented using a digital filtering approach [20]. Substituting from

Equations (2.15) and (2.17) into Equation (2.3), we get the cubic spline interpolation formula in the following form [28]:

$$\hat{f}(x) = c(x_{k-1})\left[(3+s)^3 - 4(2+s)^3 + 6(1+s)^3 - 4s^3\right]/6$$

$$+ c(x_k)\left[(2+s)^3 - 4(1+s)^3 + 6s^3\right]/6 \qquad (2.18)$$

$$+ c(x_{k+1})\left[(1+s)^3 - 4s^3\right]/6 + c(x_{k+2})s^3/6$$

2.3.3 Digital Filter Implementation of B-Spline Interpolation

For B-splines of orders 0 and 1, the basis functions are interpolating. This means that the spline coefficients are simply the pixel values. Interpolation is more complicated with higher order splines. The basis functions are non-interpolating. Therefore, a pre-processing step is required for the determination of spline coefficients prior to interpolation. This step is implemented using a digital filtering approach.

We define the B-spline kernel b_m^n obtained by sampling the B-spline basis function of degree n expanded by a factor m as follows [20]:

$$b_m^n(x_k) = \beta^n(x/m)\big|_{x=x_k} \qquad (2.19)$$

Carrying out the z-transform of this B-spline kernel gives:

$$B_m^n(z) = \sum_{x_k \in Z} b_m^n(x_k) z^{-k} \qquad (2.20)$$

Our objective is to determine the B-spline coefficients $c(x_k)$ such that we have a perfect fit at the integers, which means that:

$$\sum_{x_l \in Z} c(x_l)\beta^n(x - x_l)\big|_{x=x_k} = f(x_k) \qquad (2.21)$$

Rewriting this equation in the form of a convolution gives the following equation:

$$f(x_k) = \left(b_1^n * c\right)(x_k) \qquad (2.22)$$

Thus, the coefficients are determined using the following equation [20]:

$$c(x_k) = \left(b_1^n\right)^{-1} * f(x_k) \qquad (2.23)$$

Since b_1^n is a symmetric finite impulse response (FIR) filter, the direct B-spline filter $\left(b_1^n\right)^{-1}$ is an all-pole filter that can be implemented using a cascade of first order causal and non-causal recursive filters [20]. For the case of cubic spline interpolation, we have [20]:

$$B_1^3(z) = \frac{z + 4 + z^{-1}}{6} \tag{2.24}$$

This gives:

$$\left[B_1^3(z)\right]^{-1} = 6\left[\frac{1}{1 - z_1 z^{-1}}\right]\left[\frac{-z_1}{1 - z_1 z}\right] \tag{2.25}$$

where $z_1 = -2 + \sqrt{3}$. The above equation can be implemented with two cascaded stages as illustrated in Figure 2.4.

$$c^+(x_k) = f(x_k) + z_1 c^+(x_{k-1}) \qquad (k = 1, \ldots\ldots, N-1) \tag{2.26}$$

$$c(x_k) = z_1\left(c(x_{k+1}) - c^+(x_k)\right) \qquad (k = N-2, \ldots, 0) \tag{2.27}$$

where N is the length of the data sequence used. The initial conditions for the two stages can be calculated as follows [20]:

$$c^+(0) = \sum_{k=0}^{N-1} f(x_k) z_1^k \tag{2.28}$$

$$c(x_{N-1}) = \frac{z_1}{\left(1 - z_1^2\right)}\left(c^+(x_{N-1}) + z_1 c^+(x_{N-2})\right) \tag{2.29}$$

After the determination of the B-spline coefficients, the interpolation process is one step away. One requirement is to get an image interpolated by an up-sampling factor m. The block diagram of image interpolation by an up-sampling factor m with

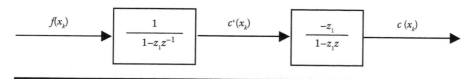

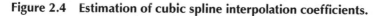

Figure 2.4   Estimation of cubic spline interpolation coefficients.

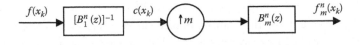

Figure 2.5   Block diagram of B-spline image interpolation.

an all-digital implementation is illustrated in Figure 2.5. The interpolated image $f_m^n(x_k)$ can be obtained using the following equation [17]:

$$f_m^n(x_k) = \sum_{l=-\infty}^{\infty} c(x_l) b_m^n(x_k - mx_l)$$
(2.30)

The above equation, which gives the interpolated signal from the obtained B-spline coefficients, is called the indirect B-spline transform. The digital filter $B_m^n(z)$ is a symmetric FIR filter. The expression for $B_m^n(z)$ is given as follows [17; Appendix A]:

$$B_m^n(z) = \frac{z^\alpha}{m^n} \left(\frac{1 - z^{-m}}{1 - z^{-1}} \right)^{n+1} B_1^n(z)$$
(2.31)

where $\alpha = (m - 1)(n + 1)/2$.

2.4 O-MOMS Interpolation

The O-MOMS term refers to a family of basis functions that achieve optimal interpolation of maximal order with minimal support. Any of these basis functions can be expressed as the weighted sum of a B-spline and its derivatives. For cubic O-MOMS image interpolation, the basis function is given by [18,19]:

$$\beta_{0-MOMS}^3(x) = \beta^3(x) + \frac{1}{42} \frac{d^2}{dx^2} \beta^3(x)$$

$$= \begin{cases} \dfrac{1}{2}|x|^3 - |x|^2 + \dfrac{1}{14}|x| + \dfrac{13}{21}, & 0 \le |x| \le 1 \\[2mm] \dfrac{-1}{6}|x|^3 + |x|^2 + \dfrac{85}{42}|x| + \dfrac{29}{21}, & 1 < |x| \le 2 \\[2mm] 0, & 2 \le |x| \end{cases}$$
(2.32)

Equation (2.3) holds for O-MOMS interpolation but with $\beta_{0-MOMS}^n(x)$ used instead of $\beta^n(x)$.

The O-MOMS basis functions are non-interpolating, and thus the coefficients in Equation (2.3) must be estimated. The estimation process can be implemented using the same digital filtering approach in B-spline image interpolation. As a result, the cubic O-MOMS interpolation formula is given by the following equation:

$$\hat{f}(x) = c(x_{k-1})\left[-\frac{1}{6}(1+s)^3 + (1+s)^2 - \frac{85}{42}(1+s) + \frac{29}{21}\right]$$

$$+ c(x_k)\left[\frac{1}{2}s^3 - s^2 + \frac{s}{14} + \frac{13}{21}\right]$$

$$+ c(x_{k+1})\left[\frac{1}{2}(1-s)^3 - (1-s)^2 + \frac{s}{14} + \frac{13}{21}\right] \qquad (2.33)$$

$$+ c(x_{k+2})\left[-\frac{1}{6}(2-s)^3 + (2-s)^2 - \frac{85}{42}(2-s) + \frac{29}{21}\right]$$

2.5 Keys' (Bicubic) Interpolation

Another family of basis functions used in interpolation is the Keys' family. Keys' interpolation is well known as bicubic interpolation. The problem encountered in the higher-order B-splines and O-MOMS interpolation is that the basis functions are non-interpolating. This means that the interpolation coefficients must be estimated before interpolation. This problem was solved by generating a cubic family of basis functions called the Keys' family that represents a compromise between ideal sync-based interpolation and B-spline interpolation. The Keys' basis function is interpolating, and it is expressed in the following form [21]:

$$\beta(x) = \begin{cases} (\alpha+2)|x|^3 - (\alpha+3)|x|^2 + 1 & 0 \leq |x| \leq 1 \\ \alpha|x|^3 - 5\alpha|x|^2 + 8\alpha|x| - 4\alpha & 1 < |x| \leq 2 \end{cases} \qquad (2.34)$$

where α is an optimization parameter that controls the rate of decay of the basis function. The Keys' basis functions with different values of α are shown in Figure 2.6.

The Keys' interpolation formula is obtained in the same manner as in the B-spline case using Equation (2.3), with pixel values replacing the interpolation coefficients. This formula is given as follows [21]:

$$\hat{f}(x) = f(x_{k-1})\left[\alpha s^3 - 2\alpha s^2 + \alpha s\right] + f(x_k)\left[(\alpha+2)s^3 - (3+\alpha)s^2 + 1\right]$$

$$+ f(x_{k+1})\left[-(\alpha+2)s^3 + (2\alpha+3)s^2 - \alpha s\right] + f(x_{k+2})\left[-\alpha s^3 + \alpha s^2\right] \qquad (2.35)$$

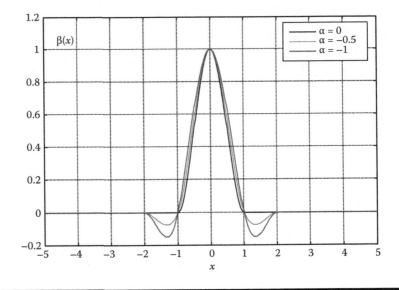

Figure 2.6   Keys' basis functions with different values of α.

2.6 Artifacts of Polynomial Image Interpolation

Some artifacts arise as a result of polynomial image interpolation. Examples are ringing, aliasing, blocking, and blurring.

2.6.1 Ringing

Ringing arises because most good synthesis functions are oscillating. Ringing is also known as the Gibbs effect. The ringing effect is not an appreciable artifact in most cases since it is possible to use different approximations for the interpolation basis function.

2.6.2 Aliasing

Unlike ringing, aliasing is a true artifact because it is never possible to perform exact recovery of the initial data from their aliased version. Aliasing is related to the discrete nature of the samples. When it is desired to represent a coarser version of the data using fewer samples, the optimal procedure is first to create a precise representation of the coarse data that uses every available sample and then down-sample only this coarse representation. In some sense, aliasing appears when this procedure is not followed or when a mismatch occurs between the coarseness of the intermediate representation and the degree of down-sampling.

2.6.3 Blocking

Blocking arises when the support of the interpolation is finite. In this case, the influence of any given pixel is limited to its surroundings, and it is sometimes possible to discern the boundary of this influence zone. Synthesis functions with sharp transitions, such as those used with nearest neighbor interpolation exacerbate this effect.

2.6.4 Blurring

Blurring is related to aliasing in the sense that it is also a mismatch between an intermediate data representation and their final down-sampled or over-sampled version. In this case, the mismatch is such that the intermediate data is too coarse for the task. This results in an image that appears out of focus.

Chapter 3

Adaptive Polynomial Image Interpolation

3.1 Introduction

Although polynomial techniques for image interpolation are the most popular methods due to their simplicity, they do not take into consideration the local activity levels of the image to be interpolated. It could be feasible if the interpolation formula of each technique is adapted from pixel to pixel to accommodate for the varying local activity levels. This adaptation reduces the blurring effect resulting from the interpolation, giving better visual quality.

The demand for high-quality interpolation results has motivated research in the field of adaptive image interpolation. Recently, a linear space-varying approach was proposed for image interpolation. This approach is based on the evaluation of a "warped-distance" between the pixel to be estimated and each of its neighbors [28]. The warping process is performed by moving the estimate of the pixel toward the more homogeneous neighboring side. This approach has succeeded to some extent for edge interpolation. It has been applied for all previously mentioned interpolation techniques.

In this chapter, we treat the problem of polynomial image interpolation adaptively to obtain better results. It is known that the efficiency of any image interpolation algorithm depends on two factors: the visual quality of the obtained image and the computational cost of the interpolation algorithm. Our objective is to make a trade-off between these two factors. Toward this objective, two adaptive algorithms are used for the implementation of polynomial image interpolation techniques.

The first algorithm is computationally simple. It is performed by weighting the pixels used in the interpolation process with different adaptive weights [30,31]. The adaptation can be used with different traditional interpolation techniques such as B-spline, O-MOMS and Keys' as well as their warped-distance implementations.

The second adaptive algorithm is much more powerful. This algorithm is based on the minimization of the squared estimation error at each pixel in the interpolated image by adaptively calculating the optimum distance of the pixel to be estimated from its neighbors. The adaptation process at each pixel can be performed iteratively to yield the best estimate of its value [32–34]. This algorithm considers the mathematical model by which an LR image is obtained from its corresponding HR one. It is also applicable to different traditional image interpolation techniques.

3.2 Low-Resolution Image Degradation Model

In the imaging process, when a scene is imaged by a high-resolution (HR) camera, the captured HR image can be named $f(n_1,n_2)$, where n_1, $n_2 = 0,1,2,....N{-}1$. If the same scene is captured by a low-resolution (LR) camera, the resulting image can be named $g(m_1,m_2)$ where m_1, $m_2 = 0,1,2,....M{-}1$. Here, $M = N/R$, and R is the ratio between the sampling rates of $f(n_1,n_2)$ and $g(m_1,m_2)$. The relationship between the LR image and the HR image, assuming no blurring, can be represented by the following mathematical model [35,36]:

$$\mathbf{g} = \mathbf{D}\mathbf{f} + \mathbf{v} \tag{3.1}$$

where $\mathbf{f}$, $\mathbf{g}$, and $\mathbf{v}$ are lexicographically ordered vectors of the unknown HR image, the captured LR image, and the additive noise values, respectively. By lexicographic ordering, we mean that the image is ordered column-by-column to form a single vector. The vector $\mathbf{f}$ is of size $N^2 \times 1$, and the vectors $\mathbf{g}$ and $\mathbf{v}$ are of size $M^2 \times 1$. The matrix $\mathbf{D}$ represents the filtering and down-sampling process that transforms the HR image to the LR one. It is of size $M^2 \times N^2$. Under the separability assumption, the matrix $\mathbf{D}$ can be written as follows [35,36]:

$$\mathbf{D} = \mathbf{D}_1 \otimes \mathbf{D}_1 \tag{3.2}$$

where $\otimes$ represents the Kronecker product, and the $M \times N$ matrix $\mathbf{D}_1$ represents the one-dimensional (1-D) filtering and down-sampling by a factor R. For $N = 2M$, $\mathbf{D}_1$ is given by [35,36]:

$$\mathbf{D}_1 = \frac{1}{2}\begin{bmatrix} 1 & 1 & 0 & 0 & \cdots & 0 & 0 \\ 0 & 0 & 1 & 1 & \cdots & 0 & 0 \\ \vdots & \vdots & \vdots & \vdots & \ddots & \vdots & \vdots \\ 0 & 0 & 0 & 0 & \cdots & 1 & 1 \end{bmatrix} \tag{3.3}$$

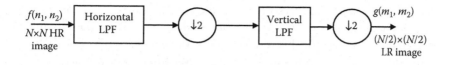

Figure 3.1   Down-sampling process from N × N HR image to N/2 × N/2 LR image.

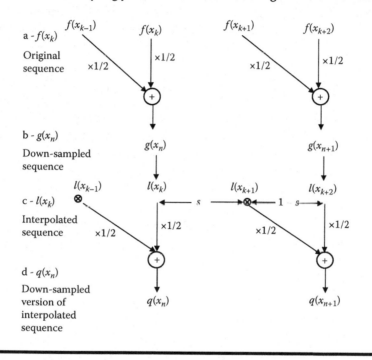

Figure 3.2   Signal down-sampling and interpolation. (a) Original data sequence. (b) Down-sampled version of original data sequence. (c) Interpolated data sequence. (d) Down-sampled version of interpolated data sequence.

The model of filtering and down-sampling is illustrated in Figure 3.1 and its mathematical implementation for a 1-D sequence is illustrated in Figure 3.2a and 3.2b. The objective of the image interpolation process is the estimation of the vector **f** given the vector **g**. Several algorithms can be used to achieve this objective. Our target is how to solve this problem adaptively.

3.3   Linear Space-Invariant Image Interpolation

The process of image interpolation aims at estimating intermediate pixel values between known pixel values. This process is performed on a 1-D basis row-by-row and then column-by-column. If we have a discrete sequence $f(x_k)$ of length N as shown in Figure 3.2a, and this sequence is filtered and down-sampled by 2, we get another sequence $g(x_n)$ of length $N/2$ as shown in Figure 3.2b. The filtering and down-sampling simulate the LR image degradation model given by Equation (3.1). The interpolation process aims at estimating a sequence $l(x_k)$ of length N as shown in Figure 3.2c, which is as close as possible to the original discrete sequence $f(x_k)$.

For interpolating the equally-spaced 1-D sampled data $g(x_n)$, all the previously-mentioned interpolation formulas can be used, but with different spatial indices for mathematical tractability. Using Equation (2.1), the value of the sample to be estimated, $l(x_{k+1})$ is, in general, given by [32–34]:

$$l(x_{k+1}) = \sum_{n=-\infty}^{\infty} c(x_n)\beta(x_{k+1} - x_n) \tag{3.4}$$

Thus, Equation (2.2) for ideal interpolation can be reformatted as follows [32–34]:

$$l(x_{k+1}) = \sum_{n=-\infty}^{\infty} g(x_n)\mathrm{sinc}(x_{k+1} - x_n) \tag{3.5}$$

Ideal interpolation is not practical, and hence alternative approximations can be used instead. The formulas of these approximations take the following forms [32–34]:

Bilinear [34]:

$$l(x_{k+1}) = (1-s)g(x_n) + sg(x_{n+1}) \tag{3.6}$$

Cubic spline [34]:

$$l(x_{k+1}) = c(x_{n-1})[(3+s)^3 - 4(2+s)^3 + 6(1+s)^3 - 4s^3]/6$$

$$+ c(x_n)[(2+s)^3 - 4(1+s)^3 + 6s^3]/6 + c(x_{n+1})[(1+s)^3 - 4s^3]/6 + c(x_{n+2})s^3/6 \tag{3.7}$$

Cubic O-MOMS [34]:

$$l(x_{k+1}) = c(x_{n-1})\left[-\frac{1}{6}(1+s)^3 + (1+s)^2 - \frac{85}{42}(1+s) + \frac{29}{21} \right]$$

$$+ c(x_n)\left[\frac{1}{2}s^3 - s^2 + \frac{s}{14} + \frac{13}{21} \right] + c(x_{n+1})\left[\frac{1}{2}(1-s)^3 - (1-s)^2 + \frac{s}{14} + \frac{13}{21} \right]$$

$$+ c(x_{n+2})\left[-\frac{1}{6}(2-s)^3 + (2-s)^2 - \frac{85}{42}(2-s) + \frac{29}{21} \right]$$

$$\tag{3.8}$$

Keys' [33,34]:

$$l(x_{k+1}) = g(x_{n-1})\left[\alpha s^3 - 2\alpha s^2 + \alpha s\right] + g(x_n)\left[(\alpha + 2)s^3 - (3 + \alpha)s^2 + 1\right]$$
$$+ g(x_{n+1})\left[-(\alpha + 2)s^3 + (2\alpha + 3)s^2 - \alpha s\right] + g(x_{n+2})\left[-\alpha s^3 + \alpha s^2\right] \tag{3.9}$$

As mentioned above, the objective is to find $l(x_{k+1})$, which is as close as possible to $f(x_{k+1})$. This process is repeated adaptively for each estimated pixel.

3.4 Warped-Distance Image Interpolation

The warped-distance algorithm is the simplest adaptive image interpolation algorithm. The idea of distance warping can be used in the implementation of any polynomial interpolation technique. This idea is based on modifying the distance s and using a new distance s' depending on the homogeneity or inhomogeneity in the neighborhood of each estimated pixel. The warped-distance s' is estimated using the following relation [28]:

$$s' = s - \tau A_n s(s - 1) \tag{3.10}$$

where A_n refers to the asymmetry of the data in the neighborhood of x_{k+1} and is defined as [28]:

$$A_n = \frac{\left|g(x_{n+1}) - g(x_{n-1})\right| - \left|g(x_{n+2}) - g(x_n)\right|}{L_{\max} - 1} \tag{3.11}$$

where $L_{max} = 256$ for 8-bit pixels. The scaling factor $L_{max} - 1$ is to keep A_n in the range of -1 to 1. The parameter τ controls the intensity of warping. It has a positive integer value. The desired effect of this warping is to avoid blurring of the edges in the interpolation process.

3.5 Weighted Image Interpolation

Instead of using the traditional image interpolation techniques, a weighting approach can be adopted. This approach depends on weighting the values of the pixels incorporated by space-varying adaptive weights. The distance s is kept fixed. The idea of weighted interpolation modifies the traditional interpolation formulas given in Equations (3.6) to (3.9).
Weighted bilinear [30]:

$$l(x_{k+1}) = (1 - s)a_0 g(x_n) + sa_1 g(x_n + 1) \tag{3.12}$$

where

$$a_0 = 1 - \gamma A_n, \qquad a_1 = 1 + \gamma A_n \qquad (3.13)$$

Weighted cubic spline [30]:

$$l(x_{k+1}) = a_{-1}c(x_{n-1})[(3+s)^3 - 4(2+s)^3 + 6(1+s)^3 - 4s^3]/6$$

$$+ a_0 c(x_n)[(2+s)^3 - 4(1+s)^3 + 6s^3]/6 + a_1 c(x_{n+1})[(1+s)^3 - 4s^3]/6 \qquad (3.14)$$

$$+ a_2 c(x_{n+2})s^3 / 6$$

Weighted cubic O-MOMS [30]:

$$l(x_{k+1}) = a_{-1}c(x_{n-1})\left[-\frac{1}{6}(1+s)^3 + (1+s)^2 - \frac{85}{42}(1+s) + \frac{29}{21} \right]$$

$$+ a_0 c(x_n)\left[\frac{1}{2}s^3 - s^2 + \frac{s}{14} + \frac{13}{21} \right] + a_1 c(x_{n+1})\left[\frac{1}{2}(1-s)^3 - (1-s)^2 + \frac{s}{14} + \frac{13}{21} \right]$$

$$+ a_2 c(x_{n+2})\left[-\frac{1}{6}(2-s)^3 + (2-s)^2 - \frac{85}{42}(2-s) + \frac{29}{21} \right]$$

$$(3.15)$$

Weighted Keys' [30]:

$$l(x_{k+1}) = a_{-1}g(x_{n-1})\left[\alpha s^3 - 2\alpha s^2 + \alpha s \right] + a_0 g(x_n)\left[(\alpha+2)s^3 - (3+\alpha)s^2 + 1 \right]$$

$$+ a_1 g(x_{n+1})\left[-(\alpha+2)s^3 + (2\alpha+3)s^2 - \alpha s \right] + a_2 g(x_{n+2})\left[-\alpha s^3 + \alpha s^2 \right] \qquad (3.16)$$

where

$$a_{-1} = a_0 = 1 - \gamma A_n, \qquad a_1 = a_2 = 1 + \gamma A_n \qquad (3.17)$$

and γ is a positive integer. The constant γ controls the intensity of weighting used for neighboring pixels. As a result, the weighting coefficients are updated at each pixel, depending on the asymmetry A_n at each pixel. There are some special cases for pixel estimation using the adaptive weighed interpolation algorithm. These cases are summarized in the following points [30]:

1. For homogeneous regions, the value of A_n tends to zero, which leads to $a_{-1} = a_0 = a_1 = a_2 = 1$. This is equivalent to the traditional image interpolation process.
2. For positive values of A_n, which means that there is an edge that is more homogeneous on the right side, the weights of the pixels on the right side

(a_1 and a_2) are increased and the weights of the pixels on the left side (a_{-1} and a_0) are decreased. As a result, the estimated pixel value is close to those of the pixels on the right side. This is expected to yield images with better visual quality.

3. For negative values of A_n, a_{-1} and a_0 are increased and a_1 and a_2 are decreased. As a result, the estimated pixel value is close to those of the pixels on the left side.

The weighted interpolation algorithm can also be applied with warping to achieve better results.

3.6 Iterative Image Interpolation

If we apply a polynomial interpolation process on the sequence $g(x_n)$ shown in Figure 3.2b, we get a sequence $l(x_k)$ of length N, which is required to be as close as possible to $f(x_k)$. This requires the minimization of the mean square error (MSE) between the estimated sequence and the original sequence. The minimization of the MSE can be performed by the minimization of the squared error between each estimated sample and its original counterpart.

If the sequence $g(x_n)$ is interpolated using bilinear interpolation to give $l(x_k)$, the squared estimation error between the estimated sample $l(x_{k+1})$ and the original sample $f(x_{k+1})$ is given by [32–34]:

$$E = [f(x_{k+1}) - l(x_{k+1})]^2 \qquad (3.18)$$

Substituting for $l(x_{k+1})$ from Equation (3.6), we get:

$$E = [f(x_{k+1}) - (1-s)g(x_n) - sg(x_{n+1})]^2 \qquad (3.19)$$

But from Figure 3.2, we have the following relations:

$$g(x_n) = \frac{1}{2}[f(x_k) + f(x_{k-1})] \qquad (3.20)$$

$$g(x_{n+1}) = \frac{1}{2}[f(x_{k+2}) + f(x_{k+1})] \qquad (3.21)$$

Substituting for $g(x_n)$ and $g(x_{n+1})$ from Equations (3.20) and (3.21) into Equation (3.19), we get:

$$E = \left[f(x_{k+1}) - \frac{(1-s)}{2}[f(x_k) + f(x_{k-1})] - \frac{s}{2}[f(x_{k+2}) + f(x_{k+1})] \right]^2 \qquad (3.22)$$

All sample values are considered constants and E is a function of s only. Differentiating Equation (3.22) with respect to s gives:

$$\frac{dE}{ds} = 2\left[f(x_{k+1}) - \frac{(1-s)}{2}[f(x_k) + f(x_{k-1})] - \frac{s}{2}[f(x_{k+1}) + f(x_{k+2})] \right]$$
$$\cdot \left[\frac{1}{2}[f(x_k) + f(x_{k-1})] - \frac{1}{2}[f(x_{k+2}) + f(x_{k+1})] \right]$$

(3.23)

Equating this derivative to zero leads to:

$$s_{opt} = \frac{2\left[f(x_{k+1}) - \frac{1}{2}[f(x_k) + f(x_{k-1})] \right]}{f(x_{k+1}) + f(x_{k+2}) - f(x_k) - f(x_{k-1})}$$

(3.24)

Substituting from Equations (3.20) and (3.21) into Equation (3.24) gives the following formula:

$$s_{opt} = \frac{f(x_{k+1}) - g(x_n)}{g(x_{n+1}) - g(x_n)}$$

(3.25)

To ensure minimum squared error at s_{opt}, we differentiate Equation (3.23) with respect to s to get:

$$\lim_{s \to s_{opt}} \frac{d^2 E}{ds^2} = \frac{1}{2}\left[[f(x_k) + f(x_{k-1})] - \frac{1}{2}[f(x_{k+1}) + f(x_{k+2})] \right]^2 = +\text{ve value} \quad (3.26)$$

Thus, at s_{opt}, there is a minimum squared error between the estimated sample $l(x_{k+1})$ and the original sample $f(x_{k+1})$ [34]. If the sequence $g(x_n)$ is interpolated using a cubic spline interpolation formula, then using Equations (3.7) and (3.18), the squared estimation error between $l(x_{k+1})$ and the original sample $f(x_{k+1})$ is given by [34]:

$$E = \left[f(x_{k+1}) - l(x_{k+1}) \right]^2 = [f(x_{k+1}) - c(x_{n-1})[(3+s)^3 - 4(2+s)^3 + 6(1+s)^3 - 4s^3]/6$$
$$- c(x_n)[(2+s)^3 - 4(1+s)^3 + 6s^3]/6$$
$$- c(x_{n+1})[(1+s)^3 - 4s^3]/6 - c(x_{n+2})s^3/6]^2$$

(3.27)

Differentiating Equation (3.27) with respect to s and equating this derivative to zero gives:

$$f(x_{k+1}) = c(x_{n-1})[(3+s_{opt})^3 - 4(2+s_{opt})^3 + 6(1+s_{opt})^3 - 4s_{opt}^3]/6$$

$$+ c(x_n)[(2+s_{opt})^3 - 4(1+s_{opt})^3 + 6s_{opt}^3]/6 + c(x_{n+1})[(1+s_{opt})^3 - 4s_{opt}^3]/6$$

$$+ c(x_{n+2})s_{opt}^3/6$$

$$(3.28)$$

If this equation is solved, it gives the value of s_{opt} for cubic spline interpolation. To ensure a minimum of the squared error at s_{opt}, we calculate the second derivative as follows:

$$\frac{d^2E}{ds^2} = 2[f(x_{k+1}) - c(x_{n-1})[(3+s)^3 - 4(2+s)^3 + 6(1+s)^3 - 4s^3]/6$$

$$- c(x_n)[(2+s)^3 - 4(1+s)^3 + 6s^3]/6 - c(x_{n+1})[(1+s)^3 - 4s^3]/6$$

$$- c(x_{n+2})s^3/6][-c(x_{n-1})[6(3+s) - 24(2+s) + 36(1+s) - 24s^3]/6$$

$$- c(x_n)[6(2+s) - 24(1+s) + 36s]/6 - c(x_{n+1})[6(1+s) - 24s]/6$$

$$- c(x_{n+2})s] + 2[-c(x_{n-1})[3(3+s)^2 - 12(2+s)^2 + 18(1+s)^2 - 12s^2]/6$$

$$- c(x_n)[3(2+s)^2 - 12(1+s)^2 + 18s^2]/6 - c(x_{n+1})[3(1+s)^2 - 12s^2]/6$$

$$- c(x_{n+2})s^2/2]^2$$

$$(3.29)$$

From Equations (3.28) and (3.29) we get:

$$\lim_{s \to s_{opt}} \frac{d^2E}{ds^2} = 2[-c(x_{n-1})[3(3+s)^2 - 12(2+s)^2 + 18(1+s)^2 - 12s^2]/6$$

$$- c(x_n)[3(2+s)^2 - 12(1+s)^2 + 18s^2]/6 - c(x_{n+1})[3(1+s)^2 - 12s^2]/6$$

$$- c(x_{n+2})s^2/2]^2$$

$$= +ve \text{ value}$$

$$(3.30)$$

The result obtained in Equation (3.30) ensures a minimum of the squared estimation error at s_{opt}[34]. Similar results can be obtained for the Keys' and cubic O-MOMS cases.

It is noted that the sequence $g(x_n)$ is available, but the sequence $f(x_k)$ is not available. Thus, one cannot directly determine s_{opt} for the above-mentioned four cases. To solve this problem, we apply a down-sampling operation on the sequence $l(x_k)$ to get a sequence $q(x_n)$ of length $N/2$ as illustrated in Figure 3.2d. The squared estimation error between $q(x_{n+1})$ and $g(x_{n+1})$ is given by:

$$E^* = [g(x_{n+1}) - q(x_{n+1})]^2 \qquad (3.31)$$

From Figure 3.2, we have:

$$l(x_{k+2}) = g(x_{n+1}) \qquad (3.32)$$

and

$$q(x_{n+1}) = \frac{1}{2}[l(x_{k+1}) + l(x_{k+2})] \qquad (3.33)$$

Substituting from Equations (3.21),(3.32), and (3.33) into Equation (3.31), we get:

$$E^* = \frac{1}{4}\left[[f(x_{k+1}) - l(x_{k+1})] + \frac{1}{2}[f(x_{k+2}) - f(x_{k+1})] \right]^2 \qquad (3.34)$$

Substituting from Equation (3.18) into Equation (3.34), we get [32–34]:

$$E^* = \frac{1}{4}\left[\sqrt{E} + K \right]^2 \qquad (3.35)$$

where

$$K = \frac{1}{2}[f(x_{k+2}) - f(x_{k+1})] \qquad (3.36)$$

It is clear that K is a constant with respect to s or with respect to s and α for the Keys' case. For the case of edge interpolation, which is the case of greatest interest, the side containing $f(x_{k+1})$ and $f(x_{k+2})$ is homogeneous and the side containing $f(x_{k-1})$ and $f(x_k)$ is also homogeneous. This means that the values of $f(x_{k+1})$ and $f(x_{k+2})$ are close to each other and the values of $f(x_{k-1})$ and $f(x_k)$ are also close to each other. Thus, the value of K is small and may be neglected. As a result, the value of s_{opt}, which minimizes E leads to the minimization of E^* regardless of the sign of K. Since E is minimum at s_{opt}, then E^* will be minimum at the same values of s_{opt}. For flat areas, the constant K is approximately equal to zero and $E^* \cong E/4$. From Equation (3.35), we can go to the following equation [32–34]:

$$E^*_{min} = \frac{1}{4}\left[\sqrt{E_{min}} + K \right]^2 \qquad (3.37)$$

To evaluate the value of s_{opt} for cases of bilinear, cubic spline or cubic O-MOMS interpolation, we follow an iterative manner as follows [34]:

$$s_{i+1} = s_i - \eta_0 \frac{dE^*}{ds}(s_i) \qquad (3.38)$$

where η_0 is the convergence parameter and $s_0 = 1/2$. In this iterative algorithm, it is required to calculate $\frac{dE^*}{ds}$. From Equation (3.31), we get:

$$\frac{dE^*}{ds} = -2\big[g(x_{n+1}) - q(x_{n+1})\big]\frac{dq(x_{n+1})}{ds} \qquad (3.39)$$

Substituting from Equation (3.33) into Equation (3.39), we get:

$$\frac{dE^*}{ds} = -\left[g(x_{n+1}) - \frac{1}{2}[l(x_{k+1}) + l(x_{k+2})]\right]\frac{dl(x_{k+1})}{ds} \qquad (3.40)$$

The term $\frac{dl(x_{k+1})}{ds}$ is calculated for each interpolation formula as follows:

Bilinear [34]:

$$\frac{dl(x_{k+1})}{ds} = -g(x_n) + g(x_{n+1}) \qquad (3.41)$$

Cubic spline [34]:

$$\frac{dl(x_{k+1})}{ds} = c(x_{n-1})[3(3+s)^2 - 12(2+s)^2 + 18(1+s)^2 - 12s^2]/6$$

$$+ c(x_n)[3(2+s)^2 - 12(1+s)^2 + 18s^2]/6 + c(x_{n+1})[3(1+s)^2 - 12s^2]/6$$

$$+ c(x_{n+2})s^2/2$$

$$(3.42)$$

Cubic O-MOMS [34]:

$$\frac{dl(x_{k+1})}{ds} = c(x_{n-1})\left[-\frac{1}{2}(1+s)^2 + 2(1+s) - \frac{85}{42}\right]$$

$$+ c(x_n)\left[\frac{3}{2}s^2 - 2s + \frac{1}{14}\right] + c(x_{n+1})\left[\frac{-3}{2}(1-s)^2 + 2(1-s) + \frac{1}{14}\right] \qquad (3.43)$$

$$+ c(x_{n+2})\left[\frac{1}{2}(2-s)^2 - 2(2-s) + \frac{85}{42}\right]$$

The iterative algorithm defined in Equation (3.38) is used with the aid of Equation (3.40) and one of the derivatives of Equations (3.41) to (3.43) for the estimation of each pixel in the interpolation process according to the used basis function.

For the case of Keys' interpolation, we have two control parameters s and α. If the sequence $g(x_n)$ is interpolated using the Keys' interpolation formula, then using Equation (3.9), the squared estimation error between $l(x_{k+1})$ and the original sample $f(x_{k+1})$ is given by [34]:

$$E = \left[f(x_{k+1}) - l(x_{k+1}) \right]^2$$

$$= \left[f(x_{k+1}) - g(x_{n-1}) \left[\alpha s^3 - 2\alpha s^2 + \alpha s \right] - g(x_n) \left[(\alpha+2)s^3 - (3+\alpha)s^2 + 1 \right] \right.$$

(3.44)

$$\left. - g(x_{n+1}) \left[-(\alpha+2)s^3 + (2\alpha+3)s^2 - \alpha s \right] - g(x_{n+2}) \left[-\alpha s^3 + \alpha s^2 \right] \right]^2$$

It is required to find the values of s_{opt} and α_{opt} that minimize the value of E in Equation (3.44). These values are then used in Equation (3.9). Unfortunately, the sequence $f(x_k)$ is unavailable. As a result, we cannot determine s_{opt} and α_{opt} from Equation (3.44). Following the same procedure as above, we get into Equation (3.37).

Note that K is a constant with respect to s and α. Thus, the values of s_{opt} and α_{opt} that minimize E lead to the minimization of E^* regardless of the sign of K. Since E is minimum at s_{opt} and α_{opt}, E^* will be minimum at the same values of s_{opt} and α_{opt} [33,34]. The error function E^* is a function of the two parameters s and α. As a result, the minimization of E^* can be performed by equating the gradient of E^* to zero [33,34].

$$\nabla E^*(s,\alpha) = \begin{bmatrix} \dfrac{\partial E^*(s,\alpha)}{\partial s} \\[2ex] \dfrac{\partial E^*(s,\alpha)}{\partial \alpha} \end{bmatrix} = 0$$

(3.45)

The values of s_{opt} and α_{opt} can be iteratively estimated as follows [58]:

$$\begin{bmatrix} s_{i+1} \\ \alpha_{i+1} \end{bmatrix} = \begin{bmatrix} s_i \\ \alpha_i \end{bmatrix} - \eta_0 \nabla E^*(s_i, \alpha_i)$$

(3.46)

where η_0 is the convergence parameter, $s_0 = 1/2$, and $\alpha_0 = -1/2$. If α is fixed at $-1/2$ and the optimization is carried out with respect to s only, Equation (3.46) can be simplified to the form [33,34]:

$$s_{i+1} = s_i - \eta_0 \frac{dE^*}{ds}(s_i)$$

(3.47)

If s is fixed at 1/2 and the optimization is carried out with respect to α only, Equation (3.46) can be simplified to the form [33,34]:

$$\alpha_{i+1} = \alpha_i - \eta_0 \frac{dE^*}{d\alpha}(\alpha_i) \tag{3.48}$$

The values of $\dfrac{\partial E^*}{\partial s}$ and $\dfrac{\partial E^*}{\partial \alpha}$ are calculated from Equation (3.31) as follows [33,34]:

$$\frac{\partial E^*}{\partial s} = -2\left[g(x_{n+1}) - q(x_{n+1})\right]\frac{\partial q(x_{n+1})}{\partial s} \tag{3.49}$$

$$\frac{\partial E^*}{\partial \alpha} = -2\left[g(x_{n+1}) - q(x_{n+1})\right]\frac{\partial q(x_{n+1})}{\partial \alpha} \tag{3.50}$$

Substituting from Equations (3.9) and (3.32) into Equation (3.33), we get [33,34]:

$$q(x_{n+1}) = \frac{1}{2}g(x_{n-1})\left[\alpha s^3 - 2\alpha s^2 + \alpha s\right] + \frac{1}{2}g(x_n)\left[(\alpha+2)s^3 - (3+\alpha)s^2 + 1\right]$$

$$+ \frac{1}{2}g(x_{n+1})\left[-(\alpha+2)s^3 + (2\alpha+3)s^2 - \alpha s + 1\right] + \frac{1}{2}g(x_{n+2})\left[-\alpha s^3 + \alpha s^2\right] \tag{3.51}$$

Using Equation (3.51), we get [33,34]:

$$\frac{\partial q(x_{n+1})}{\partial s} = \frac{1}{2}g(x_{n-1})\left[3\alpha s^2 - 4\alpha s + \alpha\right] + \frac{1}{2}g(x_n)\left[3(\alpha+2)s^2 - 2(3+\alpha)s\right]$$

$$+ \frac{1}{2}g(x_{n+1})\left[-3(\alpha+2)s^2 + 2(2\alpha+3)s - \alpha\right] + \frac{1}{2}g(x_{n+2})\left[-3\alpha s^2 + 2\alpha s\right] \tag{3.52}$$

$$\frac{\partial q(x_{n+1})}{\partial \alpha} = \frac{1}{2}g(x_{n-1})\left[s^3 - 2s^2 + s\right] + \frac{1}{2}g(x_n)\left[s^3 - s^2\right]$$

$$+ \frac{1}{2}g(x_{n+1})\left[-s^3 + 2s^2 - s\right] + \frac{1}{2}g(x_{n+2})\left[-s^3 + s^2\right] \tag{3.53}$$

The iterative algorithm defined in Equation (3.46) is used for the estimation of each pixel in the interpolation process. The optimization process of the Keys' image interpolation formula can be performed either in a single parameter (s or α) or in the two parameters. An important issue worth investigation is the computational cost of this pixel-by-pixel adaptive image interpolation process. This computational cost is studied experimentally in the following section by calculating the average number of iterations per pixel required for each interpolation process I_{av}.

3.7 Simulation Examples

To compare the different polynomial interpolation techniques and their adaptive variants, some simulation examples are included. The images used in the simulation experiments are first down-sampled using the model in Figure 3.1, and then contaminated by additive white Gaussian noise (AWGN) to simulate the LR image degradation model of Equation (3.1). After that, the LR images are interpolated to their original size and the MSE is estimated between the obtained image and the original image. The MSE is calculated as follows:

$$MSE = \frac{1}{N^2} \left\| \mathbf{f} - \hat{\mathbf{f}} \right\|^2 \tag{3.54}$$

The value of the MSE is used to calculate the peak signal-to-noise ratio (PSNR) of the obtained image, which is calculated as follows:

$$PSNR = 10 \log \left(\frac{255^2}{MSE} \right) \tag{3.55}$$

Two metrics are used for performance evaluation of any image interpolation algorithm. These metrics are the PSNR and the correlation coefficient for edge pixels between the original image and the interpolated image c_e. In applying the correlation coefficient metric, a Sobel edge detection operator with a 3×3 window is applied to both the original and the interpolated images to extract edge pixels. The correlation coefficient is estimated between edge pixels in both images. The higher the correlation coefficient, the larger is the ability of the image interpolation algorithm to preserve edges through the interpolation process.

In the first experiment, the LR 128×128 noisy woman image has been used to test all interpolation algorithms. The LR image has been obtained by down-sampling of the original image by 2 in horizontal and vertical directions, and then the obtained image is contaminated by AWGN with a signal-to-noise ratio (SNR) of 25 dB. The original image and the LR image are illustrated in Figure 3.3. This LR image has been interpolated to its original size of 256×256 using different algorithms. The MSE between the original image and the interpolated image has been estimated. The value of the MSE has been used in calculating the PSNR of each interpolated image.

All image interpolation techniques have been tested on the LR image with a support whose length is one more than the order of the interpolation kernel.

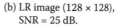

(a) Original image (256 × 256). (b) LR image (128 × 128), SNR = 25 dB.

Figure 3.3 Woman image.

The weighted and iterative image interpolation algorithms have also been tested on the same LR image for bilinear, cubic spline, cubic O-MOMS, and Keys' techniques. Two methods of implementation of the iterative algorithm have been investigated; row-by-row then column-by-column, and column-by-column then row-by-row. The average number of iterations per pixel I_{av} was estimated for the iterative interpolation algorithm to reveal the total time required for the interpolation process.

It is clear from the obtained results that the weighted interpolation algorithm gives marginal improvement as compared to traditional interpolation. It is also clear that the iterative image interpolation algorithm gives better results than the traditional implementation, the warped-distance, the weighted implementation, and the bilinear, cubic spline, cubic O-MOMS, and Keys' image interpolation techniques. The two implementations of the iterative interpolation algorithm yield approximately the same results.

Error images have been estimated between the original image and each interpolated one. If the interpolation is ideal, all the pixels of this error image must be equal to zero. The error images have been inverted and displayed. These error images reveal the ability of the adaptive algorithm to preserve edges. Results of the first experiment are given in Figure 3.4 through Figure 3.13.

A similar experiment has been carried out on the test pattern image to show edge interpolation for edges with different orientations. The results of this experiment are given in Figures 3.14 through 3.24. These results reveal the failure of the warped-distance and weighted interpolation algorithms to produce better results than traditional techniques. The iterative interpolation algorithm succeeded in getting better interpolation results.

(a) Bilinear, PSNR = 21.19 dB, c_e = 0.53.

(b) Warped-distance bilinear, PSNR = 21.27 dB, c_e = 0.55.

(c) Weighted bilinear, PSNR = 21.28 dB, c_e = 0.55.

(d) Weighted bilinear with warping, PSNR = 21.35 dB, c_e = 0.57.

(e) Iterative bilinear (row-by-row then column-by-column), PSNR = 22.29 dB, c_e = 0.69. I_{av} = 3.69.

(f) Iterative bilinear (column-by-column then row-by-row), PSNR = 22.29 dB, c_e = 0.68. I_{av} = 3.53.

Figure 3.4   Bilinear interpolation of woman image.

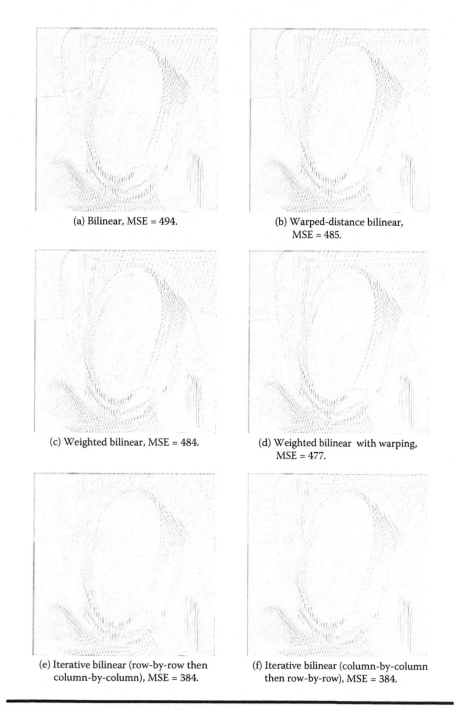

(a) Bilinear, MSE = 494.

(b) Warped-distance bilinear, MSE = 485.

(c) Weighted bilinear, MSE = 484.

(d) Weighted bilinear with warping, MSE = 477.

(e) Iterative bilinear (row-by-row then column-by-column), MSE = 384.

(f) Iterative bilinear (column-by-column then row-by-row), MSE = 384.

Figure 3.5   Error images for bilinear interpolation of woman image.

(a) Cubic spline, PSNR = 21.02 dB, $c_e = 0.5$.

(b) Warped-distance cubic spline, PSNR = 21.48 dB, $c_e = 0.59$.

(c) Weighted cubic spline, PSNR = 21.69 dB, $c_e = 0.63$.

(d) Weighted cubic spline with warping, PSNR = 21.61 dB, $c_e = 0.65$.

(e) Iterative cubic spline (row-by-row then column-by-column), PSNR = 22.23 dB, $c_e = 0.7$. $I_{av} = 3.73$.

(f) Iterative cubic spline (column-by-column then row-by-row), PSNR = 22.24 dB, $c_e = 0.71$, $I_{av} = 3.57$.

Figure 3.6   Cubic spline interpolation of woman image.

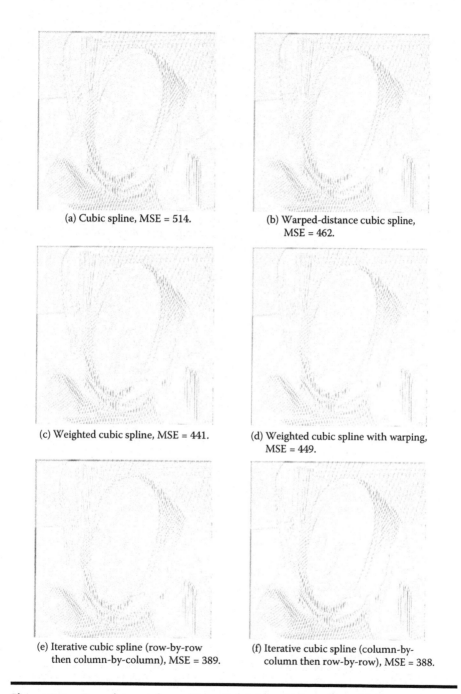

(a) Cubic spline, MSE = 514.

(b) Warped-distance cubic spline, MSE = 462.

(c) Weighted cubic spline, MSE = 441.

(d) Weighted cubic spline with warping, MSE = 449.

(e) Iterative cubic spline (row-by-row then column-by-column), MSE = 389.

(f) Iterative cubic spline (column-by-column then row-by-row), MSE = 388.

Figure 3.7   Error images for cubic spline interpolation of woman image.

(a) Cubic O-MOMS, PSNR = 21.07 dB, $c_e = 0.5$.

(b) Warped-distance cubic O-MOMS, PSNR = 21.48 dB, $c_e = 0.59$.

(c) Weighted cubic O-MOMS, PSNR = 21.73 dB, $c_e = 0.64$.

(d) Weighted cubic O-MOMS with warping, PSNR = 21.42 dB, $c_e = 0.64$.

(e) Iterative cubic O-MOMS (row-by-row then column-by-column). PSNR = 21.83 dB, $c_e = 0.67$, $I_{av} = 2.75$.

(f) Iterative cubic O-MOMS (column-by-column then row-by-row), PSNR = 21.84 dB, $c_e = 0.68$, $I_{av} = 3.56$.

Figure 3.8   Cubic O-MOMS interpolation of woman image.

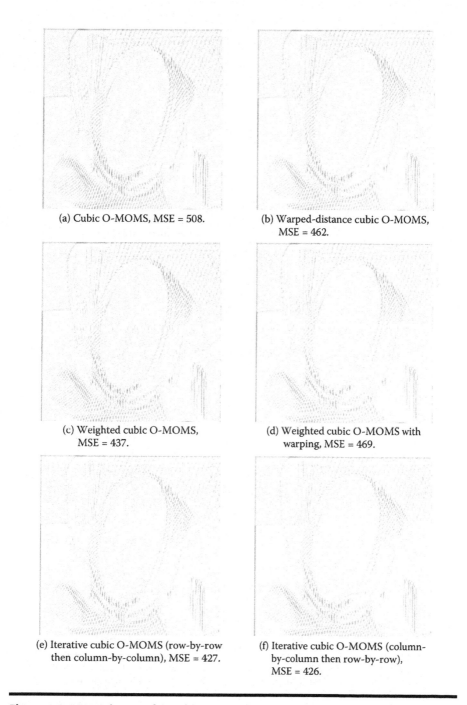

(a) Cubic O-MOMS, MSE = 508.

(b) Warped-distance cubic O-MOMS, MSE = 462.

(c) Weighted cubic O-MOMS, MSE = 437.

(d) Weighted cubic O-MOMS with warping, MSE = 469.

(e) Iterative cubic O-MOMS (row-by-row then column-by-column), MSE = 427.

(f) Iterative cubic O-MOMS (column-by-column then row-by-row), MSE = 426.

Figure 3.9   Error images for cubic O-MOMS interpolation of woman image.

(a) Keys', PSNR = 21.23 dB, c_e = 0.51.

(b) Warped-distance Keys', PSNR = 21.26 dB, c_e = 0.54.

(c) Weighted Keys', PSNR = 21.24 dB, c_e = 0.57.

(d) Weighted Keys' with warping, PSNR = 21.34 dB, c_e = 0.57.

(e) Iterative Keys' (row-by-row then column-by-column), PSNR = 22.15 dB, c_e = 0.69, I_{av} = 3.54, (Adaptive s).

(f) Iterative Keys' (column-by-column then row-by-row), PSNR = 22.15 dB, c_e = 0.69, I_{av} = 3.36, (Adaptive s).

Figure 3.10 **Keys' interpolation of woman image.**

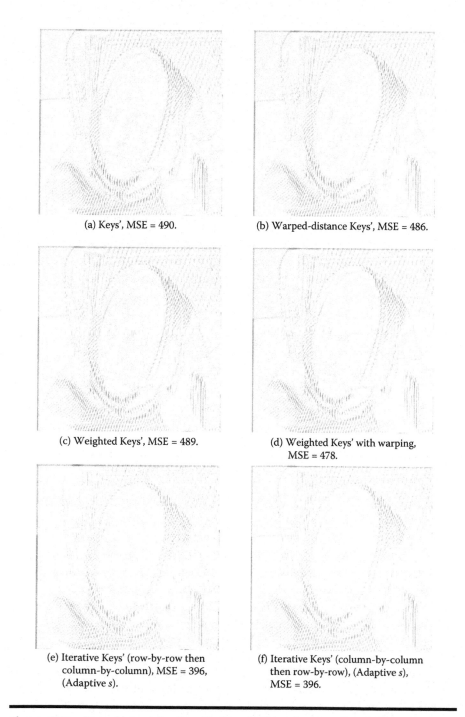

(a) Keys', MSE = 490.

(b) Warped-distance Keys', MSE = 486.

(c) Weighted Keys', MSE = 489.

(d) Weighted Keys' with warping, MSE = 478.

(e) Iterative Keys' (row-by-row then column-by-column), MSE = 396, (Adaptive s).

(f) Iterative Keys' (column-by-column then row-by-row), (Adaptive s), MSE = 396.

Figure 3.11   Error images for Keys' interpolation of woman image.

(a) Iterative Keys' (row-by-row then column-by-column), PSNR = 22.1 dB, $c_e = 0.67$, $I_{av} = 3.54$, (Adaptive α).

(b) Iterative Keys' (column-by-column then row-by-row), PSNR = 22.11 dB, $c_e = 0.67$, $I_{av} = 3.37$, (Adaptive α).

(c) Iterative Keys' (row-by-row then column-by-column), PSNR = 22.15 dB, $c_e = 0.69$, $I_{av} = 3.55$, (Adaptive s and α).

(d) Iterative Keys' (column-by-column then row-by-row), PSNR = 22.16 dB, $c_e = 0.69$, $I_{av} = 3.35$, (Adaptive s and α).

Figure 3.12   Iterative Keys' interpolation of woman image.

The performances of the traditional image interpolation techniques, their warped-distance implementation, and their iterative implementation have been studied for different SNRs on both the woman and test pattern images, and the results are given in Figure 3.25 and Figure 3.26. It is clear from the two figures that the adaptive interpolation algorithm gives the best results for noisy images with different SNRs. In all experiments with the iterative algorithm, we used $\eta_0 = 1$.

Other interpolation experiments have also been carried out on other images, and the PSNR, the c_e computation time for all interpolation techniques, and I_{av} for the iterative image interpolation algorithm are tabulated in Table 3.1 through Table 3.8. These results reveal the ability of the iterative image interpolation algorithm to preserve edges in different cases. The computation time of the iterative algorithm in all experiments is acceptable when quality is of main concern.

For Keys' interpolation, we studied the effect of adaptation of s or α, or both. The highest PSNR values can be obtained by adapting either s or s and α. The adaptation of α takes the least computation time, and the adaptation of s and α takes

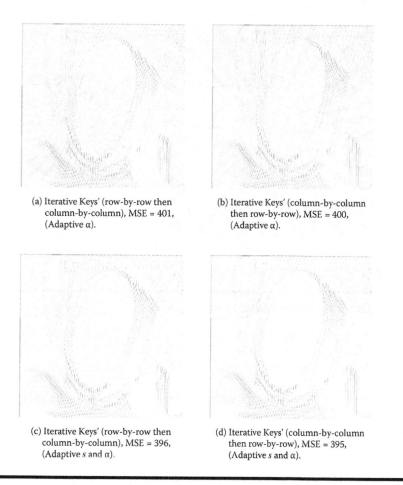

(a) Iterative Keys' (row-by-row then
column-by-column), MSE = 401,
(Adaptive α).

(b) Iterative Keys' (column-by-column
then row-by-row), MSE = 400,
(Adaptive α).

(c) Iterative Keys' (row-by-row then
column-by-column), MSE = 396,
(Adaptive s and α).

(d) Iterative Keys' (column-by-column
then row-by-row), MSE = 395,
(Adaptive s and α).

Figure 3.13   Error images for iterative Keys' interpolation of woman image.

the largest computation time. It is recommended to use the adaptation of s only in a trade-off between the computation cost and required PSNR.

Some experiments have also been carried out to compare the performance of the iterative interpolation algorithm to the commercially available image processing software products such as ACDSee and PhotoPro. Results of these experiments are tabulated in Table 3.9 through Table 3.12. These results reveal the superiority of the iterative algorithm over commercially available software, whether edge-adaptive or not.

From all the results shown, it is clear that the iterative image interpolation algorithm achieves the highest PSNR values and the highest correlation coefficient for edge pixels among all the compared algorithms. It is also clear that the iterative interpolation algorithm has a small average number of iterations per pixel in all experiments (ranging from 1.5 up to 5 iterations per pixel). This is an indication that the required time for applying this algorithm is acceptable.

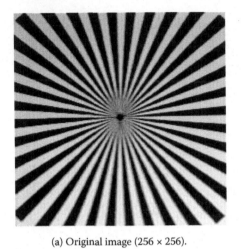

(a) Original image (256 × 256).

(b) LR image (128 × 128), SNR = 25 dB.

Figure 3.14   Test pattern image.

(a) Bilinear, PSNR = 22.78 dB, $c_e = 0.47$.

(b) Warped-distance bilinear, PSNR = 22.6 dB, $c_e = 0.46$.

(c) Weighted bilinear, PSNR = 22.6 dB, $c_e = 0.46$.

(d) Weighted bilinear with warping, PSNR = 22.48 dB, $c_e - 0.45$.

(e) Iterative bilinear (row-by-row then column-by-column), PSNR = 24.82 dB, $c_e = 0.75$, $I_{av} = 3.94$.

(f) Iterative bilinear (column-by-column then row-by-row), PSNR = 24.82 dB, $c_e = 0.75$, $I_{av} = 3.87$.

Figure 3.15   Bilinear interpolation of test pattern image.

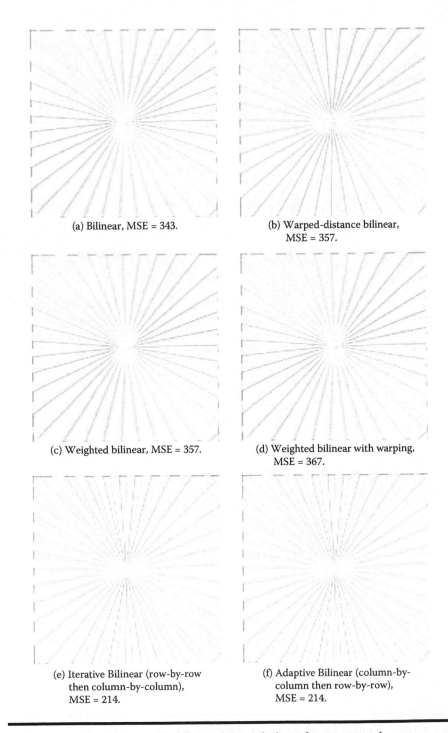

(a) Bilinear, MSE = 343.

(b) Warped-distance bilinear, MSE = 357.

(c) Weighted bilinear, MSE = 357.

(d) Weighted bilinear with warping, MSE = 367.

(e) Iterative Bilinear (row-by-row then column-by-column), MSE = 214.

(f) Adaptive Bilinear (column-by-column then row-by-row), MSE = 214.

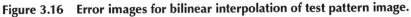

Figure 3.16   Error images for bilinear interpolation of test pattern image.

(a) Cubic spline, PSNR = 23.19 dB, $c_e = 0.46$.

(b) Warped-distance cubic spline, PSNR = 23.77 dB, $c_e = 0.56$.

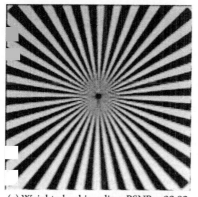 

(c) Weighted cubic spline, PSNR = 23.93 dB, $c_e = 0.58$.

(d) Weighted cubic spline with warping, PSNR = 24.12 dB, $c_e = 0.62$.

(e) Iterative cubic spline (row-by-row then column-by-column), PSNR = 25.39 dB, $c_e = 0.68$, $I_{av} = 4.02$.

(f) Iterative cubic spline (column-by-column then row-by-row), PSNR = 25.36 dB, $c_e = 0.68$, $I_{av} = 4.02$.

Figure 3.17 Cubic spline interpolation of test pattern image.

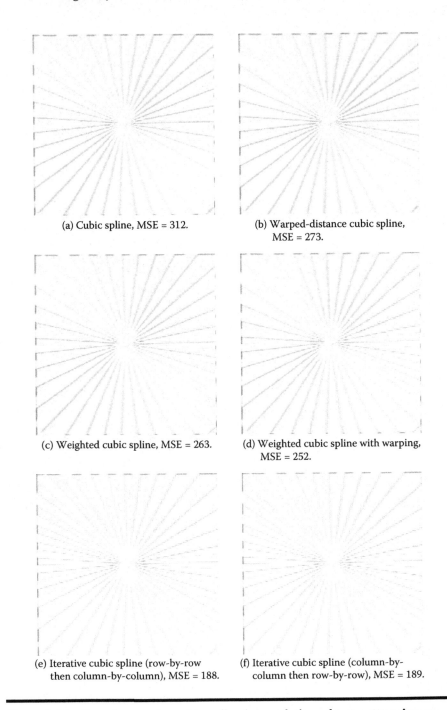

(a) Cubic spline, MSE = 312.

(b) Warped-distance cubic spline, MSE = 273.

(c) Weighted cubic spline, MSE = 263.

(d) Weighted cubic spline with warping, MSE = 252.

(e) Iterative cubic spline (row-by-row then column-by-column), MSE = 188.

(f) Iterative cubic spline (column-by-column then row-by-row), MSE = 189.

Figure 3.18   Error images for cubic spline interpolation of test pattern image.

(a) Cubic O-MOMS, PSNR = 23.19 dB, c_e = 0.47.

(b) Warped-distance cubic O-MOMS, PSNR = 23.01 dB, c_e = 0.47.

(c) Weighted cubic O-MOMS, PSNR = 23.89 dB, c_e = 0.59.

(d) Weighted cubic O-MOMS with warping, PSNR = 24 dB, c_e = 0.62.

(e) Iterative cubic O-MOMS (row-by-row then column-by-column), PSNR = 24.93 dB, c_e = 0.7, I_{av} = 3.28.

(f) Iterative cubic O-MOMS (column-by-column then row-by-row). PSNR = 24.91 dB, c_e = 0.7, I_{av} = 3.26.

Figure 3.19   Cubic O-MOMS interpolation of test pattern image.

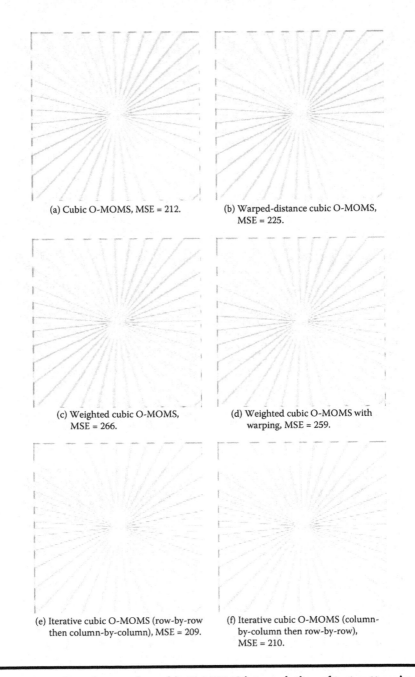

(a) Cubic O-MOMS, MSE = 212.

(b) Warped-distance cubic O-MOMS, MSE = 225.

(c) Weighted cubic O-MOMS, MSE = 266.

(d) Weighted cubic O-MOMS with warping, MSE = 259.

(e) Iterative cubic O-MOMS (row-by-row then column-by-column), MSE = 209.

(f) Iterative cubic O-MOMS (column-by-column then row-by-row), MSE = 210.

Figure 3.20   Error images for cubic O-MOMS interpolation of test pattern image.

(a) Keys', PSNR = 23.12 dB, $c_e = 0.46$.

(b) Warped-distance Keys', PSNR = 22.94 dB, $c_e = 0.46$.

(c) Weighted Keys', PSNR = 22.96 dB, $c_e = 0.46$.

(d) Weighted Keys' with warping, PSNR = 22.77 dB, $c_e = 0.45$.

(e) Iterative Keys' (row-by-row then column-by-column), PSNR = 25.28 dB, $c_e = 0.71$, $I_{av} = 3.96$, (Adaptive s).

(f) Iterative Keys' (column-by-column then row-by-row), PSNR = 25.27 dB, $c_e = 0.71$, $I_{av} = 3.97$, (Adaptive s).

Figure 3.21 Keys' interpolation of test pattern image.

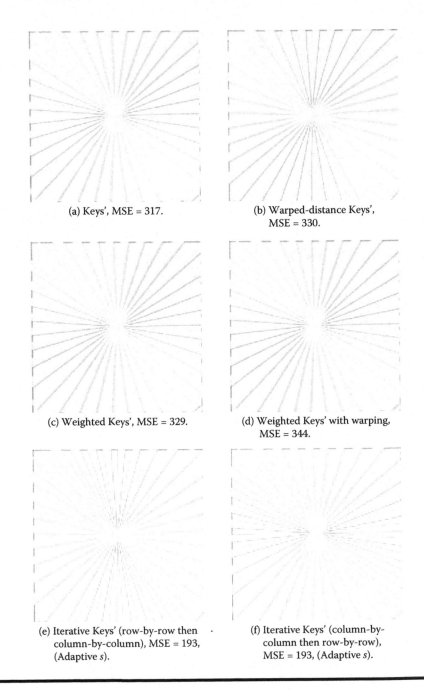

(a) Keys', MSE = 317.

(b) Warped-distance Keys',
MSE = 330.

(c) Weighted Keys', MSE = 329.

(d) Weighted Keys' with warping,
MSE = 344.

(e) Iterative Keys' (row-by-row then
column-by-column), MSE = 193,
(Adaptive *s*).

(f) Iterative Keys' (column-by-
column then row-by-row),
MSE = 193, (Adaptive *s*).

Figure 3.22   Error images for Keys' interpolation of test pattern image.

(a) Iterative Keys' (row-by-row then column-by-column), PSNR = 25.27 dB, $c_e = 0.72$, $I_{av} = 3.99$, (Adaptive α).

(b) Iterative Keys' (column-by-column then row-by-row), PSNR = 25.27 dB, $c_e = 0.72$, $I_{av} = 3.96$, (Adaptive α).

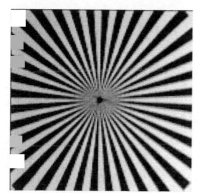

(c) Iterative Keys' (row-by-row then column-by-column), PSNR = 25.31 dB, $c_e = 0.73$, $I_{av} = 3.98$, (Adaptive s and α).

(d) Iterative Keys' (column-by-column then row-by-row). PSNR = 25.32 dB, $c_e = 0.73$, $I_{av} = 3.96$, (Adaptive s and α).

Figure 3.23 Iterative Keys' interpolation of test pattern image.

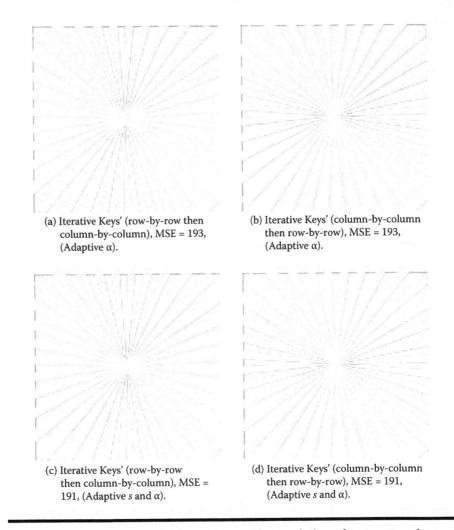

(a) Iterative Keys' (row-by-row then column-by-column), MSE = 193, (Adaptive α).

(b) Iterative Keys' (column-by-column then row-by-row), MSE = 193, (Adaptive α).

(c) Iterative Keys' (row-by-row then column-by-column), MSE = 191, (Adaptive s and α).

(d) Iterative Keys' (column-by-column then row-by-row), MSE = 191, (Adaptive s and α).

Figure 3.24 **Error images for iterative Keys' interpolation of test pattern image.**

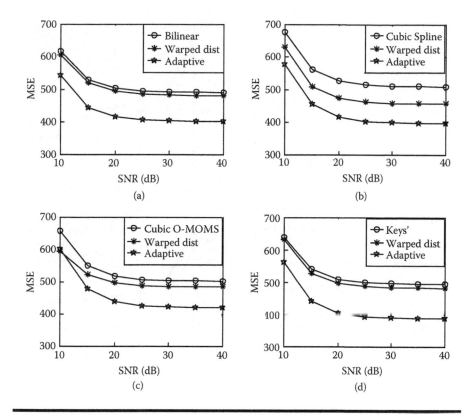

Figure 3.25 MSE versus SNR for woman image. Adaptive interpolation was carried out using iterative algorithm.

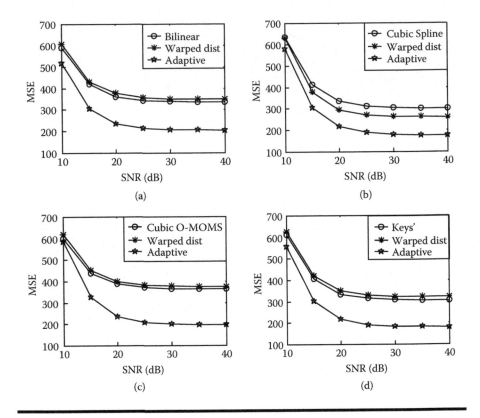

Figure 3.26 **MSE versus SNR for test pattern image. Adaptive interpolation was carried out using iterative algorithm.**

Table 3.1 Bilinear Interpolation Results of Noise-Free Images

Image	Bilinear	Bilinear (Warped-Distance)	Iterative Bilinear
Cameraman (128 × 128)	PSNR = 24.05 $c_e = 0.61$ CPU = 1.1 s	PSNR = 24.12 $c_e = 0.62$ CPU = 2.6 s	PSNR = 25.37 $c_e = 0.77$ $I_{av} = 2.18$, CPU = 10 s
Lenna (64 × 64)	PSNR = 22.76 $c_e = 0.58$ CPU = 0.3 s	PSNR = 22.77 $c_e = 0.58$ CPU = 0.7 s	PSNR = 23.91 $c_e = 0.76$ $I_{av} = 3.24$, CPU = 3.71 s
Mandrill (128 × 128)	PSNR = 18.22 $c_e = 0.47$ CPU = 1.1 s	PSNR = 18.2 $c_e = 0.48$ CPU = 2.6 s	PSNR = 19.01 $c_e = 0.67$ $I_{av} = 4.1$, CPU = 18.8 s
Building (64 × 64)	PSNR = 17.25 $c_e = 0.23$ CPU = 0.3 s	PSNR = 17.28 $c_e = 0.26$ CPU = 0.7 s	PSNR = 18.03 $c_e = 0.65$ $I_{av} = 3.62$, CPU = 4.15 s
Plane (64 × 64)	PSNR = 26.67 $c_e = 0.79$ CPU = 0.3 s	PSNR = 24.88 $c_e = 0.74$ CPU = 0.7 s	PSNR = 28.09 $c_e = 0.88$ $I_{av} = 2.17$, CPU = 2.5 s

Table 3.2 Bilinear Interpolation Results of Noisy Images with SNR = 20 dB

Image	Bilinear	Bilinear (Warped-Distance)	Iterative Bilinear
Cameraman (128 × 128)	PSNR = 23.72 c_e = 0.61 CPU = 1.1 s	PSNR = 23.78 c_e = 0.65 CPU = 2.6 s	PSNR = 24.85 c_e = 0.76 I_{av} = 3.14, CPU = 14.4 s
Lenna (64 × 64)	PSNR = 22.6 c_e = 0.57 CPU = 0.3 s	PSNR = 22.61 c_e = 0.58 CPU = 0.7 s	PSNR = 23.7 c_e = 0.75 I_{av} = 3.65, CPU = 4.18 s
Mandrill (128 × 128)	PSNR = 18.13 c_e = 0.47 CPU = 1.1 s	PSNR = 18.1 c_e = 0.48 CPU = 2.6 s	PSNR = 18.88 c_e = 0.67 I_{av} = 4.38, CPU = 20 s
Building (64 × 64)	PSNR = 17.21 c_e = 0.24 CPU = 0.3 s	PSNR = 17.25 c_e = 0.26 CPU = 0.7 s	PSNR = 17.97 c_e = 0.65 I_{av} = 3.81, CPU = 4.37 s
Plane (64 × 64)	PSNR = 26.05 c_e = 0.78 CPU = 0.3 s	PSNR = 24.47 c_e = 0.73 CPU = 0.7 s	PSNR = 27.16 c_e = 0.87 I_{av} = 3.11, CPU = 3.56 s

Table 3.3 Cubic Spline Interpolation Results of Noise-Free Images

Image	Cubic Spline	Cubic Spline (Warped-Distance)	Iterative Cubic Spline
Cameraman (128 × 128)	PSNR = 24.08 c_e = 0.6 CPU = 3.31 s	PSNR = 24.29 c_e = 0.63 CPU = 7 s	PSNR = 25.3 c_e = 0.74 I_{av} = 2.31, CPU = 22.9 s
Lenna (64 × 64)	PSNR = 22.83 c_e = 0.57 CPU = 0.82 s	PSNR = 22.89 c_e = 0.58 CPU = 1.5 s	PSNR = 24.14 c_e = 0.73 I_{av} = 3.4, CPU = 8.44 s
Mandrill (128 × 128)	PSNR = 18.13 c_e = 0.45 CPU = 3.31 s	PSNR = 18.09 c_e = 0.45 CPU = 7 s	PSNR = 18.95 c_e = 0.62 I_{av} = 4.15, CPU = 10.3 s
Building (64 × 64)	PSNR = 16.97 c_e = 0.4 CPU = 0.82 s	PSNR = 17.02 c_e = 0.39 CPU = 1.5 s	PSNR = 17.61 c_e = 0.57 I_{av} = 3.74, CPU = 9.29 s
Plane (64 × 64)	PSNR = 24.98 c_e = 0.71 CPU = 0.82 s	PSNR = 25.1 c_e = 0.73 CPU = 1.5 s	PSNR = 26.25 c_e = 0.81 I_{av} = 2.56, CPU = 6.35 s

Table 3.4 Cubic Spline Interpolation Results of Noisy Images with SNR = 20 dB

Image	Cubic Spline	Cubic Spline (Warped-Distance)	Iterative Cubic Spline
Cameraman (128 × 128)	PSNR = 23.63 c_e = 0.6 CPU = 3.31 s	PSNR = 23.81 c_e = 0.63 CPU = 7 s	PSNR = 24.62 c_e = 0.73 I_{av} = 3.21, CPU = 31.86 s
Lenna (64 × 64)	PSNR = 22.62 c_e = 0.57 CPU = 0.82 s	PSNR = 22.69 c_e = 0.58 CPU = 1.5 s	PSNR = 23.82 c_e = 0.72 I_{av} = 3.79, CPU = 9.4 s
Mandrill (128 × 128)	PSNR = 17.99 c_e = 0.44 CPU = 3.31 s	PSNR = 17.97 c_e = 0.45 CPU = 7 s	PSNR = 18.77 c_e = 0.61 I_{av} = 4.42, CPU = 43.8 s
Building (64 × 64)	PSNR = 16.92 c_e = 0.4 CPU = 0.82 s	PSNR = 17.97 c_e = 0.4 CPU = 1.5 s	PSNR = 17.54 c_e = 0.57 I_{av} = 4, CPU = 9.93 s
Plane (64 × 64)	PSNR = 24.38 c_e = 0.71 CPU = 0.82 s	PSNR = 24.55 c_e = 0.73 CPU = 1.5 s	PSNR = 25.42 c_e = 0.8 I_{av} = 3.39, CPU = 8.4 s

Table 3.5 Cubic O-MOMS Interpolation Results of Noise-Free Images

Image	Cubic O-MOMS	Cubic O-MOMS (Warped-Distance)	Iterative Cubic O-MOMS
Cameraman (128 × 128)	PSNR = 24.06 $c_e = 0.6$ CPU = 3.31 s	PSNR = 24.25 $c_e = 0.63$ CPU = 7.3 s	PSNR = 24.7 $c_e = 0.74$ $I_{av} = 1.81$, CPU = 19.82 s
Lenna (64 × 64)	PSNR = 22.85 $c_e = 0.57$ CPU = 0.82 s	PSNR = 22.91 $c_e = 0.58$ CPU = 1.9 s	PSNR = 23.74 $c_e = 0.74$ $I_{av} = 2.49$, CPU = 6.82 s
Mandrill (128 × 128)	PSNR = 18.14 $c_e = 0.45$ CPU = 3.31 s	PSNR = 18.11 $c_e = 0.45$ CPU = 7.3 s	PSNR = 18.79 $c_e = 0.61$ $I_{av} = 3.34$, CPU = 36.57s
Building (64 × 64)	PSNR = 16.99 $c_e = 0.41$ CPU = 0.82 s	PSNR = 17.04 $c_e = 0.4$ CPU = 1.9 s	PSNR = 17.41 $c_e = 0.56$ $I_{av} = 3.02$, CPU = 8.27 s
Plane (64 × 64)	PSNR = 24.97 $c_e = 0.71$ CPU = 0.82 s	PSNR = 25.08 $c_e = 0.73$ CPU = 1.9 s	PSNR = 25.43 $c_e = 0.82$ $I_{av} = 2.01$, CPU = 5.5 s

Table 3.6 Cubic O-MOMS Interpolation Results of Noise Images with SNR = 20 dB

Image	Cubic O-MOMS	Cubic O-MOMS (Warped-Distance)	Iterative Cubic O-MOMS
Cameraman (128 × 128)	PSNR = 23.63 $c_e = 0.6$ CPU = 3.31 s	PSNR = 23.76 $c_e = 0.62$ CPU = 7.3 s	PSNR = 24.12 $c_e = 0.74$ $I_{av} = 2.12$, CPU = 23.21 s
Lenna (64 × 64)	PSNR = 22.66 $c_e = 0.57$ CPU = 0.82 s	PSNR = 22.75 $c_e = 0.6$ CPU = 1.9 s	PSNR = 23.48 $c_e = 0.73$ $I_{av} = 2.58$, CPU = 7.06 s
Mandrill (128 × 128)	PSNR = 18.01 $c_e = 0.45$ CPU = 3.31 s	PSNR = 17.97 $c_e = 0.45$ CPU = 7.3 s	PSNR = 18.6 $c_e = 0.6$ $I_{av} = 3.5$, CPU = 38.32 s
Building (64 × 64)	PSNR = 16.94 $c_e = 0.4$ CPU = 0.82 s	PSNR = 16.98 $c_e = 0.4$ CPU = 1.9 s	PSNR = 17.35 $c_e = 0.56$ $I_{av} = 3.03$, CPU = 8.29 s
Plane (64 × 64)	PSNR = 24.44 $c_e = 0.7$ CPU = 0.82 s	PSNR = 24.53 $c_e = 0.72$ CPU = 1.9 s	PSNR = 24.77 $c_e = 0.82$ $I_{av} = 2.26$, CPU = 6.19 s

Table 3.7 Keys' Interpolation Results of Noise-Free Images

Image	Keys'	Keys' (Warped-Distance)	Iterative Keys' (Adaptive s)	Iterative Keys' (Adaptive α)	Iterative Keys' (Adaptive s and α)
Cameraman (128 × 128)	PSNR = 24.24 c_e = 0.6 CPU = 1.5 s	PSNR = 24.33 c_e = 0.62 CPU = 5 s	PSNR = 25.65 c_e = 0.75 I_{av} = 2.17 CPU = 20.52 s	PSNR = 25.53 c_e = 0.75 I_{av} = 2.17 CPU = 19.8 s	PSNR = 25.65 c_e = 0.75 I_{av} = 2.17 CPU = 27.24 s
Lenna (64 × 64)	PSNR = 22.87 c_e = 0.57 CPU = 0.41 s	PSNR = 22.88 c_e = 0.57 CPU = 1.3 s	PSNR = 24.12 c_e = 0.74 I_{av} = 3.24 CPU = 7.66 s	PSNR = 24.05 c_e = 0.73 I_{av} = 3.24 CPU = 7.4 s	PSNR = 24.12 c_e = 0.73 I_{av} = 3.24 CPU = 10.17 s
Mandrill (128 × 128)	PSNR = 18.28 c_e = 0.47 CPU = 1.5 s	PSNR = 18.26 c_e = 0.48 CPU = 5 s	PSNR = 19.17 c_e = 0.66 I_{av} = 4 CPU = 37.8 s	PSNR = 19.12 c_e = 0.65 I_{av} = 4 CPU = 36.52 s	PSNR = 19.17 c_e = 0.66 I_{av} = 4 CPU = 50.2 s

(Continued)

Table 3.7 (Continued) Keys' Interpolation Results of Noise-Free Images

Image	Keys'	Keys' (Warped-Distance)	Iterative Keys' (Adaptive s)	Iterative Keys' (Adaptive α)	Iterative Keys' (Adaptive s and α)
Building (64 × 64)	PSNR = 17.23 c_e = 0.21 CPU = 0.41 s	PSNR = 17.27 c_e = 0.25 CPU = 1.3 s	PSNR = 18.08 c_e = 0.62 I_{av} = 3.52 CPU = 8.32 s	PSNR = 18.02 c_e = 0.6 I_{av} = 3.53 CPU = 8.06 s	PSNR = 18.08 c_e = 0.62 I_{av} = 3.52 CPU = 11.05 s
Plane (64 × 64)	PSNR = 25.03 c_e = 0.7 CPU = 0.41 s	PSNR = 25.07 c_e = 0.74 CPU = 1.3 s	PSNR = 26.4 c_e = 0.82 I_{av} = 2.39 CPU = 5.65 s	PSNR = 26.34 c_e = 0.82 I_{av} = 2.4 CPU = 5.48 s	PSNR = 26.4 c_e = 0.82 I_{av} = 2.39 CPU = 7.5 s

Table 3.8 Keys' Interpolation Results of Noisy Images with SNR = 20 dB

Image	Keys'	Keys' (Warped-Distance)	Iterative Keys' (Adaptive s)	Iterative Keys' (Adaptive α)	Iterative Keys' (Adaptive s and α)
Cameraman (128 × 128)	PSNR = 23.83 $c_e = 0.61$ CPU = 1.5 s	PSNR = 23.92 $c_e = 0.62$ CPU = 5 s	PSNR = 25 $c_e = 0.74$ $I_{av} = 2.88$ CPU = 27.24 s	PSNR = 24.9 $c_e = 0.74$ $I_{av} = 2.88$ CPU = 26.29 s	PSNR = 25 $c_e = 0.75$ $I_{av} = 2.88$ CPU = 36.15 s
Lenna (64 × 64)	PSNR = 22.71 $c_e = 0.56$ CPU = 0.41 s	PSNR = 22.72 $c_e = 0.57$ CPU = 1.3 s	PSNR = 23.88 $c_e = 0.74$ $I_{av} = 3.53$ CPU = 8.35 s	PSNR = 23.9 $c_e = 0.73$ $I_{av} = 3.52$ CPU = 8.03 s	PSNR = 23.85 $c_e = 0.73$ $I_{av} = 3.51$ CPU = 11.01 s
Mandrill (128 × 128)	PSNR = 18.17 $c_e = 0.47$ CPU = 1.5 s	PSNR = 18.14 $c_e = 0.48$ CPU = 5 s	PSNR = 19 $c_e = 0.66$ $I_{av} = 4.25$ CPU = 40.19 s	PSNR = 18.95 $c_e = 0.65$ $I_{av} = 4.22$ CPU = 38.53 s	PSNR = 19.01 $c_e = 0.65$ $I_{av} = 4.24$ CPU = 53.22 s

(Continued)

Table 3.8 (Continued) Keys' Interpolation Results of Noisy Images with SNR = 20 dB

Image	Keys'	Keys' (Warped-Distance)	Iterative Keys' (Adaptive s)	Iterative Keys' (Adaptive α)	Iterative Keys' (Adaptive s and α)
Building (64 × 64)	PSNR = 17.18 c_e = 0.22 CPU = 0.41 s	PSNR = 17.22 c_e = 0.26 CPU = 1.3 s	PSNR = 18.02 c_e = 0.61 I_{av} = 3.73 CPU = 8.82 s	PSNR = 17.96 c_e = 0.59 I_{av} = 3.69 CPU = 8.42 s	PSNR = 18.02 c_e = 0.62 I_{av} = 3.68 CPU = 11.55 s
Plane (64 × 64)	PSNR = 24.55 c_e = 0.71 CPU = 0.41 s	PSNR = 24.64 c_e = 0.74 CPU = 1.3 s	PSNR = 25.58 c_e = 0.81 I_{av} = 3.09 CPU = 7.3 s	PSNR = 25.58 c_e = 0.82 I_{av} = 3.03 CPU = 6.92 s	PSNR = 25.65 c_e = 0.82 I_{av} = 3.02 CPU = 9.48 s

Table 3.9   Interpolation Results of Noise-Free Images Using ACDSee Program

Image	Lanczos	Mitchell	Bell
Cameraman (128 × 128)	PSNR = 21.44 $c_e = 0.44$	PSNR = 21.75 $c_e = 0.47$	PSNR = 21.89 $c_e = 0.49$
Lenna (64 × 64)	PSNR = 20.66 $c_e = 0.41$	PSNR = 20.98 $c_e = 0.44$	PSNR = 21.05 $c_e = 0.48$
Mandrill (128 × 128)	PSNR = 15.74 $c_e = 0.23$	PSNR = 16.18 $c_e = 0.29$	PSNR = 16.49 $c_e = 0.32$
Building (64 × 64)	PSNR = 15.11 $c_e = -0.13$	PSNR = 15.51 $c_e = -0.12$	PSNR = 15.85 $c_e = -0.12$
Plane (64 × 64)	PSNR = 20.84 $c_e = 0.41$	PSNR = 21.26 $c_e = 0.44$	PSNR = 21.49 $c_e = 0.46$

Table 3.10   Interpolation Results of Noisy Images Using ACDSee Program (SNR = 20 dB)

Image	Lanczos	Mitchell	Bell
Cameraman (128 × 128)	PSNR = 21.15 $c_e = 0.44$	PSNR = 21.56 $c_e = 0.47$	PSNR = 21.76 $c_e = 0.49$
Lenna (64 × 64)	PSNR = 20.55 $c_e = 0.41$	PSNR = 20.91 $c_e = 0.44$	PSNR = 21.02 $c_e = 0.48$
Mandrill (128 × 128)	PSNR = 15.66 $c_e = 0.44$	PSNR = 16.14 $c_e = 0.29$	PSNR = 16.46 $c_e = 0.33$
Building (64 × 64)	PSNR = 15.09 $c_e = -0.13$	PSNR = 15.5 $c_e = -0.11$	PSNR = 15.84 $c_e = -0.11$
Plane (64 × 64)	PSNR = 20.6 $c_e = 0.41$	PSNR = 21.1 $c_e = 0.45$	PSNR = 21.38 $c_e = 0.47$

Table 3.11   Interpolation Results of Noise-Free Images Using PhotoPro Program

Image	Lanczos with Unsharp Masking (Edge Adaptive)	Mitchell with Unsharp Masking (Edge Adaptive)	Bell with Unsharp Masking (Edge Adaptive)
Cameraman (128 × 128)	PSNR = 22.66 $c_e = 0.6$	PSNR = 22.87 $c_e = 0.62$	PSNR = 22.85 $c_e = 0.65$
Lenna (64 × 64)	PSNR = 20.49 $c_e = 0.52$	PSNR = 20.69 $c_e = 0.55$	PSNR = 20.65 $c_e = 0.58$
Mandrill (128 × 128)	PSNR = 18.12 $c_e = 0.44$	PSNR = 18.3 $c_e = 0.47$	PSNR = 18.17 $c_e = 0.51$
Building (64 × 64)	PSNR = 16.27 $c_e = -0.2$	PSNR = 16.49 $c_e = -0.19$	PSNR = 16.44 $c_e = -0.13$
Plane (64 × 64)	PSNR = 20.29 $c_e = 0.69$	PSNR = 20.5 $c_e = 0.71$	PSNR = 20.55 $c_e = 0.74$

Table 3.12   Interpolation Results of Noisy Images Using PhotoPro Program (SNR = 20 dB)

Image	Lanczos with Unsharp Masking (Edge Adaptive)	Mitchell with Unsharp Masking (Edge Adaptive)	Bell with Unsharp Masking (Edge Adaptive)
Cameraman (128 × 128)	PSNR = 22.15 $c_e = 0.59$	PSNR = 22.54 $c_e = 0.62$	PSNR = 22.65 $c_e = 0.65$
Lenna (64 × 64)	PSNR = 20.32 $c_e = 0.51$	PSNR = 20.59 $c_e = 0.55$	PSNR = 20.59 $c_e = 0.57$
Mandrill (128 × 128)	PSNR = 17.94 $c_e = 0.44$	PSNR = 18.18 $c_e = 0.46$	PSNR = 18.09 $c_e = 0.5$
Building (64 × 64)	PSNR = 16.22 $c_e = -0.2$	PSNR = 16.45 $c_e = -0.19$	PSNR = 16.42 $c_e = -0.13$
Plane (64 × 64)	PSNR = 19.99 $c_e = 0.68$	PSNR = 20.32 $c_e = 0.7$	PSNR = 20.44 $c_e = 0.73$

Chapter 4

Neural Modeling of Polynomial Image Interpolation

4.1 Introduction

Artificial neural networks (ANNs) are efficient optimization tools that have emerged to simulate the methods of human thinking. These networks have certain structures based on neurons and activation functions. Researchers have shown that the human brain stores information in the form of patterns. ANNs are massively parallel networks designed to do the same task. They are used in modeling to save certain patterns in databases that are further used for pattern matching [37].

In addition, ANNs can be used as powerful tools for modeling general input–output relationships [37]. Applications of ANNs have been reported in areas such as control, telecommunications, remote sensing, pattern recognition, and manufacturing. However, in recent years, they have been used more and more in the area of image processing. They can be taught the formulas of polynomial image interpolation and their variants to carry them out.

4.2 Fundamentals of ANNs

Concepts related to ANNs give enough details to provide some understanding of what can be accomplished with neural network models and how these models are developed for image interpolation. The basic concepts of a neural network will be defined in the following subsections.

4.2.1 Cells

A cell (or unit) is an autonomous processing element that models a neuron. The cell can be thought of as a very simple computer. The purpose of a cell is to receive information from other cells. It performs a relatively simple processing task on the combined information and sends the result to one or more other cells [38].

4.2.2 Layers

A layer is a collection of cells that can be thought of as performing a common function. It is generally assumed that no cell is connected to another cell in the same layer. All ANNs have an input layer and an output layer to interface with the external environment [38]. Each layer has at least one cell. Any cell between the input layer and the output layer is said to be in a hidden layer. ANNs are often classified as single- or multi-layer networks.

4.2.3 Arcs

An arc or connection is a one-way communication link between two cells. A feed-forward network is one in which the information flows from the input layer through some hidden layers to the output layer. A feedback network, by contrast, also permits backward communication [38].

4.2.4 Weights

A weight w_{ij} is a real number that indicates the influence that cell u_j has on cell u_i. The weights are often combined into a weight matrix **W**. The weights may be initialized as zeros or as random numbers and they can be altered during the training phase [37].

4.2.5 Activation Rules

An activation rule is a network rule that applies to all cells, and specifies how outputs from cells are combined into an overall network input to cell u_i. The term net_i in Equation (4.1) indicates this combination. The most common is the weighted-sum rule, wherein a sum is formed by adding the products of the inputs and their corresponding weights as follows [39,40]:

$$net_i = \sum_{j=1}^{n} w_{ij} x_j + \theta_i \tag{4.1}$$

The term x_j represents the input to cell u_i from cell u_j. The w_{ij} represents the associated weight to cell u_i from cell u_j, and θ_i represents a bias associated with cell u_i.

Adding one or more special cell(s) having a constant input of unity often simulates these cell biases. In practice, each bias value θ may be treated as a weight ($w_{io} = -\theta$) with a constant unit input. The vector form of Equation (4.1) can then be written as [40]:

$$\mathbf{net} = \mathbf{WX} \tag{4.2}$$

where **W** is the weight matrix that corresponds to all cells in a given layer and **X** is the input vector that corresponds to inputs in a given layer.

4.2.6 Activation Functions

The activation function $\Lambda(x)$ is used to produce the neuron output signal. Most frequently, the same function is used for all of the cells. Several different functions have been used in neural network simulations [38].

4.2.6.1 Identity Function

$$\Lambda(x) = x \text{ for all } x \tag{4.3}$$

This activation function gives the value of the combined inputs.

4.2.6.2 Step Function

$$\Lambda(x) = \begin{cases} 1 & \text{for } x \geq \theta \\ 0 & \text{for } x < \theta \end{cases} \tag{4.4}$$

The output is zero until the activation reaches a threshold θ; then it jumps to 1.

4.2.6.3 Sigmoid Function

$$\Lambda(x) = \frac{1}{1 + e^{-x}} \tag{4.5}$$

The sigmoid (meaning S-shaped) function [37] is bounded within a specific range of [0,1]. It is often used as an activation function for ANNs in which the desired output values are either binary or in the interval between 0 and 1.

4.2.6.4 Piecewise-Linear Function [38]:

$$\Lambda(x) = \begin{cases} 1 & \text{for } x \geq \dfrac{1}{2} \\ x & \text{for } -\dfrac{1}{2} < x < \dfrac{1}{2} \\ 0 & \text{for } x \leq -\dfrac{1}{2} \end{cases} \tag{4.6}$$

4.2.6.5 Arc Tangent Function [38]:

$$\Lambda(x) = \left(\frac{2}{\pi}\right)\arctan(x) \tag{4.7}$$

4.2.6.6 Hyperbolic Tangent Function [38]:

$$f(x) = \frac{\left(e^x - e^{-x}\right)}{\left(e^x + e^{-x}\right)} \tag{4.8}$$

4.2.7 Outputs

The output O_i of a cell is a function of its activation level.

4.2.8 Learning Rules

The learning rule is applied to all the connections, and it specifies how the weights w_{ij} are updated on the basis of experience. Two types of learning or training algorithms can be distinguished: supervised and unsupervised.

4.2.8.1 Supervised Learning

During supervised training, every problem presented to a neural network is accompanied with an expected response. The network must modify a set of free parameters in order to yield an output that matches the desired response [37].

4.2.8.2 Unsupervised Learning

Unsupervised learning allows a neural network to extract useful information only from the redundancy of the patterns that are presented to it [37].

4.3 Neural Network Structures

There are several types of neural structures, such as multi-layer perceptrons (MLPs), radial basis function (RBF) networks [37], wavelet neural networks, and recurrent neural networks. A brief discussion of these structures follows.

4.3.1 Multi-Layer Perceptrons

MLPs are the most common ANNs. They belong to the feed-forward group in which information processing follows in one direction from input to output. MLPs consist of neurons grouped into an input layer, several hidden layers, and an output layer. Each neuron from a layer is connected with all neurons in the next layer, but there are no connections between same-layer neurons. A standard MLP is shown in Figure 4.1.

4.3.2 Radial Basis Function Networks

ANNs with single hidden layers that use radial basis activation functions for hidden neurons are classified as radial basis function (RBF) networks. They belong to the feed-forward ANNs. Any function Λ that satisfies the property $\Lambda(x) = \Lambda(\|x\|)$ is a radial function. Commonly used radial basis activation functions are the Gaussian and the multi-quadratic functions. The Gaussian function is given by [37]:

$$\Lambda(x) = \exp(-x^2) \tag{4.9}$$

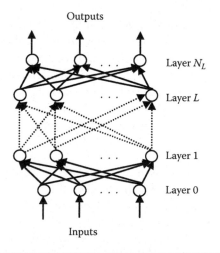

Figure 4.1   MLP neural network.

The multi-quadratic function is given by [38]:

$$\Lambda(x) = \frac{1}{(c^2 + x^2)^\alpha}, \quad \alpha > 0 \tag{4.10}$$

4.3.3 Wavelet Neural Network

A wavelet network is a feed-forward network with one hidden layer. The hidden neuron activation functions are wavelet functions [37]. Let $\Psi(x) = \eta(\|x\|)$, where η is a function of a single variable. A radial function $\Psi(x)$ is a wavelet function, if

$$C_\Psi = (2\pi)^n \int_0^\infty \frac{|\eta(\xi)|^2}{\xi} d\xi < \infty \tag{4.11}$$

4.3.4 Recurrent ANNs

Recurrent neural networks (RNNs) form a class of neural networks in which connections between units form directed cycles. This creates an internal state of a network that allows it to exhibit a dynamic temporal behavior. Unlike feed-forward ANNs, RNNs can use their internal memories to process arbitrary sequences of inputs [37].

4.4 Training Algorithm

Error back propagation (EBP) is one of the most commonly used training algorithms for ANNs. The EBP networks are widely used because of their robustness that allows them to be applied for a wide range of tasks. The EBP algorithm is a way of using known input–output pairs of a target function to find the coefficients that make a certain mapping function to approximate the target function as closely as possible [41].

The EBP is a systematic method for training a multi-layer ANN, the computations of which have been cleverly organized to reduce their complexity in time and memory space. Given P training pairs $\{x_1,d_1,x_2,d_2,\ldots, x_i,d_i\}$, where x_i is an $I \times 1$ vector, d_i is an $I \times 1$ vector, and $i = 1,2,\ldots, I$. The hidden layer has outputs $\mathbf{z}$ of size $J \times 1$, and $j = 1,2,\ldots J$. The output layer has outputs $\mathbf{o}$ of size $I \times 1$. Note that the 0th input of x_i is of value 1, since input vectors have been augmented. The 0th component of z_j also is of value 1, since hidden layer outputs have also been augmented. The EBP algorithm is organized as follows [39–41]:

1. Let I be the number of units in the input layer, as determined by the length of the training input vectors. I is the number of units in the output layer. Now, choose J, the number of units in the hidden layer. As shown in Figure 4.2, the input and hidden layers have an extra unit used for thresholding.
2. Initialize the weights in the network. Each weight should be set randomly to a number between −0.1 and 0.1. Take a learning rate $\eta > 0$ and choose E_{min}, which is the value for target error.

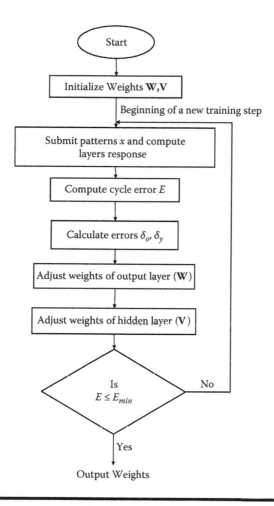

Figure 4.2   Flowchart of EBP algorithm.

3. Initialize the thresholds for activation functions. The values of these thresholds should never change, i.e., $x_o = 1$, $z_o = 1$.
4. Set $q = 1$ and $p = 1$, where q is an integer that denotes the number of training steps and p is an integer denoting the counter within the training cycle.
5. Propagate the activations from the units in the input layer to the units in the hidden layer using the activation function.

$$z_j = \frac{1}{1 + e^{-\left(\sum_{i=0}^{I} v_{ji} x_i\right)}} \quad \text{for } j = 1, 2, \dots, J \qquad (4.12)$$

6. Propagate the activations from the units in the hidden layer to the units in the output layer using the activation function.

$$o_i = \frac{1}{1 + e^{-\left(\sum_{j=0}^{J} w_{ij} z_j\right)}} \quad \text{for } i = 1, 2, \ldots, I \tag{4.13}$$

7. Compute the error value.

$$E_i = \frac{1}{2}(d_i - o_i)^2 \quad \text{for } i = 1, 2, \ldots I \tag{4.14}$$

8. The errors of the units in the output layer are denoted by δ_{oi}. Errors are based on the network actual output o_i and the target output d_i.

$$\delta_{oi} = (d_i - o_i)(1 - o_i) o_i \quad \text{for } i = 1, 2, \ldots, I. \tag{4.15}$$

9. The errors of the units in the hidden layer denoted by δ_{zj} are calculated as follows:

$$\delta_{zj} = z_j (1 - z_j) \sum_{i=1}^{I} \delta_{oi} w_{ij} \quad \text{for } j = 1, 2, \ldots, J. \tag{4.16}$$

10. Adjust the weights between the hidden and output layers as follows:

$$w_{ij} = w_{ij} + \eta \delta_{oi} z_j \quad \text{for } i = 1, 2, \ldots, I \text{ and } j = 1, 2, \ldots, J \tag{4.17}$$

11. Adjust the weights between the input and hidden layers as follows:

$$v_{ji} = v_{ji} + \eta \delta_{zj} x_i \quad \text{for } i = 1, 2, \ldots, I \text{ and } j = 1, 2, \ldots, J \tag{4.18}$$

12. If $p < I$, then $p = p + 1$, $q = q + 1$, and go to step 4. Otherwise go to step 13.
13. If $E < E_{min}$, terminate the training session. Output **W**, **V**, q, and E, otherwise $p = 1$ and initiate a new training cycle by going to step 4.

Figure 4.2 is a flowchart of the EBP algorithm.

4.5 Neural Image Interpolation

The steps of the neural implementation of polynomial interpolation techniques are summarized below [42]. In the training phase:

1. A set of images is interpolated using a certain polynomial interpolation technique.
2. The input points of the region of support used for each pixel estimation are sorted in a vector form (two points for bilinear and four points for Keys' and cubic spline interpolation).
3. All vectors are used as inputs to the neural network with the interpolation results as outputs.
4. The neural network is trained with this available dataset.

In the testing phase:

1. In the image to be interpolated, the input points of the region of support used for each pixel estimation are sorted in a vector form.
2. This vector is used as an input to the neural network to estimate the required pixel value.

A neural network of any fixed size can be used with all types of interpolation. Training of the neural network is accomplished by adjusting its weights using a training algorithm. The training algorithm adapts the weights by attempting to minimize the sum of the squared error between a desired output and the actual outputs of the output neurons.

Each weight in the neural network is adjusted by adding an increment to reduce the error between the actual and desired outputs as rapidly as possible. The adjustment is carried out over several training iterations, until a satisfactorily small value of the error is obtained or a given number of epochs is reached. The EBP algorithm can be used for this task [42].

4.6 Simulation Examples

A number of simulation experiments have been carried out to test neural image interpolation. The images used in these experiments are first down-sampled, and then contaminated by additive white Gaussian noise (AWGN) to simulate the low-resolution (LR) image degradation model. The LR image is then interpolated to its original size and the peak signal-to-noise ratio (PSNR) is estimated between the obtained image and the original image. In the simulation experiments, the traditional image interpolation techniques (bilinear, Keys', and cubic spline) with their adaptive variants and the neural implementation of these techniques have been tested on the LR 128 × 128 noisy woman image. The results of these experiments are shown in Figures 4.3 through 4.8 and Tables 4.1 through 4.3.

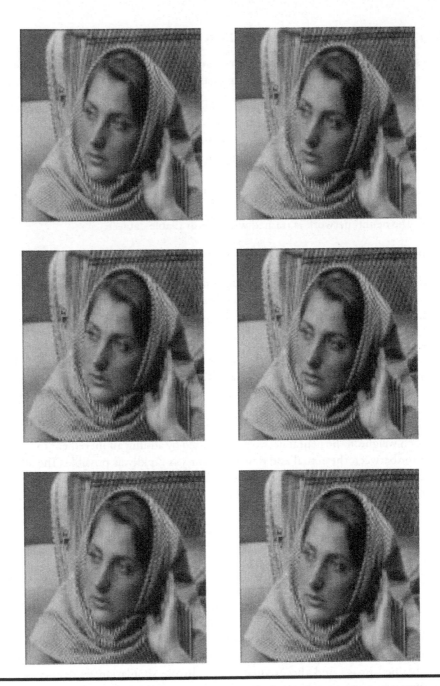

Figure 4.3 Bilinear interpolation with neural implementation. (a) Bilinear interpolation, noise free. (b) Neural implementation, 5 neurons. (c) Neural implementation, 2 neurons. (d) Bilinear interpolation, SNR = 20 dB. (e) Neural implementation, 5 neurons. (f) Neural implementation, 2 neurons.

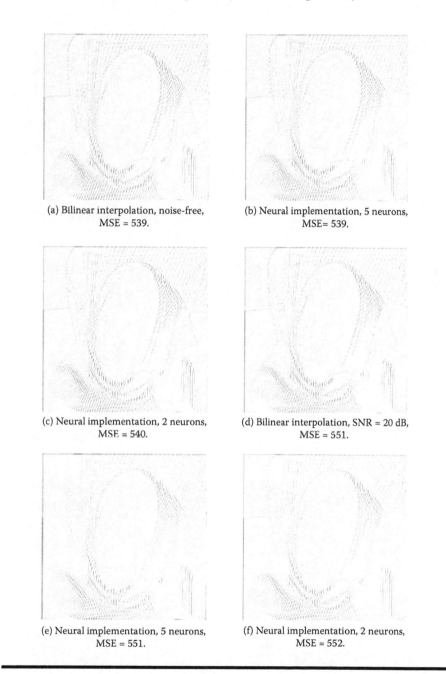

(a) Bilinear interpolation, noise-free, MSE = 539.

(b) Neural implementation, 5 neurons, MSE= 539.

(c) Neural implementation, 2 neurons, MSE = 540.

(d) Bilinear interpolation, SNR = 20 dB, MSE = 551.

(e) Neural implementation, 5 neurons, MSE = 551.

(f) Neural implementation, 2 neurons, MSE = 552.

Figure 4.4   Error images for bilinear interpolation with neural implementation. (a) Bilinear interpolation, noise free, MSE = 539. (b) Neural implementation, 5 neurons, MSE = 539. (c) Neural implementation, 2 neurons, MSE = 540. (d) Bilinear interpolation, SNR = 20 dB, MSE = 551. (e) Neural implementation, 5 neurons, MSE = 551. (f) Neural implementation, 2 neurons, MSE = 552.

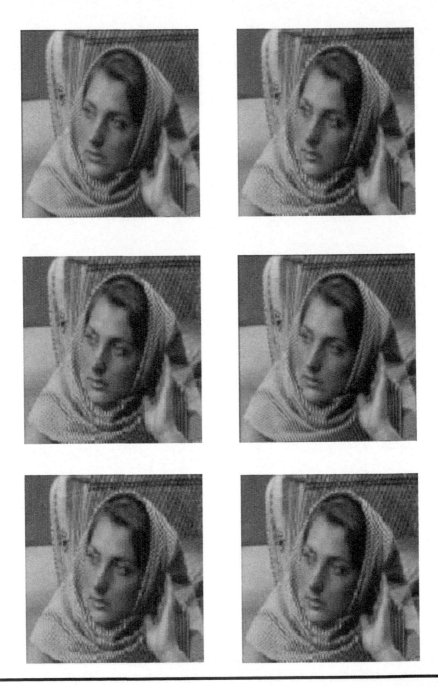

Figure 4.5 Keys' interpolation with neural implementation. (a) Keys' interpolation, noise free. (b) Neural implementation, 5 neurons. (c) Neural implementation, 2 neurons. (d) Keys' interpolation, SNR = 20 dB. (e) Neural implementation, 5 neurons. (f) Neural implementation, 2 neurons.

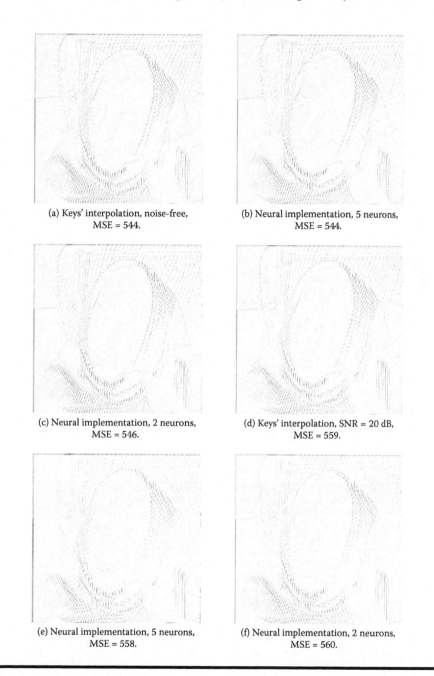

(a) Keys' interpolation, noise-free, MSE = 544.

(b) Neural implementation, 5 neurons, MSE = 544.

(c) Neural implementation, 2 neurons, MSE = 546.

(d) Keys' interpolation, SNR = 20 dB, MSE = 559.

(e) Neural implementation, 5 neurons, MSE = 558.

(f) Neural implementation, 2 neurons, MSE = 560.

Figure 4.6 Error images for Keys' interpolation with neural implementation. (a) Keys' interpolation, noise free, MSE = 544. (b) Neural implementation, 5 neurons, MSE = 544. (c) Neural implementation, 2 neurons, MSE = 546. (d) Keys' interpolation, SNR = 20 dB, MSE = 559. (e) Neural implementation, 5 neurons, MSE = 558. (f) Neural implementation, 2 neurons, MSE = 560.

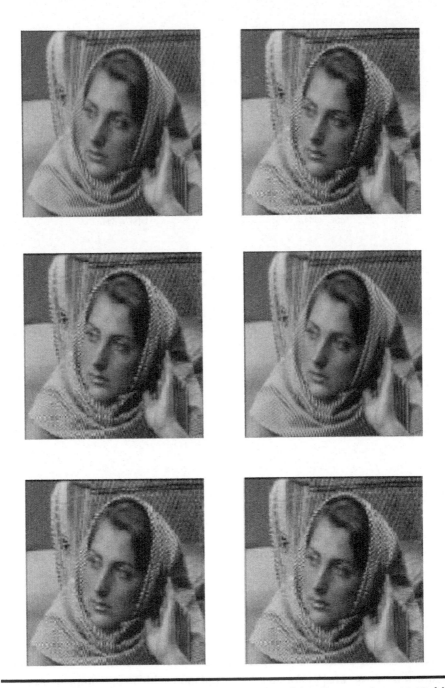

Figure 4.7 Cubic spline interpolation with neural implementation. (a) Cubic spline interpolation, noise free. (b) Neural implementation, 2 neurons. (c) Neural implementation, 2 neurons. (d) Cubic spline interpolation, SNR = 20 dB. (d) Neural implementation, 5 neurons. (f) Neural implementation, 2 neurons.

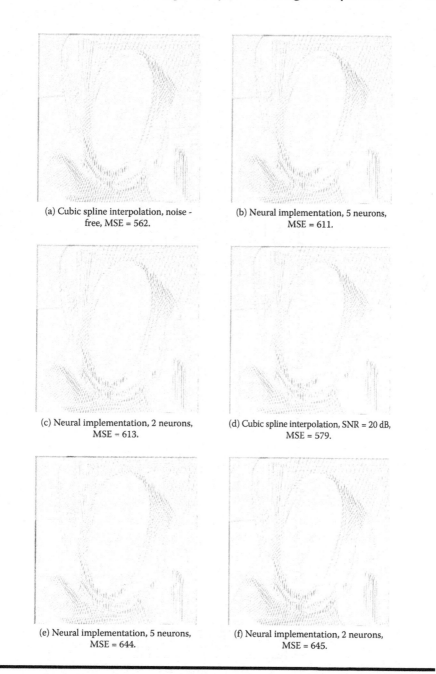

(a) Cubic spline interpolation, noise - free, MSE = 562.

(b) Neural implementation, 5 neurons, MSE = 611.

(c) Neural implementation, 2 neurons, MSE = 613.

(d) Cubic spline interpolation, SNR = 20 dB, MSE = 579.

(e) Neural implementation, 5 neurons, MSE = 644.

(f) Neural implementation, 2 neurons, MSE = 645.

Figure 4.8 Error images for cubic spline interpolation with neural implementation. (a) Cubic spline interpolation, noise free, MSE = 562. (b) Neural implementation, 5 neurons, MSE = 611. (c) Neural implementation, 2 neurons, MSE = 613. (d) Cubic spline interpolation, SNR = 20 dB, MSE = 579. (e) Neural implementation, 5 neurons, MSE = 644. (f) Neural implementation, 2 neurons, MSE = 645.

Table 4.1 PSNR Values for Space-Invariant Interpolation of Woman Image

Type	Noise-Free			SNR = 20 dB		
	Traditional Method	Neural Method		Traditional Method	Neural method	
		5 Neurons	2 Neurons		5 Neurons	2 Neurons
Bilinear	20.8139	20.8139	20.7997	20.7175	20.7166	20.7048
Keys'	20.7725	20.7725	20.7551	20.6571	20.6581	20.6425
Cubic Spline	20.6265	20.2655	20.2506	20.5019	20.0419	20.0346

Table 4.2 PSNR Values for Warped-Distance Interpolation of Woman Image

Type	Noise-Free			SNR = 20 dB		
	Traditional Method	*Neural Method*		*Traditional Method*	*Neural Method*	
		5 Neurons	*2 Neurons*		*5 Neurons*	*2 Neurons*
Bilinear	20.8943	20.8896	20.8891	20.7883	20.7916	20.7750
Keys'	20.8876	20.8861	20.8845	20.7661	20.7670	20.7487
Cubic Spline	21.0870	20.4146	20.4047	20.9297	20.1807	20.1816

Table 4.3 PSNR Values for Weighted Interpolation of Woman Image

Type	Noise-Free			SNR = 20 dB		
	Traditional Method	Neural Method		Traditional Method	Neural Method	
		5 Neurons	2 Neurons		5 Neurons	2 Neurons
Bilinear	20.8943	20.8907	20.8891	20.7801	20.7840	20.7809
Keys'	20.8673	20.8657	20.8637	20.7441	20.7387	20.7513
Cubic Spline	21.2874	20.8547	20.8348	21.1210	20.6755	20.6451

Experiments were carried out with a neural network having five neurons in a single hidden layer, which gives better results close to traditional techniques, or a neural network having two neurons to reduce complexity. As the number of neurons is increased, the complexity is increased and the accuracy is also slightly increased so there is no need for the complexity to be increased. The experiments proved that the neural implementation of polynomial interpolation techniques has a very close performance to the traditional implementation with a fixed computational complexity.

Chapter 5

Color Image Interpolation

5.1 Introduction

Color image interpolation or demosaicking is a process by which a raw image generated by a digital still camera with the help of a color filter array (CFA) is converted to a full color image by estimating the missing color components of each pixel from its neighbors. In order to reduce the cost of digital still cameras used to capture color images, each camera uses a single charge-coupled device (CCD) instead of three CCDs [43,44]. The CFA consists of a set of spectrally selective filters arranged in a certain interleaved pattern so that each sensor pixel samples only one of the three primary color components. In digital still cameras, color images are encoded by the CFA pattern, and a subsequent interpolation process produces full color images.

The Bayer CFA pattern is the one most frequently used [43,44]. Since there is only one color element available at each pixel position, the two missing color elements must be estimated from the adjacent pixels. In the Bayer CFA pattern, half of the pixels are assigned to the green (G) channel and the other half are divided between the red (R) and blue (B) channels that represent the chrominance signal. This chapter presents only the concepts of color image interpolation in order to use this idea robustly in pattern recognition applications as will be explained in Chapter 6.

5.2 Color Filter Arrays

Early color films were made by placing color filters over black-and-white films. Starch particles dyed red, green, and blue were used. An alternative technology used yellow stripes crossed by blue and red stripes, so that half the area was devoted to yellow, and the area between the yellow stripes was half red and half blue. Modern color film is made by adding successive layers sensitive to each color. The principle of combining layers of filters that successively remove components of the incoming light with successively sensitive layers of photographic emulsion is illustrated by Figure 5.1.

The Bayer pattern shown in Figure 5.2 is used in most digital cameras. A Bayer pattern uses the subtractive primaries. Half of the total number of pixels are green, while a quarter of the total is assigned to red and a quarter to blue. In order to obtain this color information, the color image sensor is covered with a red, green, or blue filter in a repeating pattern. This pattern or sequence of filters can vary, but the widely-adopted Bayer pattern invented at Kodak is a repeating 2 × 2 arrangement [43,44].

When the image sensor is read out, line by line, the pixel sequence is *GRGRGR*, etc., and then the alternate line sequence is *BGBGBG*, etc. This output is called sequential *RGB* (or *sRGB*). Since each pixel has been made sensitive only to one color, the overall sensitivity of a color image sensor is lower than that of a monochrome

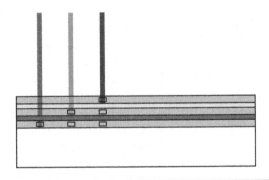

Figure 5.1   Layers of filters used for color imaging.

G	R	G	R
B	G	B	G
G	R	G	R
B	G	B	G

Figure 5.2   Pattern of Bayer CFA.

sensor. As a result, monochrome sensors are better for low-light applications such as security cameras.

White balance and color correction are processing operations performed to ensure proper color fidelity in a captured digital camera image. The sensors do not detect light exactly as the human eye does, and so some processing or correction of the detected image is necessary to ensure that the final image realistically represents the colors of the original scene.

5.2.1 White Balance

The first step in processing raw pixel data is to perform a white balance operation. A white object will have equal values of reflectivity for each primary color, i.e., $R = G = B$. An image of a white object can be captured and its histogram analyzed. The color channel that has the highest level is set as the target mean and the remaining two channels are increased with a gain multiplier to match. For example, if the green channel has the highest mean, a gain a is applied to the red channel and a gain b is applied to the blue channel.

The white balance varies based on the color lighting source applied to the object and the amount of each color component within it. A full-color natural scene can also be processed in the same fashion. This gray world method assumes that the world is gray and the distribution of primary colors is equal. The white patch method attempts to locate the objects that are truly white within the scene by assuming the white pixels are also the brightest ($I = R + G + B$). Then, only the top percentage intensity pixels are included in the calculation of means.

5.2.2 Bayer Interpolation

To convert an image from the Bayer format to an RGB-per-pixel format, we need to interpolate the two missing color values at each pixel position. As suggested in Reference 45, R and B values are interpolated linearly from the nearest neighbors of the same color. There are four possible cases, as shown in Figure 5.3. When interpolating the missing values of R and B on a green pixel, as in Figure 5.3(a) and (b), we take the average values of the two nearest neighbors of the same color. For example, in Figure 5.3(a), the value for the blue component at the central G pixel will be the average of the blue pixels above and below the G pixel, while the value for the red component will be the average of the two red pixels to the left and right of the G pixel.

Figure 5.3(c) shows a case in which the value of the blue component is to be estimated at the central R pixel. In such case, we take the average of the four nearest blue pixels surrounding the R pixel. Similarly, to determine the value of the red component at the central B pixel in Figure 5.3(d), we take the average of the four nearest red pixels surrounding the B pixel. The green component is adaptively interpolated from a pair of nearest neighbors. To illustrate the procedure, consider two

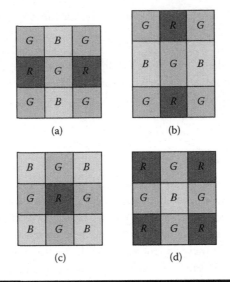

Figure 5.3   Four possible cases for interpolating *R* and *B* components.

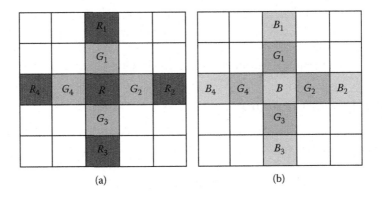

Figure 5.4   Two possible cases for interpolating *G* component.

possible cases in Figure 5.4. In Figure 5.4(a), the value of the green component is to be interpolated at an *R* pixel as follows [45]:

$$G(R) = \begin{cases} (G_1 + G_3)/2, & \text{if } |R_1 - R_3| < |R_2 - R_4| \\ (G_2 + G_4)/2, & \text{if } |R_1 - R_3| > |R_2 - R_4| \\ (G_1 + G_2 + G_3 + G_4)/4, & \text{if } |R_1 - R_3| = |R_2 - R_4| \end{cases} \quad (5.1)$$

We take into account the correlation in the red component to interpolate the green component. If the difference between R_1 and R_3 is smaller than the difference between R_2 and R_4, indicating that the correlation is stronger in the vertical direction, we use the average of the vertical neighbors G_1 and G_3 to interpolate the required value. If the horizontal correlation is larger, we use horizontal neighbors. If neither direction dominates the correlation, we use all four neighbors. Similarly, from Figure 5.4(b), we will have

$$G(B) = \begin{cases} (G_1 + G_3)/2 & \text{if } |B_1 - B_3| < |B_2 - B_4| \\ (G_2 + G_4)/2 & \text{if } |B_1 - B_3| > |B_2 - B_4| \\ (G_1 + G_2 + G_3 + G_4)/4 & \text{if } |B_1 - B_3| = |B_2 - B_4| \end{cases} \tag{5.2}$$

If the speed of execution is of major concern, one can safely use simple linear interpolation of the green component from the four nearest neighbors, without any adaptation as follows:

$$G = (G_1 + G_2 + G_3 + G_4)/4 \tag{5.3}$$

According to Reference 45, this method of interpolation executes twice as fast as the adaptive method and achieves only slightly worse performance on real images. For even fast updates, only two of the four green values are averaged. However, this method displays false color on edges or zipper artifacts.

5.3 Linear Interpolation with Laplacian Second Order Correction

A gradient-based algorithm has been developed for color image interpolation to enhance the visual quality of color images [45]. The missing green color components at the pixel locations containing a red or blue color component are first estimated. In Figure (5.5), the objective is to estimate the missing green component G_5 at pixel location B_5. The horizontal and vertical gradients at this pixel location are defined as follows [45]:

$$\delta H = |G_4 - G_6| + |(B_5 - B_3) - (B_7 - B_5)| \tag{5.4}$$

$$\delta V = |G_2 - G_8| + |(B_5 - B_1) - (B_9 - B_5)| \tag{5.5}$$

Intuitively, δH and δV above can be considered as combinations of the luminance gradient and the chrominance gradient. In the expression of δH above, as an example, the first term $|G_4 - G_6|$ is the first order difference of the neighboring green values considered to be the luminance gradient. The second term $|(B_5 - B_3) - (B_7 - B_5)|$ is

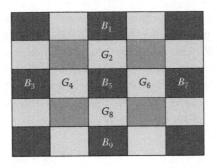

Figure 5.5 Reference Bayer CFA sample.

the second order derivative of the neighboring blue values considered as the chrominance gradient. Using these two gradients, the missing green component G_5 at the pixel location B_5 is estimated as [45]:

$$
G_5 = \begin{cases}
(G_4 + G_6)/2 + (-B_3 + 2B_5 - B_7)/4 & \text{if } \delta H < \delta V \\
(G_2 + G_8)/2 + (-B_1 + 2B_5 - B_9)/4 & \text{if } \delta H > \delta V \\
(G_2 + G_4 + G_6 + G_8)/4 + (-B_1 - B_3 + 4B_5 - B_7 - B_9)/8 & \text{if } \delta H = \delta V
\end{cases}
$$

$$(5.6)$$

The interpolation step for G_5 has two parts. The first part is the linear average of the neighboring green values, and the second part can be considered a second order correction term based on the neighboring blue (red) values. The missing red (or blue) color components are estimated at every pixel location after the estimation of the missing green components at every pixel location using information of the reconstructed green channel. For interpolation of red and blue pixel values, the gradient value is utilized.

5.4 Adaptive Color Image Interpolation

To determine the missing green values in Bayer CFA images, the conventional method of Laplacian second order correction can be used with a refinement. This method consists of three steps. The first is an initial determination of interpolation direction using an edge indicator. The second step applies the interpolation algorithm to the result of the initial interpolation direction decision. The third step is the interpolation of the green color channel.

To refine the interpolation directions, we modify the edge directions by comparing the adjacent edge indicator. We define a numeric map of edge directions at missing pixel locations as shown in Figure 5.6. The values of $Eh_{i,j}$ and $Ev_{i,j}$ at indices i,j are defined as [45]:

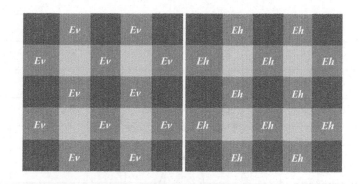

Figure 5.6   Numeric edge maps.

$$Eh_{i,j} = \begin{cases} 1 & \delta H_{i,j} < \delta V_{i,j} \\ 0.5 & \delta H_{i,j} = \delta V_{i,j} \\ 0 & otherwise \end{cases} \tag{5.7}$$

$$Ev_{i,j} = \begin{cases} 1 & \delta H_{i,j} > \delta V_{i,j} \\ 0.5 & \delta H_{i,j} = \delta V_{i,j} \\ 0 & otherwise \end{cases} \tag{5.8}$$

Eh is set to 1 if the edge direction is horizontal, and *Ev* is set to 1 if the direction is vertical. For an edge that shows equal horizontal and vertical gradients, the values for *Eh* and *Ev* are set to 0.5. A compensation process is performed on the initial edge direction to obtain a more correct edge direction. The compensated edge direction [horizontal (*Eh'*) and vertical (*Ev'*)] values can be defined as [45]:

$$Eh'_{i,j} = Eh_{i,j-2} + Eh_{i,j} + Eh_{i,j+2} \tag{5.9}$$

$$Ev'_{i,j} = Ev_{i,j-2} + Ev_{i,j} + Ev_{i,j+2} \tag{5.10}$$

We can interpolate a missing pixel of the green channel along the edge direction as follows [45]:

$$G_5 = \begin{cases} (G_4 + G_6)/2 + (-B_3 + 2B_5 - B_7)/4 & \text{if } Eh'_7 \geq 2.5 \\ (G_2 + G_8)/2 + (-B_1 + 2B_5 - B_9)/4 & \text{if } Ev'_7 \geq 2.5 \\ (G_2 + G_4 + G_6 + G_8)/4 + (-B_1 - B_3 + 4B_5 - B_7 - B_9)/8 & Otherwise \end{cases} \tag{5.11}$$

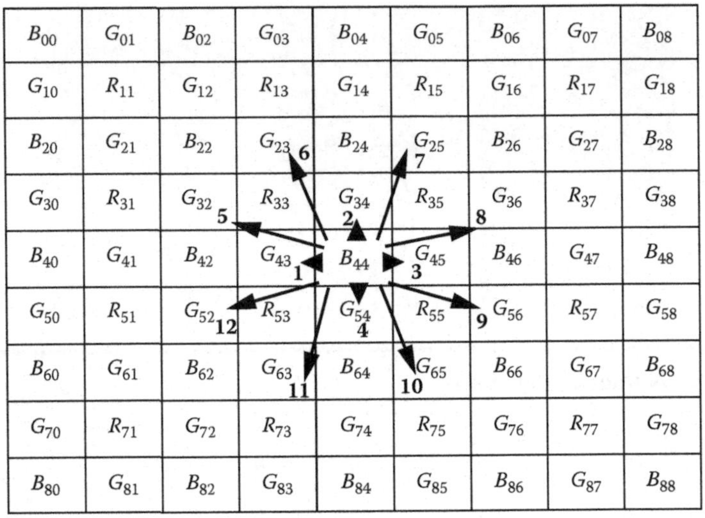

B_{00}	G_{01}	B_{02}	G_{03}	B_{04}	G_{05}	B_{06}	G_{07}	B_{08}
G_{10}	R_{11}	G_{12}	R_{13}	G_{14}	R_{15}	G_{16}	R_{17}	G_{18}
B_{20}	G_{21}	B_{22}	G_{23} 6	B_{24}	G_{25} 7	B_{26}	G_{27}	B_{28}
G_{30}	R_{31}	G_{32} 5	R_{33}	G_{34} 2	R_{35}	G_{36} 8	R_{37}	G_{38}
B_{40}	G_{41}	B_{42}	G_{43} 1	B_{44}	G_{45} 3	B_{46}	G_{47}	B_{48}
G_{50}	R_{51}	G_{52} 12	R_{53}	G_{54} 4	R_{55}	G_{56} 9	R_{57}	G_{58}
B_{60}	G_{61}	B_{62}	G_{63} 11	B_{64}	G_{65} 10	B_{66}	G_{67}	B_{68}
G_{70}	R_{71}	G_{72}	R_{73}	G_{74}	R_{75}	G_{76}	R_{77}	G_{78}
B_{80}	G_{81}	B_{82}	G_{83}	B_{84}	G_{85}	B_{86}	G_{87}	B_{88}

Figure 5.7   Interpolation directions.

Table 5.1   Positions of Nearby Samples for Interpolation

N	1	2	3	4
v_n	−1	−1	1	1
h_n	−1	1	1	−1

To increase the edge sensitivity, 12 samples can be considered for interpolation as illustrated in Figure 5.7 [45]. For notational convenience, we number the directions from 1 to 12 as shown in the figure. For a given sample, the nearest sample in the same color plane is either 2 or 2.5 pixels away in each edge direction.

Here we consider only the four nearest samples of the same color in the diagonal directions, because there are no similar samples in the other eight directions as shown in Figure 5.7. The vertical and horizontal positions of these four samples (indexed from $n = 1$ to $n = 4$) relative to the sample to be interpolated are listed in Table 5.1. Let $E_n(i, j)$ be the edge indicator for direction n.

$$E_n(i, j) = |P(i + v_n, j + h_n) - P(i - v_n, j - h_n)| + |P(i + 2\, v_n, j + 2h_n) - P(i, j)| \quad (5.12)$$

$P(i,j)$ denotes the sample at the position (i, j). Both h_n and v_n are listed in Table (5.1). They denote the horizontal and vertical positions, respectively, of a nearest sample relative to the sample to be interpolated.

$$w_n(i,j) = \left(\sum_{n=1}^{4} \frac{1}{1+E_n(i,j)} \right)^{-1} \cdot \frac{1}{1+E_n(i,j)} \qquad (5.13)$$

The missing red value of the blue sample $B(i,j)$ is obtained by [45]:

$$R(i,j) = G(i,j) - \sum_{n=1}^{4} w_n(i,j) \cdot K_{r,n}(i+v_n, j+h_n) \qquad (5.14)$$

where

$$K_{r,n}(i,j) = G(i+v_n, j+h_n) - R(i+v_n, j+h_n) \qquad (5.15)$$

is the color difference along direction n.
The missing red value of a green sample $G(i,j)$ is obtained by [45]:

$$R(i,j) = G(i,j) - \sum_{n=1}^{4} w'_n(i,j) \cdot K_{r,n}(i+v'_n, j+h'_n) \qquad (5.16)$$

where

$$w'_n(i,j) = \left(\sum_{n=1}^{12} \frac{1}{1+E'_n(i,j)} \right)^{-1} \cdot \frac{1}{1+E'_n(i,j)} \qquad (5.17)$$

The edge indicator $E'_n(i,j)$ is computed as:

$$E'_n(i,j) = \gamma_n E_n(i,j) \qquad (5.18)$$

The stochastic adjustment γ_n is set to:

$$\gamma_n = \begin{cases} 1, & 1 \le n \le 4 \\ 0.6 & 5 \le n \le 12 \end{cases} \qquad (5.19)$$

The interpolation of the remaining blue pixels is performed similarly.

Post processing can be performed to reduce the visible artifacts in the interpolated images. The green values of red and blue pixels are interpolated again to reduce aliasing and false color. The missing green value $G(i,j)$ of a blue sample $B(i,j)$ is determined by [45]:

$$G(i, j) = B(i, j) - \sum_{n=1}^{12} w'_n(i, j) \cdot K_{b,n}(i + v'_n, j + h'_n) \qquad (5.20)$$

where

$$K_{b,n}(i, j) = G(i + v'_n, j + h'_n) - B(i + v'_n, j + h'_n) \qquad (5.21)$$

is the color difference along direction n. Finally, the red and blue values are interpolated again as described above.

Chapter 6

Image Interpolation for Pattern Recognition

6.1 Introduction

A new application for image interpolation is in pattern recognition. The basis of pattern recognition is storing a large number of images in databases to compare them to any new one to decide whether it is already contained in the database. The main problem with databases is the huge storage size. This problem can be solved by making a database containing down-sampled versions of the original images in smaller sizes than the sizes of the original images. The original images could be retrieved when needed through image interpolation. The idea of saving down-sampled images rather than the original ones can be used with different types of databases such as fingerprint and landmine databases with gray-scale images and flower and retinal databases with color images.

Fingerprints are biometric signs that can be utilized for identification and authentication purposes in biometric systems. Among all the biometric indicators, fingerprints have one of the highest levels of reliability [46]. The main reasons for the popularity of fingerprint-based identification are the uniqueness and permanence of fingerprints. It has been claimed that no two individuals, including identical twins, have the exact same fingerprints. Another claim is that the fingerprint of an individual does not change throughout his lifetime, with the exception of a significant injury to the finger that creates a permanent scar [47].

Fingerprints are graphical patterns of locally parallel ridges and valleys with well defined orientations on the surfaces of fingertips. Ridges are the lines on the tips of fingers. A unique pattern of lines can form a loop, whorl, or arch pattern.

Figure 6.1   Examples of minutiae points. (a) Ridge ending. (b) Bifurcation.

Valleys are the spaces or gaps on either side of a ridge. The most important features in fingerprints are called the minutiae, usually defined as ridge endings and ridge bifurcations. A ridge bifurcation is the point where a ridge forks into a branch ridge [48]. Examples of minutiae are shown in Figure 6.1.

A full fingerprint normally contains 50 to 80 minutiae. A partial fingerprint may contain fewer than 20 minutiae. According to the Federal Bureau of Investigation, it suffices to identify a fingerprint by matching 12 minutiae, but it has been reported that in most cases, 8 matched minutiae are enough.

Landmines are small explosive objects buried under the earth surface. They are classified as anti-personnel (AP) landmines, used to kill persons, and anti-tank (AT) landmines, used to attack vehicles and their occupants. There are about 100 million buried landmines covering more than 200,000 square kilometers of the world surface and affecting about 70 countries [49].

Many obstacles are faced in removing buried landmines, such as the lack of maps or information about the mines or where they are buried, changes of locations due to climatic and physical factors, the large variety of types of AP and AT landmines, and the high costs of removal. The production cost of a landmine is very low (perhaps $3 per mine), but the detection and removal costs are high (more than $1,000 per mine).

Several techniques have been proposed for demining (detecting and clearing) buried mines. One of the promising detection techniques is the acoustic-to-seismic (A/S) technique that detects landmines by vibrating them with acoustic or seismic waves that are generated and received by acoustic or seismic transducers, respectively [49–53]. An acoustic or a seismic wave is excited by a source at a known position. It travels through the soil to interact with underground objects. This method involves a transmission system that generates an acoustic or seismic wave into the area under test and a receiving system that senses changes in the mechanical properties of the area. The mechanical changes produce acoustic or seismic images for the areas under test.

Detection of landmines from acoustic images can be accomplished by traditional shape-based techniques. These techniques begin by intensity thresholding of images to reject the dark background, and then the detection of landmines based on their dimensions [49–53]. The drawbacks of these techniques lie in their inability to detect landmines with small dimensions in the images and their inability to reject background noise in the intensity thresholding approach. In most of these

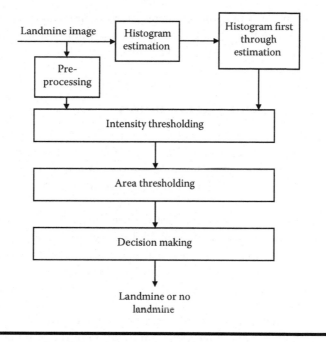

Figure 6.2   Steps of traditional landmine detection technique.

techniques, pre-processing steps like morphological operations are required. In spite of the ability of morphological operations to smooth objects in images, they can disclose small clutter shapes to give false alarms.

The detection of a landmine can be achieved by the intensity thresholding of the landmine image with a certain threshold to remove the dark background. This process helps eliminate the background and reveals objects. After that, an area thresholding process is performed based on the areas of the expected objects to remove unwanted small-area clutters. A pre-processing step such as the use of morphological operations may be required prior to thresholding. Figure 6.2 shows a block diagram of a traditional landmine detection technique.

Traditional landmine detection based on geometrical information has several limitations. The intensity thresholding may not remove all unwanted noise or clutter in images. The area thresholding process requires certain thresholds that may differ from image to image leading to either false alarms or missed landmines. Without pre-processing, the detection probability of landmines is about 90%. Morphological pre-processing can increase this detection probability to about 97% with a small false-alarm probability, but all these probabilities are in the absence of any type of noise [49–53]. The issue of noise effect is rarely studied by researchers in this area.

A robust cepstral approach has been presented in the literature for both fingerprint recognition and landmine detection [54–58]. This approach can be used successfully with noise. This approach is presented in this chapter and its sensitivity to synthetic pixels obtained through interpolation is studied. The objective of this study is to achieve success in reducing database sizes.

6.2 Cepstral Pattern Recognition

A pattern recognition system is composed of feature extraction and feature matching for the purpose of classification [58]. It operates in two modes: training and recognition. Both modes include feature extraction. A feature extractor converts a digital one-dimensional (1-D) signal into a sequence of numerical descriptors called feature vectors.

Several feature extraction techniques are used in signal recognition systems such as linear prediction coefficients (LPCs), linear predictive cepstral coefficients (LPCCs), perceptual linear predictive (PLP) analysis, and mel frequency cepstral coefficients (MFCCs). Classification is a process that has two phases: image modeling and pattern matching. For successful classification, each image is modeled using a set of data samples in the training mode, from which a set of feature vectors is generated and saved in a database. Features are extracted from the training data, essentially stripping away all unnecessary information in the training samples and leaving only the characteristic information with which image models can be constructed [54–58].

When a sample of data from some unknown pattern arrives, pattern matching techniques are used to map the features from the input sample to a model corresponding to a known pattern [54–58]. In this chapter, pattern recognition is based on MFCCs. Figure 6.3 is a diagram of the steps of the cepstral recognition system. The steps of the feature extraction process from an image can be summarized as follows:

1. The image is converted to a 1-D signal.
2. The obtained 1-D signal can be used in time domain or in another discrete transform domain. The discrete cosine transform (DCT), discrete sine transform (DST), and discrete wavelet transform (DWT) can be used for this purpose.
3. MFCCs and polynomial shape coefficients are extracted from either the 1-D signal, the discrete transform of the signal, or both of them.

6.3 Feature Extraction

The concept of feature extraction using MFCCs is widely known in speaker identification. It contributes to the goal of identifying speakers based on low level properties. Pattern images after conversion to a 1-D signal are treated in this chapter like speech signals.

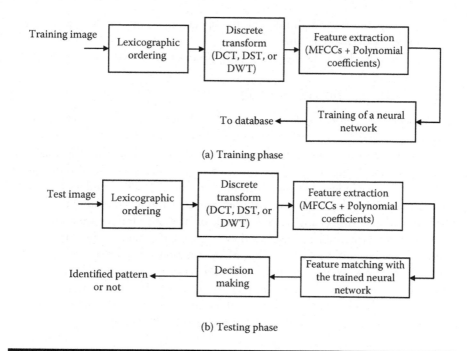

(a) Training phase

(b) Testing phase

Figure 6.3 **Diagram of cepstral pattern recognition system.**

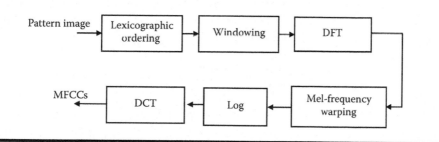

Figure 6.4 **Extraction of MFCCs from image.**

6.3.1 Extraction of MFCCs

MFCCs are coefficients used to represent signal distribution. They are commonly used as features in speaker identification systems. They are derived through cepstral analysis. The difference between the cepstrum and the mel frequency cepstrum is that the frequency bands are divided using a mel scale [54–58]. The steps of MFCC extraction from an image are shown in Figure 6.4 and summarized as follows:

1. Lexicographic ordering
2. Slicing of the obtained signal with a pre-determined window
3. Performing the fast Fourier transform (FFT) on the sliced signal

4. Mapping the log amplitudes of the spectrum onto the mel scale using triangular overlapping filters

5. Performing a DCT on the mel log amplitudes to yield MFCCs

6.3.1.1 Framing and Windowing

After lexicographic ordering of an image into a 1-D signal, we obtain a slowly time-varying signal partitioned into short time segments called frames. To make the frame parameters vary smoothly, there is normally a 50% overlap between each two adjacent frames. Windowing is performed on each frame with one of the popular signal processing windows like the Hamming window [54–58]. Windowing is often applied to increase the continuity between adjacent frames and smooth the end points such that abrupt changes between ends of successive frames are minimized.

As a frame is multiplied by a window, most of the data at the edges of the frame become insignificant, causing loss of information. An approach to tackle this problem is to allow overlapping in the sections between frames; this allows adjacent frames to include portions of data in the current frame. This means that the edges of the current frame are included as the center data of adjacent frames. Typically, around 50% overlapping is sufficient to embrace the lost information.

6.3.1.2 Discrete Fourier Transform

Fourier analysis provides a way of analyzing the spectral properties of a given signal in the frequency domain. The Fourier transform converts a discrete signal $s(n)$ from a time domain into a frequency domain with the equation [54–58]:

$$S(k) = \sum_{n=0}^{N-1} s(n)e^{-j2\pi nk/N}, \qquad 0 \le k \le N-1 \tag{6.1}$$

where $n = 0,1,\ldots,N-1$ and N is the number of samples in the signal $s(n)$; k represents the discrete frequency index and j is equal to $\sqrt{-1}$. The result of the discrete Fourier transform (DFT) is a complex valued sequence of length N. The inverse discrete Fourier transform (IDFT) is defined as:

$$s(n) = \frac{1}{N} \sum_{k=0}^{N-1} S(k)e^{j2\pi nk/N}, \qquad 0 \le n \le N-1 \tag{6.2}$$

6.3.1.3 Mel Filter Bank

The main advantage of the MFCCs method is that it uses mel frequency scaling, which is defined as [54–58]:

$$Mel(f) = 2595 \log\left(1 + \frac{f}{700}\right) \tag{6.3}$$

where *mel* is the mel frequency scale and *f* is the frequency on the linear frequency scale.

6.3.1.4 Discrete Cosine Transform

The final stage involves performing a discrete cosine transform (DCT) on the log of the mel spectrum. If the power at output of the mth mel filter is $\tilde{S}(m)$, the MFCCs are given as [54–58]:

$$c_g = \sqrt{\frac{2}{N_f}} \sum_{m=1}^{N_f} \log(\tilde{S}(m)) \cos\left(\frac{g\pi}{N_f}(m-0.5)\right) \tag{6.4}$$

where $g = 0,1,\ldots N_f - 1$, N_f is the number of mel filters, and c_g is the gth MFCC. The number of the resulting MFCCs is chosen between 12 and 20, since most of the signal information is represented by the first few coefficients. The 0th coefficient represents the mean value of the input signal.

6.3.2 Polynomial Coefficients

The MFCCs are sensitive to mismatches between training and testing data, and they are also pattern dependent. Polynomial coefficients are added to the MFCCs to solve this problem. They help in increasing the similarity between the training and testing data [54–58]. The importance of these coefficients arises from the fact that they can preserve valuable information (mean, slope, and curvature) about the shape of the time function of each cepstral coefficient of the training and testing data.

To calculate the polynomial coefficients, the time waveforms of the cepstral coefficients are expanded by orthogonal polynomials. The following two orthogonal polynomials can be used [54–58]:

$$P_1(i) = i-5 \tag{6.5}$$

$$P_2(i) = i^2 - 10i + 55/3 \tag{6.6}$$

To model the shapes of the MFCC time functions, a nine-element window at each MFCC is used. Based on this windowing assumption, the polynomial coefficients can be calculated as follows [54–58]:

$$a_g(t) = \frac{\displaystyle\sum_{i=1}^{9} P_1(i) c_g(t+i+1)}{\displaystyle\sum_{i=1}^{9} P_1^2(i)} \tag{6.7}$$

$$b_g(t) = \frac{\sum_{i=1}^{9} P_2(i)c_g(t+i+1)}{\sum_{i=1}^{9} P_2^2(i)} \tag{6.8}$$

where $a_g(t)$ and $b_g(t)$ are the slope and the curvature of the MFCC time functions at each c_g. The vectors containing all c_g, a_g, and b_g are concatenated together for each frame of the signal corresponding to an image.

6.4 Feature Extraction from Discrete Transforms

Discrete transforms can be used for extraction of robust MFCCs. The DWT, the DCT, and the DST have been investigated in the literature for this purpose [54–58].

6.4.1 Discrete Wavelet Transform

It is known that the DFT considers the analysis of a signal separately in the time and frequency domains and does not provide temporal information about frequencies. Although the DFT may be a good tool for analyzing a stationary signal, non-stationary signals need another tool. When analyzing a non-stationary signal, in addition to the frequency content of the signal, we need to know how the frequency content of the signal changes with time.

To overcome this deficiency, a modified transform called the short time Fourier transform (STFT) has been adopted because it allows the representation of the signal in both time and frequency domains through time windowing functions. The window length determines a constant time and frequency resolution. The main idea behind the STFT is to have localization in time domain. A drawback of the STFT is its small and fixed window, so that the STFT cannot capture the rapid changes in the signal. Moreover, it does not give information about the slowly changing parts of the signal.

Wavelet analysis provides an exciting alternative to Fourier analysis for signal processing. Wavelet transform allows a variable time–frequency resolution that leads to locality in both the time and frequency domains. The locality of the transform of a signal is important in two ways for pattern recognition. First, different parts of the signal may convey different amounts of information. Second, when the signal is corrupted by local noise in a time and/or frequency domain, the noise affects only a few coefficients if the coefficients represent local information in the time and frequency domains.

In fact, the wavelet transform is a mathematical operation used to divide a given signal into different sub-bands of different scales to study each scale separately.

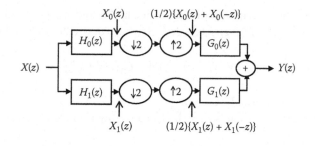

Figure 6.5 **Two-band decomposition–reconstruction wavelet filter bank.**

The idea of the discrete wavelet transform (DWT) is to represent a signal as a series of approximations (low-pass version) and details (high-pass version) at different resolutions. The signal is low-pass filtered to give an approximation signal and high-pass filtered to give a detail signal. Both filters can be used to model the signal. The wavelet decomposition and reconstruction process is illustrated in Figure 6.5.

The multi-level DWT can be regarded as equivalent to filtering the signal with a bank of band-pass filters whose impulse responses are all approximately given by scaled versions of a mother wavelet. The scaling factor between adjacent filters is usually 2, leading to octave bandwidths and center frequencies that are one octave apart [59–75]. The outputs of the filters are usually maximally decimated so that the number of DWT output samples equals the number of input samples, and thus no redundancy occurs in this transform.

The art of finding a good wavelet lies in the design of the set of filters, H_0, H_1, G_0, and G_1 to achieve various trade-offs between spatial and frequency domain characteristics, while satisfying the perfect reconstruction (PR) condition [35]. In Figure 6.5, the process of decimation and interpolation by 2 at the outputs of H_0 and H_1 effectively sets all odd samples of these signals to zero. For the low-pass branch, this is equivalent to multiplying $x_0(n)$ by $\frac{1}{2}(1+(-1)^n)$. Hence, $X_0(z)$ is converted to $\frac{1}{2}\{X_0(z)+X_0(-z)\}$. Similarly, $X_1(z)$ is converted to $\frac{1}{2}\{X_1(z)+X_1(-z)\}$. As a result, the expression for $Y(z)$ is given by [73]:

$$Y(z) = \frac{1}{2}\{X_0(z)+X_0(-z)\}G_0(z)+\frac{1}{2}\{X_1(z)+X_1(-z)\}G_1(z)$$

$$= \frac{1}{2}X(z)\{H_0(z)G_0(z)+H_1(z)G_1(z)\} \tag{6.9}$$

$$+\frac{1}{2}X(-z)\{H_0(-z)G_0(z)+H_1(-z)G_1(z)\}$$

The first PR condition requires aliasing cancellation and forces the above term in $X(-z)$ to be zero. Hence $\{H_0(-z)G_0(z) + H_1(-z)G_1(z)\} = 0$, which can be achieved if [73]:

$$H_1(z) = z^{-k}G_0(-z) \quad \text{and} \quad G_1(z) = z^k H_0(-z) \tag{6.10}$$

where k must be odd (usually $k = \pm 1$).

The second PR condition is that the transfer function from $X(z)$ to $Y(z)$ should be unity:

$$\{H_0(z)G_0(z) + H_1(z)G_1(z)\} = 2 \tag{6.11}$$

If we define a product $P(z) = H_0(z)G_0(z)$ and substitute from Equation (6.10) into (6.11), then the PR condition becomes [73]:

$$H_0(z)G_0(z) + H_1(z)G_1(z) = P(z) + P(-z) = 2 \tag{6.12}$$

This needs to be true for all z, and since the odd powers of z in $P(z)$ cancel with those in $P(-z)$, it requires that $p_0 = 1$ and $p_n = 0$ for all n, even and non-zero. The polynomial $P(z)$ should be a zero-phase polynomial to minimize distortion. In general, $P(z)$ is of the following form [73]:

$$P(z) = \cdots + p_5 z^5 + p_3 z^3 + p_1 z + 1 + p_1 z^{-1} + p_3 z^{-3} + p_5 z^{-5} + \cdots \tag{6.13}$$

The design method for the PR filters can be summarized in the following steps [73]:

1. Choose $p_1, p_3, p_5, \cdots$ to give a zero-phase polynomial $P(z)$ with good characteristics.
2. Factorize $P(z)$ into $H_0(z)$ and $G_0(z)$ with similar low-pass frequency response.
3. Calculate $H_1(z)$ and $G_1(z)$ from $H_0(z)$ and $G_0(z)$.

To simplify this procedure, we can use the following relation:

$$P(z) = P_t(Z) = 1 + P_{t,1}Z + P_{t,3}Z^3 + P_{t,5}Z^5 + \cdots \tag{6.14}$$

where

$$Z = \frac{1}{2}\left(z + z^{-1}\right) \tag{6.15}$$

The Haar wavelet is the simplest type of wavelet. In the discrete form, Haar wavelets are related to a mathematical operation called the Haar transform that serves as a prototype for all other wavelet transforms [73]. Like all wavelet transforms, the Haar decomposes a discrete signal into two sub-signals of half its length. One is a running average or trend; the other is a running difference or fluctuation. This uses the simplest possible $P_t(Z)$ with a single zero at $Z = -1$. It is represented as follows [73]:

$$P_t(Z) = 1 + Z \quad \text{and} \quad Z = \frac{1}{2}\left(z + z^{-1}\right) \tag{6.16}$$

Thus

$$P(z) = \frac{1}{2}\left(z + 2 + z^{-1}\right) = \frac{1}{2}(z+1)\left(1 + z^{-1}\right) = G_0(z)H_0(z) \tag{6.17}$$

We can find $H_0(z)$ and $G_0(z)$ as follows:

$$H_0(z) = \frac{1}{2}\left(1 + z^{-1}\right) \tag{6.18}$$

$$G_0(z) = (z + 1) \tag{6.19}$$

Using Equation (6.17) with $k = 1$:

$$G_1(z) = zH_0(-z) = \frac{1}{2}z\left(1 - z^{-1}\right) = \frac{1}{2}(z - 1)$$

$$H_1(z) = z^{-1}G_0(-z) = z^{-1}(-z + 1) = (z^{-1} - 1) \tag{6.20}$$

The two outputs of $H_0(z)$ and $H_1(z)$ are concatenated to form a single vector of the same length as the original signal. The features are extracted from this vector and used for pattern recognition.

6.4.2 Discrete Cosine Transform

The discrete cosine transform (DCT) is a 1-D transform with excellent energy compaction property. For a signal $x(n)$, the DCT is represented by [73]:

$$X(k) = \alpha(k) \sum_{n=0}^{N-1} x(n) \cos\left(\frac{\pi(2n+1)k}{2N}\right), \quad k = 0,1,2......, N-1 \tag{6.21}$$

where

$$\alpha(0) = \sqrt{\frac{1}{N}}, \quad \alpha(k) = \sqrt{\frac{2}{N}}$$

The inverse discrete cosine transform (IDCT) is given by:

$$x(n) = \sum_{K=0}^{N-1} \alpha(k) X(k) \cos\left(\frac{\pi(2n+1)}{2N}\right), \quad n = 0,1,2,\ldots, N-1 \qquad (6.22)$$

The features are extracted from $X(k)$ and used for pattern recognition.

6.4.3 Discrete Sine Transform

The discrete sine transform (DST) is another triangular transform that has common properties with the DCT. The mathematical representation of the DST is given by [73]:

$$X(k) = \sum_{n=0}^{N-1} x(n) \sin\left(\frac{\pi}{N+1}(n-1)(k+1)\right), \quad k = 0, \ldots, N-1 \qquad (6.23)$$

The features are extracted from $X(k)$ and used for pattern recognition.

6.5 Feature Matching Using ANNs

The classification step in the cepstral recognition method is in fact a matching process of the features of a new image and the features saved in the database. Neural networks are widely used for feature matching. Multi-layer perceptrons (MLPs) consisting of an input layer, one or more hidden layers, and an output layer can be used for this purpose [54–58].

For an input vector $\mathbf{X}$, the neural network output vector $\mathbf{Y}$ can be obtained according to the following matrix equation [54–58]:

$$\mathbf{Y} = \mathbf{W}_2 * F(\mathbf{W}_1 * \mathbf{X} + \mathbf{B}_1) + \mathbf{B}_2 \qquad (6.24)$$

where $\mathbf{W}_1$ and $\mathbf{W}_2$ are the weight matrices between the input and the hidden layer and between the hidden and the output layers, respectively, and $\mathbf{B}_1$ and $\mathbf{B}_2$ are bias matrices for the hidden and the output layers, respectively.

6.6 Simulation Examples

Several experiments have been carried out to test the performance of the cepstral pattern recognition method after several types of image interpolation to retain the original image sizes. Time and transform domains have been used for feature extraction after interpolation. The degradations considered are additive white Gaussian noise (AWGN), impulsive noise, and speckle noise with and without blurring.

In the training phase of the cepstral recognition approach, a database is composed first. Twenty images have been used to generate this database. The MFCCs and polynomial coefficients have been estimated to form the feature vectors of the database. In the testing phase, similar features to those used in the training have been extracted from the degraded pattern images after interpolation with different interpolation methods: bilinear, Keys', warped-distance bilinear, warped-distance Keys', neural bilinear, and neural Keys'.

The features used in all experiments are 13 MFCCs and 26 polynomial coefficients forming feature vectors of 39 coefficients for each frame of the pattern signals. Seven methods have been used for feature extraction. In the first method, the MFCCs and polynomial coefficients have been extracted from the time domain signals only. In the second method, the features have been extracted from the DWTs of these signals. In the third method, the features have been extracted from both the original signals and the DWTs of these signals and concatenated together. In the fourth method, the features have been extracted from the DCTs of the time domain signals. In the fifth method, the features have been extracted from both the original signals and the DCTs of these signals and concatenated together. In the sixth method, the features have been extracted from the DSTs of the time domain signals. In the last method, the features have been extracted from both the original signals and the DSTs of these signals and concatenated together.

Samples of the fingerprint and landmine images used in creating the databases are shown in Figures 6.6 and 6.7, respectively. The results of the experiments on interpolated fingerprint images appear in Figures 6.8 through 6.49. The results of experiments on interpolated landmine images are given in Figures 6.50 through 6.91. As seen from these figures, it is clear that feature extractions from transform domains like the DCT and DST are not sensitive to synthetic pixels obtained through all types of interpolations. This is attributed to the averaging effect of the transformation equation that cancels the effects of pixel synthesis errors. The results also reveal that neural interpolation is feasible with the used pattern recognition approach.

In color image interpolation, we have used flower and retinal image databases. Samples of these databases are shown in Figure 6.92 and Figure 6.93, respectively. A comparison study of the flower and retinal images was conducted for all these extraction methods for the above mentioned degradation cases. Features were extracted from the intensity components of the color images after red, green, blue (RGB) to intensity, hue, and saturation (IHS) transformation. Figures 6.94 through 6.99 show results for flower images and Figures 6.100 through 6.103 show results for retinal images.

From this comparison for both the flower and retinal images, it is clear that the features extracted from both the original signals and the DCTs of these signals achieve the highest recognition rates. As in gray scale images, this is attributed to the energy compaction property of the DCT that enables accurate feature extraction from the first frames of the 1-D signals after the DCT that can characterize each signal. Features extracted from DCTs are less sensitive to synthetic pixels obtained from color image interpolation.

Figure 6.6   Samples of fingerprint images used in training phase.

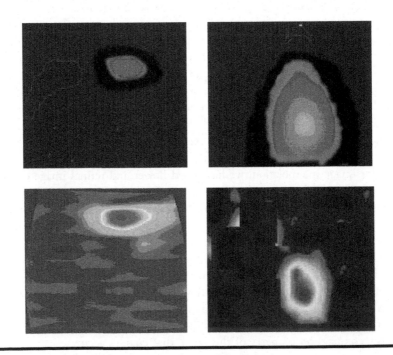

Figure 6.7   Samples of acoustic landmine images.

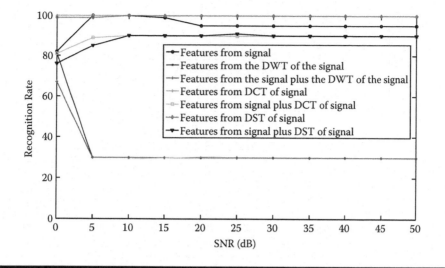

Figure 6.8 **Recognition rate versus signal-to-noise ratio for different feature extraction methods from fingerprint images contaminated by AWGN.**

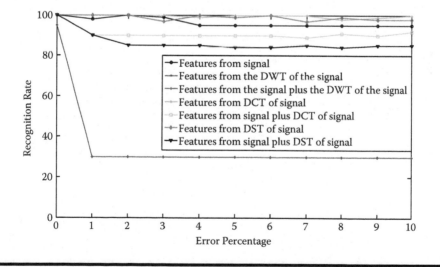

Figure 6.9 **Recognition rate versus error percentage for different feature extraction methods from fingerprint images contaminated by impulsive noise.**

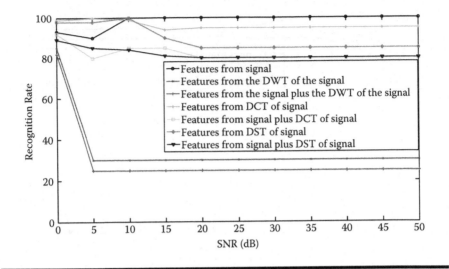

Figure 6.10 **Recognition rate versus signal-to-noise ratio for different feature extraction methods from blurred fingerprint images contaminated by AWGN.**

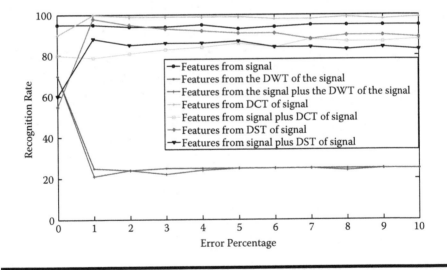

Figure 6.11 **Recognition rate versus error percentage for different feature extraction methods from blurred fingerprint images contaminated by impulsive noise.**

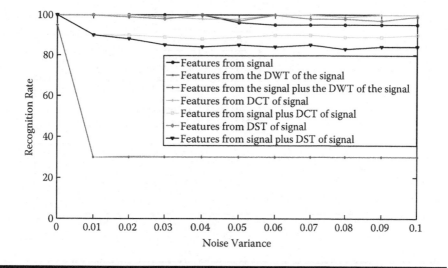

Figure 6.12 Recognition rate versus noise variance for different feature extraction methods from fingerprint images contaminated by speckle noise.

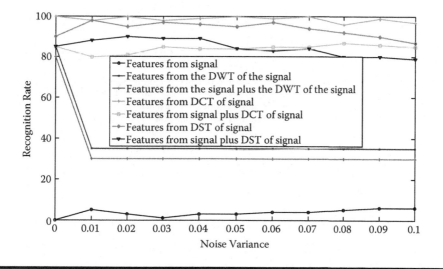

Figure 6.13 Recognition rate versus noise variance for different feature extraction methods from blurred fingerprint images contaminated by speckle noise.

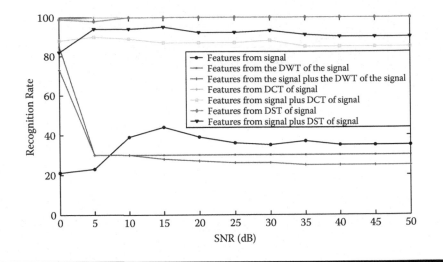

Figure 6.14 **Recognition rate versus signal-to-noise ratio for different feature extraction methods from fingerprint images contaminated by AWGN and interpolated with bilinear method.**

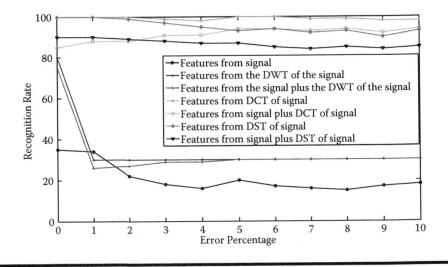

Figure 6.15 **Recognition rate versus error percentage for different feature extraction methods from fingerprint images contaminated by impulsive noise and interpolated with bilinear method.**

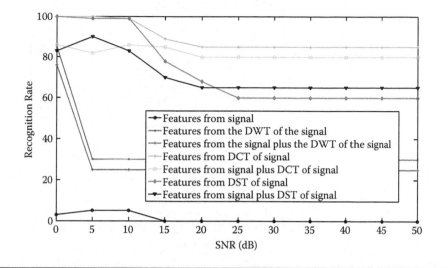

Figure 6.16 Recognition rate versus signal-to-noise ratio for different feature extraction methods from blurred fingerprint images contaminated by AWGN and interpolated with bilinear method.

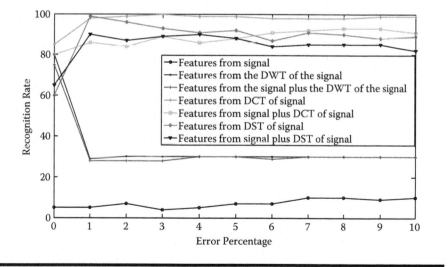

Figure 6.17 Recognition rate versus error percentage for different feature extraction methods from blurred fingerprint images contaminated by impulsive noise and interpolated with bilinear method.

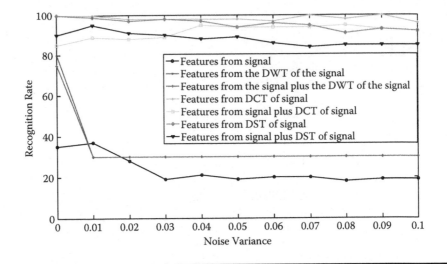

Figure 6.18 Recognition rate versus noise variance for different feature extraction methods from fingerprint images contaminated by speckle noise and interpolated with bilinear method.

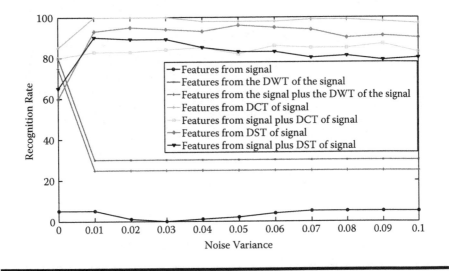

Figure 6.19 Recognition rate versus noise variance for different feature extraction methods from blurred fingerprint images contaminated by speckle noise and interpolated with bilinear method.

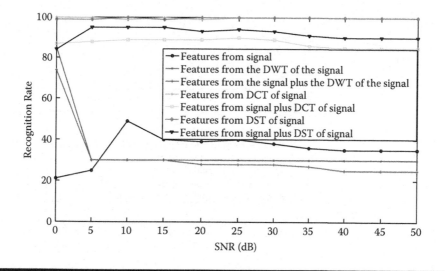

Figure 6.20 Recognition rate versus signal-to-noise ratio for different feature extraction methods from fingerprint images contaminated by AWGN and interpolated with Keys' method.

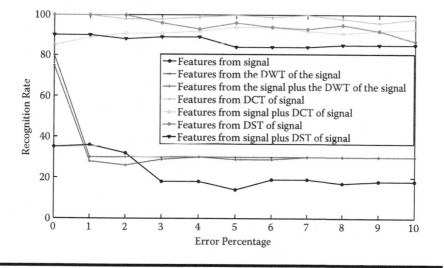

Figure 6.21 Recognition rate versus error percentage for different feature extraction methods from fingerprint images contaminated by impulsive noise and interpolated with Keys' method.

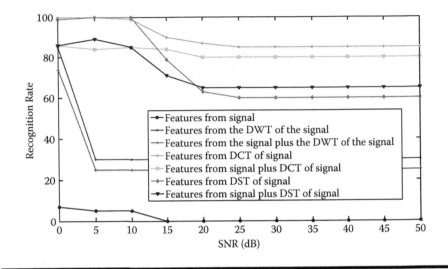

Figure 6.22 Recognition rate versus signal-to-noise ratio for different feature extraction methods from blurred fingerprint images contaminated by AWGN and interpolated with Keys' method.

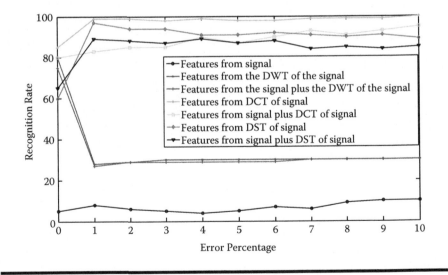

Figure 6.23 Recognition rate versus error percentage for different feature extraction methods from blurred fingerprint images contaminated by impulsive noise and interpolated with Keys' method.

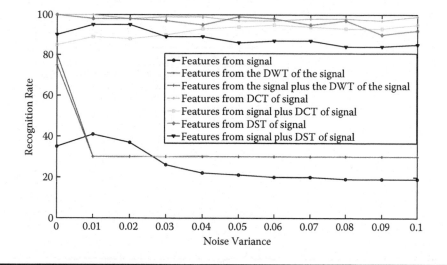

Figure 6.24 **Recognition rate versus noise variance for different feature extraction methods from fingerprint images contaminated by speckle noise and interpolated with Keys' method.**

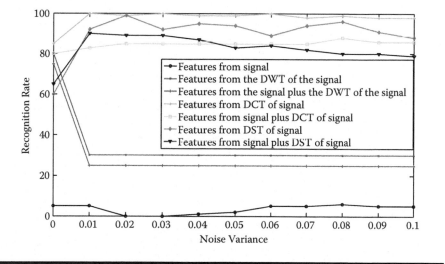

Figure 6.25 **Recognition rate versus noise variance for different feature extraction methods from blurred fingerprint images contaminated by speckle noise and interpolated with Keys' method.**

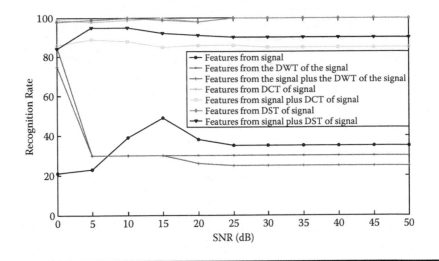

Figure 6.26 Recognition rate versus signal-to-noise ratio for different feature extraction methods from fingerprint images contaminated by AWGN and interpolated with warped-distance bilinear method.

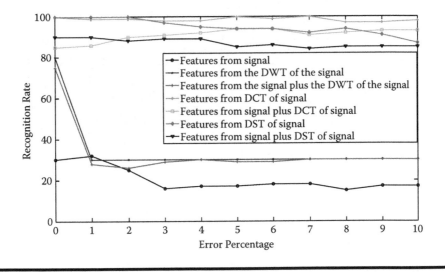

Figure 6.27 Recognition rate versus error percentage for different feature extraction methods from fingerprint images contaminated by impulsive noise and interpolated with warped-distance bilinear method.

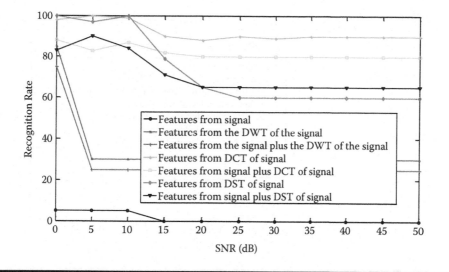

Figure 6.28 Recognition rate versus signal-to-noise ratio for different feature extraction methods from blurred fingerprint images contaminated by AWGN and interpolated with warped-distance bilinear method.

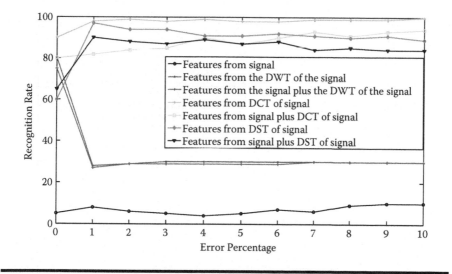

Figure 6.29 Recognition rate versus error percentage for different feature extraction methods from blurred fingerprint images contaminated by impulsive noise and interpolated with warped-distance bilinear method.

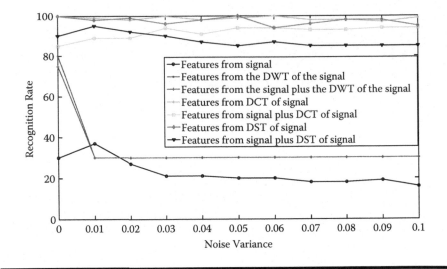

Figure 6.30 **Recognition rate versus noise variance for different feature extraction methods from fingerprint images contaminated by speckle noise and interpolated with warped-distance bilinear method.**

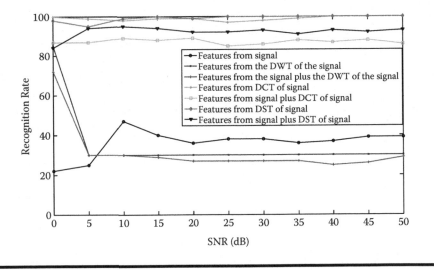

Figure 6.31 **Recognition rate versus noise variance for different feature extraction methods from blurred fingerprint images contaminated by speckle noise and interpolated with warped-distance bilinear method.**

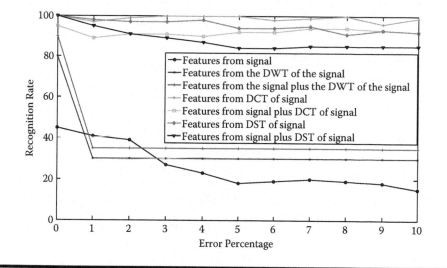

Figure 6.32 Recognition rate versus signal-to-noise ratio for different feature extraction methods from fingerprint images contaminated by AWGN and interpolated with warped Keys' method.

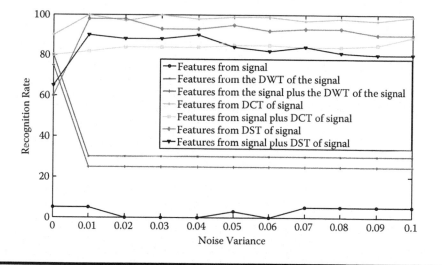

Figure 6.33 Recognition rate versus error percentage for different feature extraction methods from fingerprint images contaminated by impulsive noise and interpolated with warped-distance Keys' method.

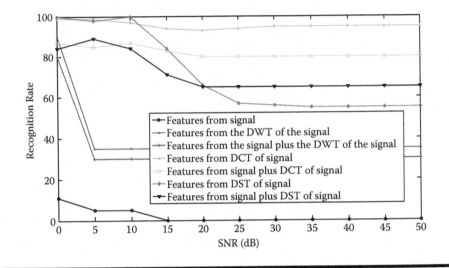

Figure 6.34 **Recognition rate versus signal-to-noise ratio for different feature extraction methods from blurred fingerprint images contaminated by AWGN and interpolated with warped-distance Keys' method.**

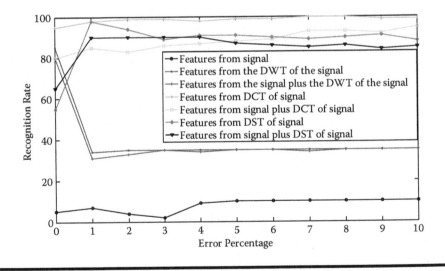

Figure 6.35 **Recognition rate versus error percentage for different feature extraction methods from blurred fingerprint images contaminated by impulsive noise and interpolated with warped-distance Keys' method.**

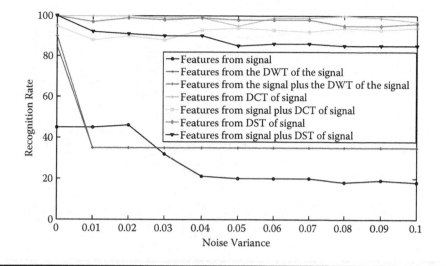

Figure 6.36 **Recognition rate versus noise variance for different feature extraction methods from fingerprint images contaminated by speckle noise and interpolated with warped-distance Keys' method.**

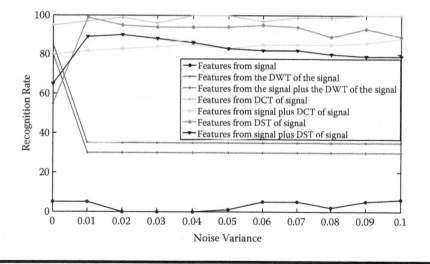

Figure 6.37 **Recognition rate versus noise variance for different feature extraction methods from blurred fingerprint images contaminated by speckle noise and interpolated with warped-distance Keys' method.**

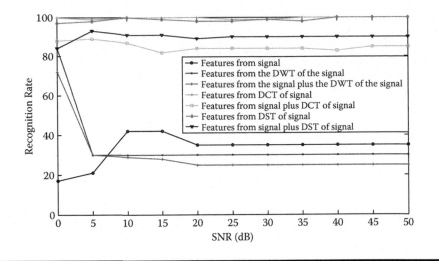

Figure 6.38 **Recognition rate versus signal-to-noise ratio for different feature extraction methods from fingerprint images contaminated by AWGN and interpolated with neural bilinear method.**

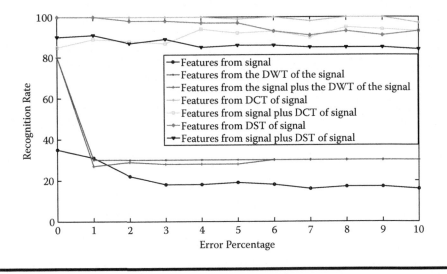

Figure 6.39 **Recognition rate versus error percentage for different feature extraction methods from fingerprint images contaminated by impulsive noise and interpolated with neural bilinear method.**

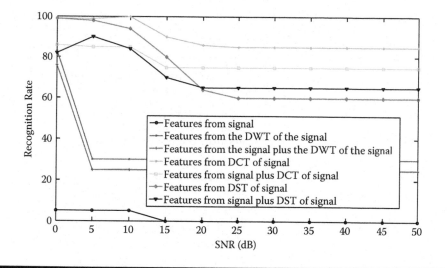

Figure 6.40 **Recognition rate versus signal-to-noise ratio for different feature extraction methods from blurred fingerprint images contaminated by AWGN and interpolated with neural bilinear method.**

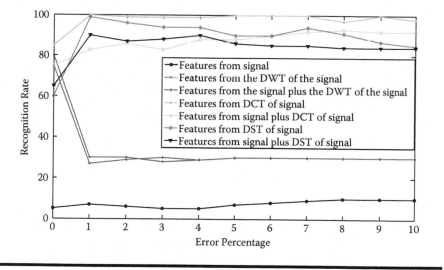

Figure 6.41 **Recognition rate versus error percentage for different feature extraction methods from blurred fingerprint images contaminated by impulsive noise and interpolated with neural bilinear method.**

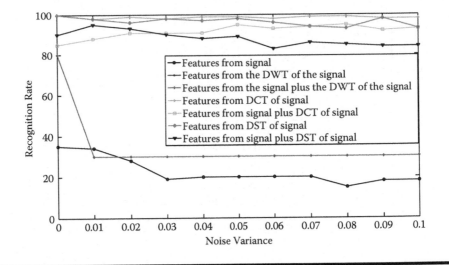

Figure 6.42 Recognition rate versus noise variance for different feature extraction methods from fingerprint images contaminated by speckle noise and interpolated with neural bilinear method.

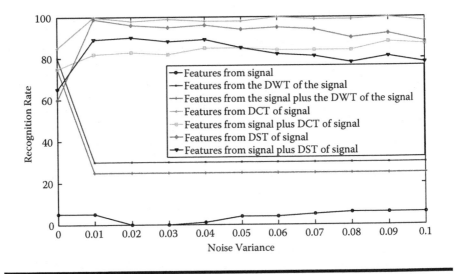

Figure 6.43 Recognition rate versus noise variance for different feature extraction methods from blurred fingerprint images contaminated by speckle noise and interpolated with neural bilinear method.

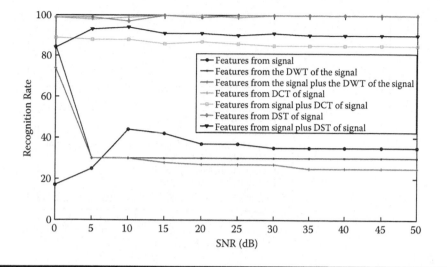

Figure 6.44 Recognition rate versus signal-to-noise ratio for different feature extraction methods from fingerprint images contaminated by AWGN and interpolated with neural Keys' method.

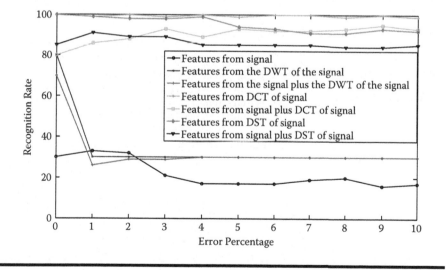

Figure 6.45 Recognition rate versus error percentage for different feature extraction methods from fingerprint images contaminated by impulsive noise and interpolated with neural Keys' method.

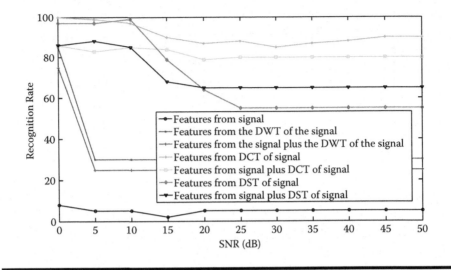

Figure 6.46 **Recognition rate versus signal-to-noise ratio for different feature extraction methods from blurred fingerprint images contaminated by AWGN and interpolated with neural Keys' method.**

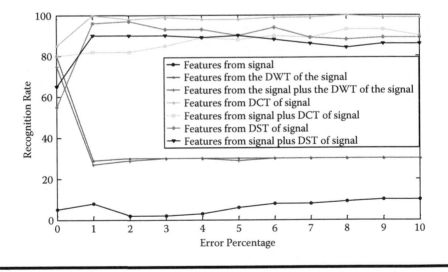

Figure 6.47 **Recognition rate versus error percentage for different feature extraction methods from blurred fingerprint images contaminated by impulsive noise and interpolated with neural Keys' method.**

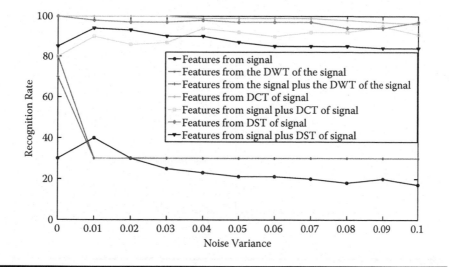

Figure 6.48 Recognition rate versus noise variance for different feature extraction methods from fingerprint images contaminated by speckle noise and interpolated with neural Keys' method.

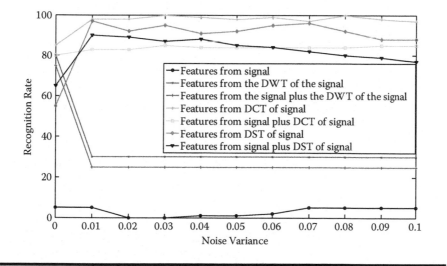

Figure 6.49 Recognition rate versus noise variance for different feature extraction methods from blurred fingerprint images contaminated by speckle noise and interpolated with neural Keys' method.

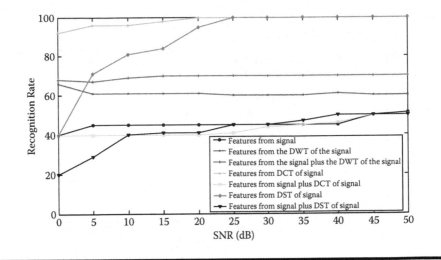

Figure 6.50 **Recognition rate versus signal-to-noise ratio for different feature extraction methods from landmine images contaminated by AWGN.**

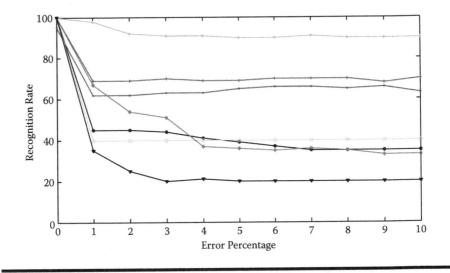

Figure 6.51 **Recognition rate versus error percentage for different feature extraction methods from landmine images contaminated by impulsive noise.**

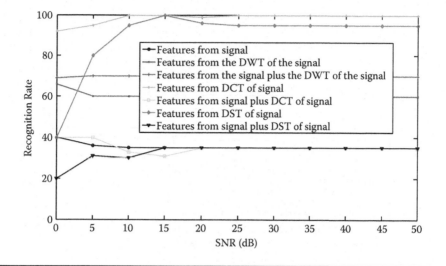

Figure 6.52 **Recognition rate versus signal-to-noise ratio for different feature extraction methods from blurred landmine images contaminated by AWGN**

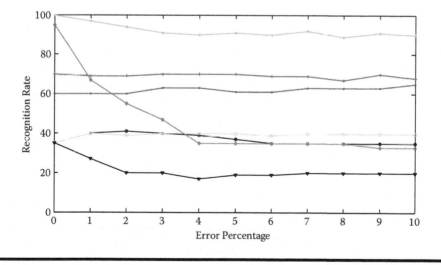

Figure 6.53 **Recognition rate versus error percentage for different feature extraction methods from blurred landmine images contaminated by impulsive noise.**

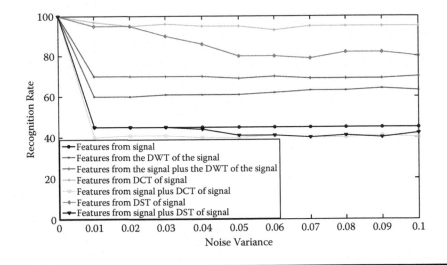

Figure 6.54 **Recognition rate versus noise variance for different feature extraction methods from landmine images contaminated by speckle noise.**

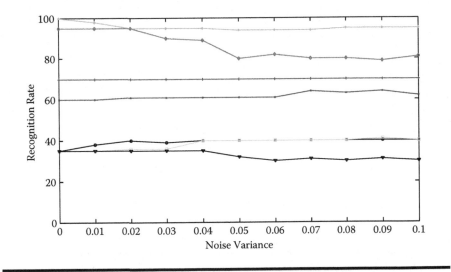

Figure 6.55 **Recognition rate versus noise variance for different feature extraction methods from blurred landmine images contaminated by speckle noise.**

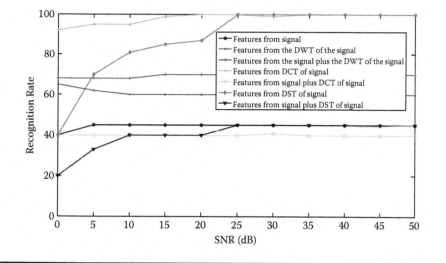

Figure 6.56 Recognition rate versus signal-to-noise ratio for different feature extraction methods from landmine images contaminated by AWGN and interpolated with bilinear method.

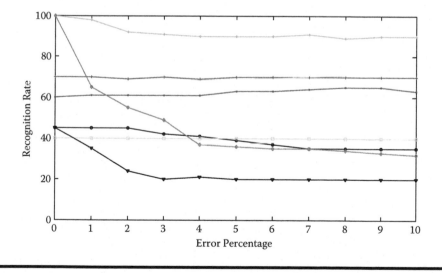

Figure 6.57 Recognition rate versus error percentage for different feature extraction methods from landmine images contaminated by impulsive noise and interpolated with bilinear method.

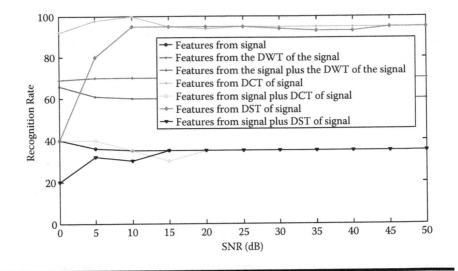

Figure 6.58 **Recognition rate versus signal-to-noise ratio for different feature extraction methods from blurred landmine images contaminated by AWGN and interpolated with bilinear method.**

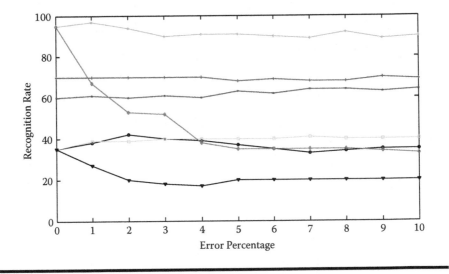

Figure 6.59 **Recognition rate versus error percentage for different feature extraction methods from blurred landmine images contaminated by impulsive noise and interpolated with bilinear method.**

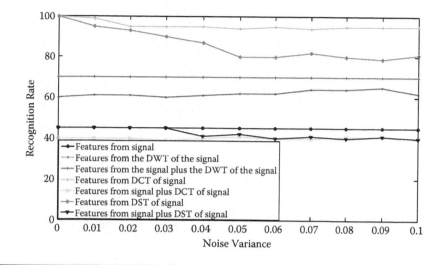

Figure 6.60 Recognition rate versus noise variance for different feature extraction methods from landmine images contaminated by speckle noise and interpolated with bilinear method.

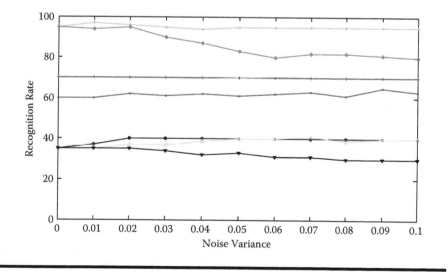

Figure 6.61 Recognition rate versus noise variance for different feature extraction methods from blurred landmine images contaminated by speckle noise and interpolated with bilinear method.

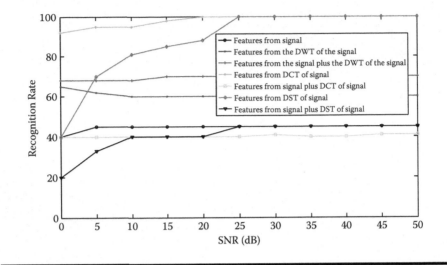

Figure 6.62 **Recognition rate versus SNR for different feature extraction methods from landmine images contaminated by AWGN and interpolated with Keys' method.**

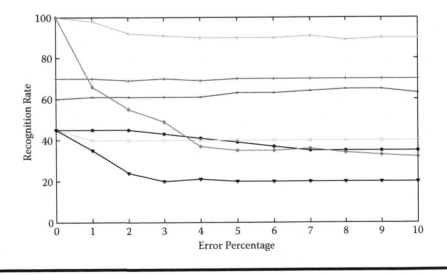

Figure 6.63 **Recognition rate versus error percentage for different feature extraction methods from landmine images contaminated by impulsive noise and interpolated with Keys' method.**

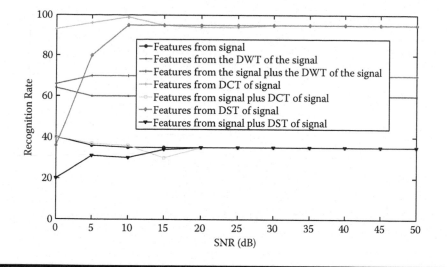

Figure 6.64 Recognition rate versus SNR for different feature extraction methods from blurred landmine images contaminated by AWGN and interpolated with Keys' method.

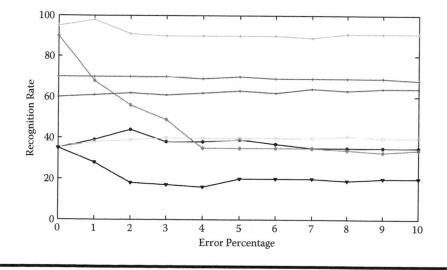

Figure 6.65 Recognition rate versus error percentage for different feature extraction methods from blurred landmine images contaminated by impulsive noise and interpolated with Keys' method.

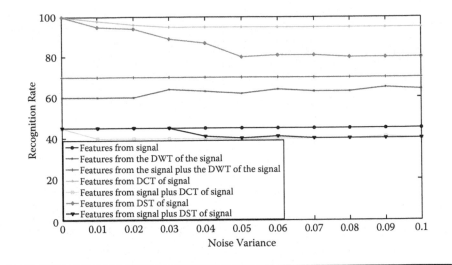

Figure 6.66 Recognition rate versus noise variance for different feature extraction methods from landmine images contaminated by speckle noise and interpolated with Keys' method.

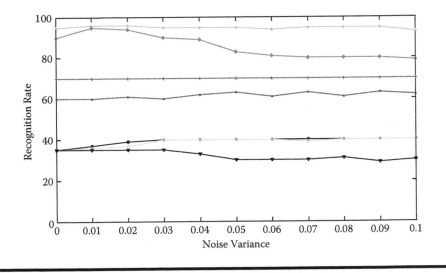

Figure 6.67 Recognition rate versus noise variance for different feature extraction methods from blurred landmine images contaminated by speckle noise and interpolated with Keys' method.

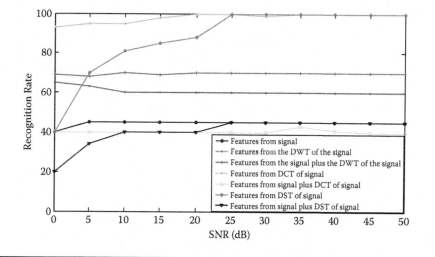

Figure 6.68 Recognition rate versus signal-to-noise ratio for different feature extraction methods from landmine images contaminated by AWGN and interpolated with warped-distance bilinear method.

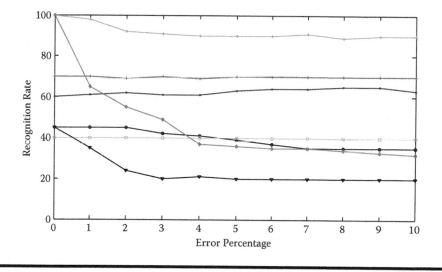

Figure 6.69 Recognition rate versus error percentage for different feature extraction methods from landmine images contaminated by impulsive noise and interpolated with warped-distance bilinear method.

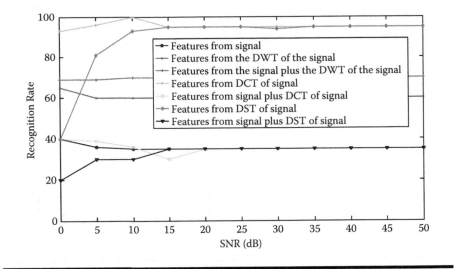

Figure 6.70 **Recognition rate versus signal-to-noise ratio for different feature extraction methods from blurred landmine images contaminated by AWGN and interpolated with warped-distance bilinear method.**

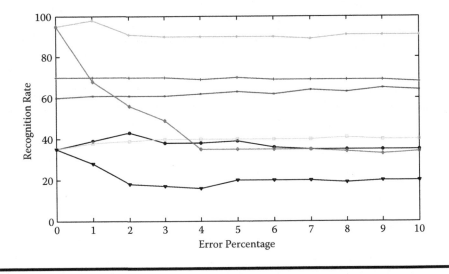

Figure 6.71 **Recognition rate versus error percentage for different feature extraction methods from blurred landmine images contaminated by impulsive noise and interpolated with warped-distance bilinear method.**

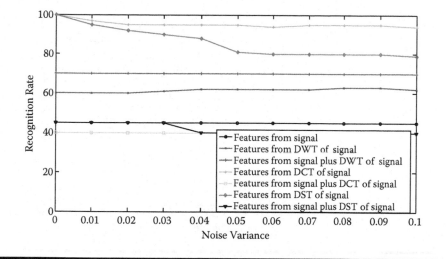

Figure 6.72 **Recognition rate versus noise variance for different feature extraction methods from landmine images contaminated by speckle noise and interpolated with warped-distance bilinear method.**

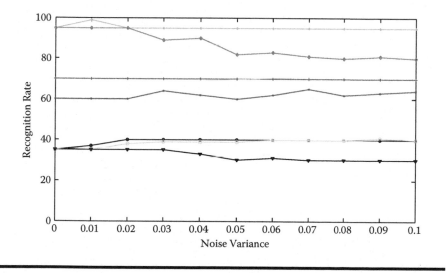

Figure 6.73 **Recognition rate versus noise variance for different feature extraction methods from blurred landmine images contaminated by speckle noise and interpolated with warped-distance bilinear method.**

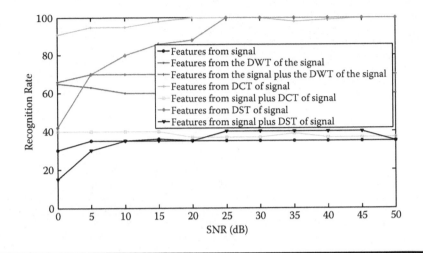

Figure 6.74 **Recognition rate versus SNR for different feature extraction methods from landmine images contaminated by AWGN and interpolated with warped-distance Keys' method.**

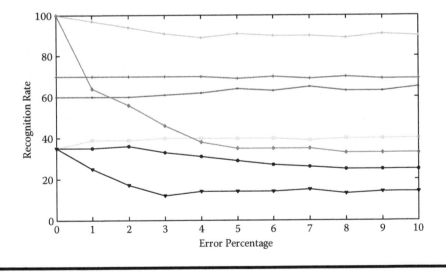

Figure 6.75 **Recognition rate versus error percentage for different feature extraction methods from landmine images contaminated by impulsive noise and interpolated with warped-distance Keys' method.**

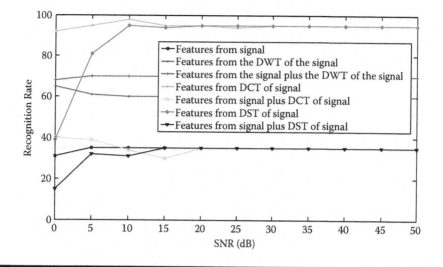

Figure 6.76 Recognition rate versus signal-to-noise ratio for different feature extraction methods from blurred landmine images contaminated by AWGN and interpolated with warped-distance Keys' method.

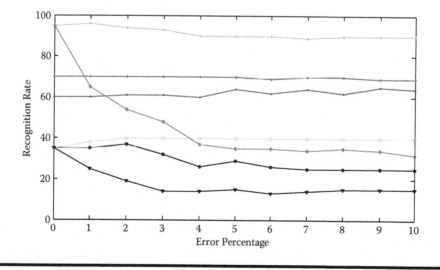

Figure 6.77 Recognition rate versus error percentage for different feature extraction methods from blurred landmine images contaminated by impulsive noise and interpolated with warped-distance Keys' method.

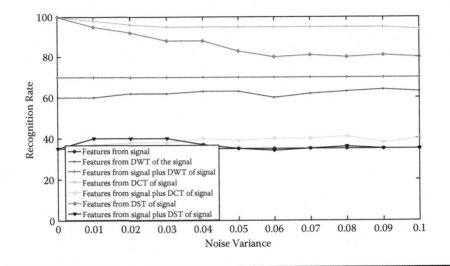

Figure 6.78 **Recognition rate versus noise variance for different feature extraction methods from landmine images contaminated by speckle noise and interpolated with warped-distance Keys' method.**

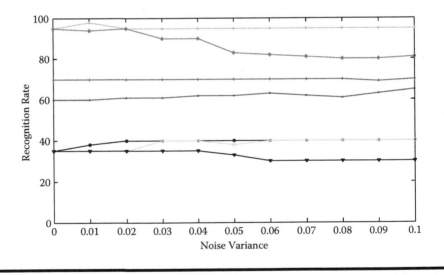

Figure 6.79 **Recognition rate versus noise variance for different feature extraction methods from blurred landmine images contaminated by speckle noise and interpolated with warped-distance Keys' method.**

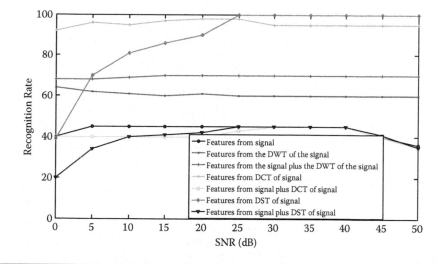

Figure 6.80 Recognition rate versus signal-to-noise ratio for different feature extraction methods from landmine images contaminated by AWGN and interpolated with neural bilinear method.

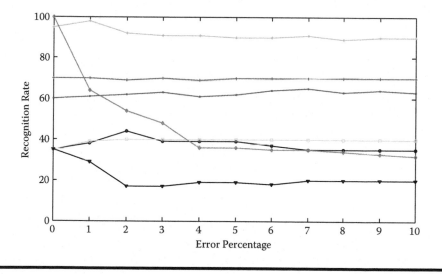

Figure 6.81 Recognition rate versus error percentage for different feature extraction methods from landmine images contaminated by impulsive noise and interpolated with neural bilinear method.

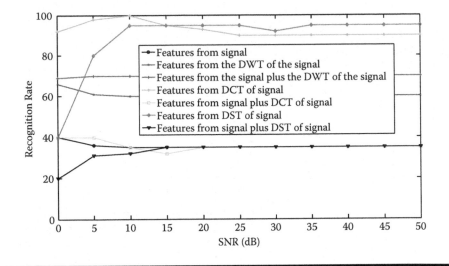

Figure 6.82 **Recognition rate versus signal-to-noise ratio for different feature extraction methods from blurred landmine images contaminated by AWGN and interpolated with neural bilinear method.**

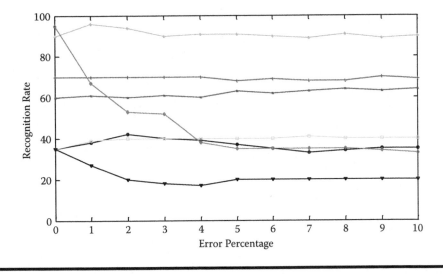

Figure 6.83 **Recognition rate versus error percentage for different feature extraction methods from blurred landmine images contaminated by impulsive noise and interpolated with neural bilinear method.**

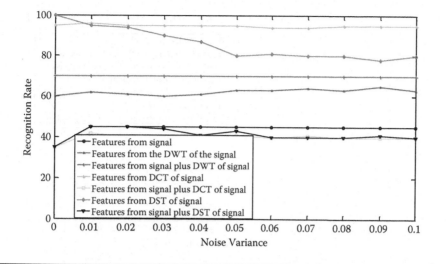

Figure 6.84 Recognition rate versus noise variance for different feature extraction methods from landmine images contaminated by speckle noise and interpolated with neural bilinear method.

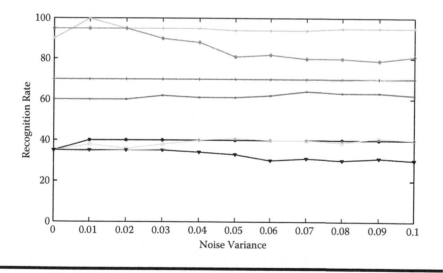

Figure 6.85 Recognition rate versus noise variance for different feature extraction methods from blurred landmine images contaminated by speckle noise and interpolated with neural bilinear method.

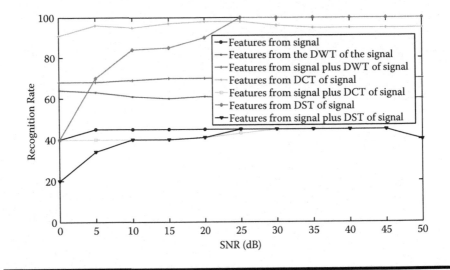

Figure 6.86 Recognition rate versus signal-to-noise ratio for different feature extraction methods from landmine images contaminated by AWGN and interpolated with neural Keys' method.

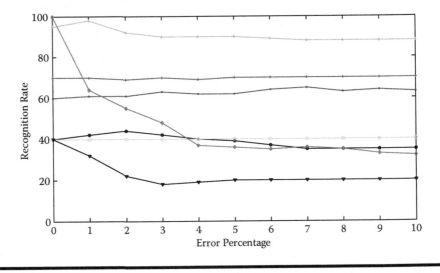

Figure 6.87 Recognition rate versus error percentage for different feature extraction methods from landmine images contaminated by impulsive noise and interpolated with neural Keys' method.

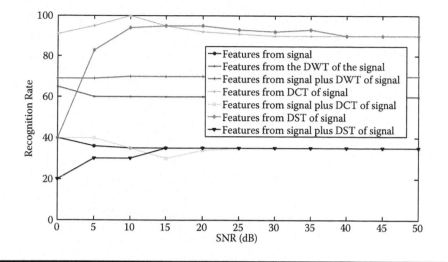

Figure 6.88   Recognition rate versus signal-to-noise ratio for different feature extraction methods from blurred landmine images contaminated by AWGN and interpolated with neural Keys' method.

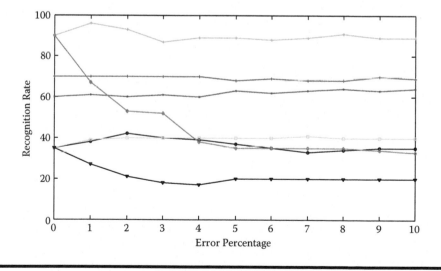

Figure 6.89   Recognition rate versus error percentage for different feature extraction methods from blurred landmine images contaminated by impulsive noise and interpolated with neural Keys' method.

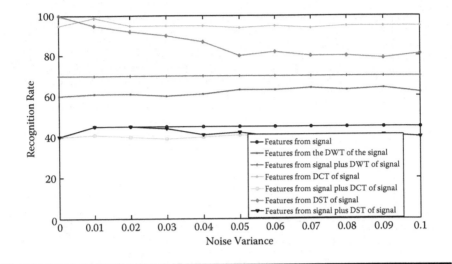

Figure 6.90 Recognition rate versus noise variance for different feature extraction methods from landmine images contaminated by speckle noise and interpolated with neural Keys' method.

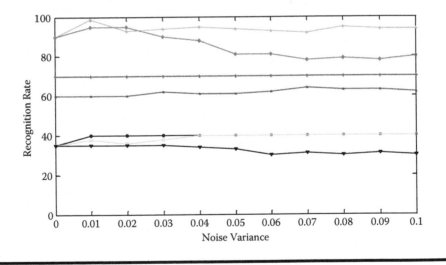

Figure 6.91 Recognition rate versus noise variance for different feature extraction methods from blurred landmine images contaminated by speckle noise and interpolated with neural Keys' method.

Figure 6.92 Samples of flower images used in experiments.

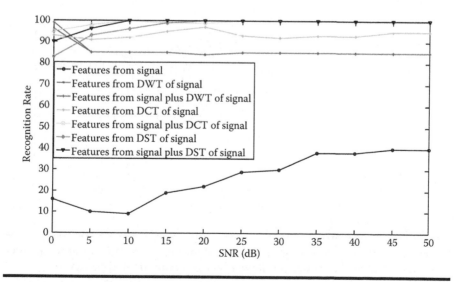

Figure 6.93 Samples of retinal images used in experiments.

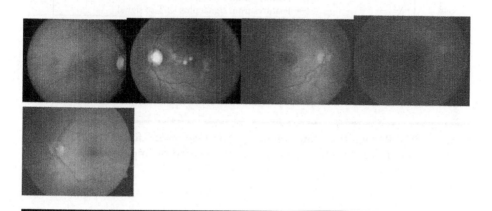

Figure 6.94 Recognition rate versus signal-to-noise ratio for different feature extraction methods from flower images contaminated by AWGN.

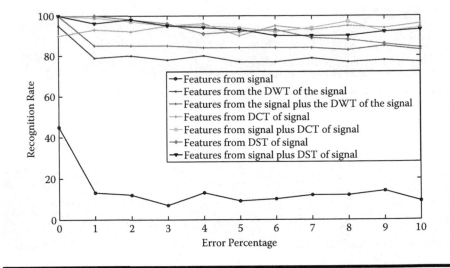

Figure 6.95 Recognition rate versus percentage error for different feature extraction methods from flower images contaminated by impulsive noise.

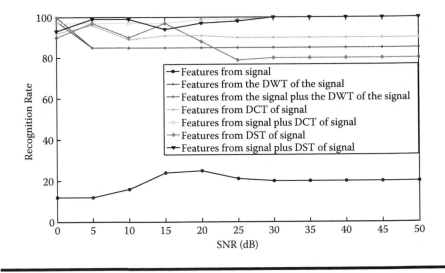

Figure 6.96 Recognition rate versus signal-to-noise ratio for different feature extraction methods from blurred flower images contaminated by AWGN.

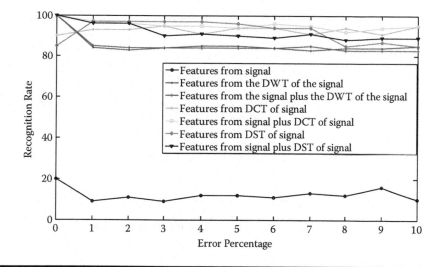

Figure 6.97 **Recognition rate versus percentage error for different feature extraction methods from blurred flower images contaminated by impulsive noise,**

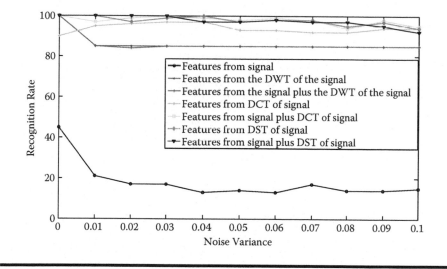

Figure 6.98 **Recognition rate versus noise variance for different feature extraction methods from flower images contaminated by speckle noise.**

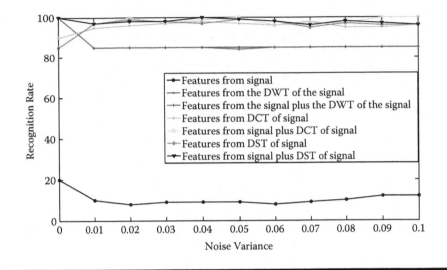

Figure 6.99 **Recognition rate versus noise variance for different feature extraction methods from blurred flower images contaminated by speckle noise.**

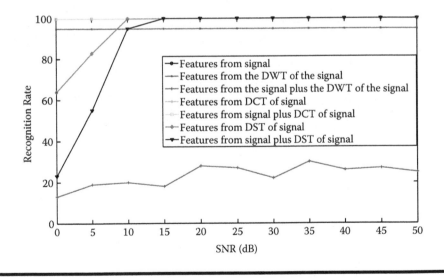

Figure 6.100 **Recognition rate versus signal-to-noise ratio for different feature extraction methods from retinal images contaminated by AWGN.**

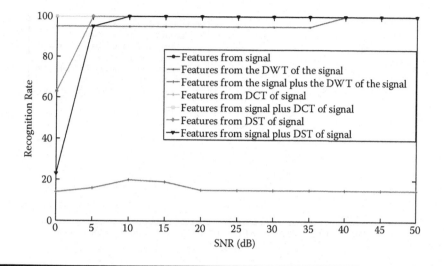

Figure 6.101 **Recognition rate versus signal-to-noise ratio for different feature extraction methods from blurred retinal images contaminated by AWGN.**

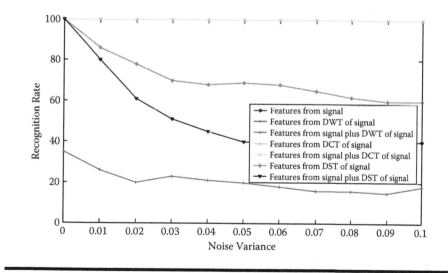

Figure 6.102 **Recognition rate versus noise variance for different feature extraction methods from retinal images contaminated by speckle noise.**

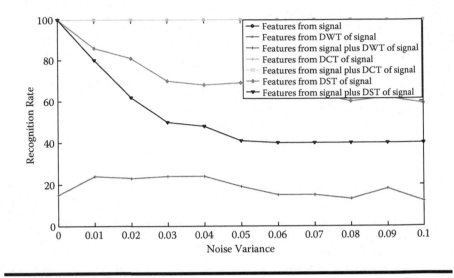

Figure 6.103 **Recognition rate versus noise variance for different feature extraction methods from blurred retinal images contaminated by speckle noise.**

Chapter 7

Image Interpolation as Inverse Problem

7.1 Introduction

In Chapters 2 and 3, we introduced a study of polynomial image interpolation. We have dealt with both space-invariant and space-varying (adaptive) interpolation. The previously mentioned techniques of image interpolation are signal synthesis techniques. They are based on synthesizing unkown pixel values using their known neighborhoods. In most of the work in the field of interpolation, ideal sampling of the source low resolution (LR) images is assumed. The LR image degradation model is not considered. We have considered this degradation model in deriving the iterative interpolation algorithm in Chapter 3, but there are many more requirements to consider.

According to the modern sampling theory, image interpolation requires a prefiltering step in the reconstruction process [76]. Formally speaking, the interpolation basis functions such as the bilinear, Keys', cubic spline, and cubic O-MOMS basis functions generate corresponding approximation spaces. When the image acquisition model is non-ideal, which is our case, a correction filter is required prior to interpolation to compensate for the non-ideality of the image acquisition model. This correction filter is obtained as the inverse of the cross-correlation sequence between the acquisition model filter and the reconstruction filter. Unfortunately, the estimation of this correction filter in our case is difficult or even impossible, because ours is an ill-posed inverse problem [76].

An inverse problem is characterized as ill-posed when there is no guarantee for the existence, uniqueness, and stability of the solution based on direct inversion. The solution of an inverse problem is not guaranteed stable if a small perturbation of the data may produce a large effect on the solution [68–70]. Image interpolation belongs to a general class of problems that were rigorously classified as ill-posed. The treatment of an ill-posed inverse problem in the presence of noise is performed using different techniques such as regularization and linear minimum mean square error (LMMSE) filtering [76,77].

This chapter presents four solutions for the image interpolation problem developed in a general framework based on dealing with image interpolation as an inverse problem rather than a signal synthesis problem. Based on the observation model given by Equation (3.1), our objective is to obtain a high-resolution (HR) image that is as close as possible to the original image subject to certain constraints.

In the first solution, an adaptive least squares interpolation algorithm is implemented [78–80]. In the second solution, an LMMSE approach is utilized. The necessary assumptions required to reduce the computational complexity of the LMMSE solution are presented [81,82]. In the third solution, the concept of *a priori* entropy maximization in the required HR image is used to derive a closed-form solution for the image interpolation problem, namely maximum entropy interpolation [82]. Finally, the regularization theory is used to find a regularized image interpolation formula. An efficient sectioned implementation of this formula is presented to avoid the large computational cost required [82,83].

The performances of all the above-mentioned algorithms are studied and compared to each other and to traditional polynomial interpolation techniques and the previously-mentioned iterative algorithms. The suitability of each solution to interpolating different images is also studied.

7.2 Adaptive Least-Squares Image Interpolation

In this adaptive algorithm, the image to be interpolated is divided into small overlapping blocks of size $M \times M$, and the objective is to obtain an interpolated version of each block of size $N \times N$. We assume that the relation between the available LR and the estimated HR block is given by [78–80]:

$$\hat{\mathbf{f}}_{i,j} = \mathbf{W}\mathbf{g}_{i,j} \tag{7.1}$$

where $\mathbf{g}_{i,j}$ and $\hat{\mathbf{f}}_{i,j}$, are the $M^2 \times 1$ and $N^2 \times 1$ lexicographically ordered LR and estimated HR blocks at position (i,j), respectively. $\mathbf{W}$ is the $N^2 \times M^2$ weight matrix required to obtain the HR block from the LR block. This matrix is required to be adaptive from block to block to accommodate the local activity levels of each block.

The first insight at Equation (7.1) leads to the least-squares solution obtained by minimizing the mean square error (MSE) of estimation as follows [78–80]:

$$\psi = \left\| \mathbf{f}_{i,j} - \hat{\mathbf{f}}_{i,j} \right\|^2 = \left\| \mathbf{f}_{i,j} - \mathbf{W}\mathbf{g}_{i,j} \right\|^2 \tag{7.2}$$

Differentiating both sides of Equation (7.2) with respect to **W** gives:

$$\frac{\partial \psi}{\partial \mathbf{W}} = -2(\mathbf{f}_{i,j} - \hat{\mathbf{f}}_{i,j})(\mathbf{g}_{i,j})^t \qquad (7.3)$$

This minimization leads directly to the following solution for **W** [78–80]:

$$\mathbf{W}^{k+1} = \mathbf{W}^k - \eta \left[\frac{\partial \psi}{\partial \mathbf{W}}\right]^k = \mathbf{W}^k + \mu(\mathbf{f}_{i,j} - \hat{\mathbf{f}}_{i,j}^k)(\mathbf{g}_{i,j}^k)^t \qquad (7.4)$$

where η is a constant and μ is the convergence parameter.

Using the above equation in estimating the weight matrix **W** requires knowledge of the samples of the original HR block $\mathbf{f}_{i,j}$, which is not practical. This problem has been previously solved by deriving the weights from another HR image, and using these weights to interpolate the available LR image. This approach is expected to give poor visual quality of the interpolated image.

An alternative approach is to consider the model that relates the available LR block to the original HR block. This model is given for each block by the following relation [78–80]:

$$\mathbf{g}_{i,j} = \mathbf{D}\mathbf{f}_{i,j} \qquad (7.5)$$

The matrix **D** is of size $M^2 \times N^2$. Thus, we can minimize the following cost function [78–80]:

$$\Omega = \left\| \mathbf{D}\left(\mathbf{f}_{i,j} - \hat{\mathbf{f}}_{i,j}\right)\right\|^2 \qquad (7.6)$$

The above equation means minimizing the MSE between the available LR block and a down-sampled version of the estimated HR block. This leads to [78–80]:

$$\Omega = \left\| \mathbf{g}_{i,j} - \mathbf{D}\hat{\mathbf{f}}_{i,j}\right\|^2 = \left\| \mathbf{g}_{i,j} - \mathbf{D}\mathbf{W}\mathbf{g}_{i,j}\right\|^2 \qquad (7.7)$$

Differentiating Equation (7.7) with respect to **W** and using Equation (7.1) leads to:

$$\frac{\partial \Omega}{\partial \mathbf{W}} = -2\mathbf{D}^t (\mathbf{g}_{i,j} - \mathbf{D}\hat{\mathbf{f}}_{i,j})(\mathbf{g}_{i,j})^t \qquad (7.8)$$

Using Equation (7.8), the weight matrix can be adapted using the following equation [63]:

$$\mathbf{W}^{k+1} = \mathbf{W}^k - \eta \left[\frac{\partial \Omega}{\partial \mathbf{W}}\right]^k = \mathbf{W}^k + \mu \mathbf{D}^t (\mathbf{g}_{i,j} - \mathbf{D}\hat{\mathbf{f}}_{i,j}^k)(\mathbf{g}_{i,j}^k)^t \qquad (7.9)$$

The adaptation of Equation (7.9) can be easily performed since it does not require the original HR block to be known *a priori*.

7.3 LMMSE Image Interpolation

Considering the image observation model given by Equation (3.1), the LMMSE criterion requires the MSE of estimation to be minimum over the entire ensemble of all possible estimates of the image. The optimization problem here is given by [81,82]:

$$\min_{\hat{\mathbf{f}}} E\left[\mathbf{e}^t\mathbf{e}\right] = E\left[Tr(\mathbf{ee}^t)\right] \tag{7.10}$$

with

$$\mathbf{e} = \mathbf{f} - \hat{\mathbf{f}} \tag{7.11}$$

where $\hat{\mathbf{f}}$ is the estimate of the required HR image. Since the down-sampling and filtering matrix $\mathbf{D}$ is linear, the estimate of $\mathbf{f}$ will be linear. That is an estimate of $\mathbf{f}$ can be derived by a linear operation on the degraded image such that [81,82]:

$$\hat{\mathbf{f}} = \mathbf{Tg} \tag{7.12}$$

where $\mathbf{T}$ is derived subject to solving Equation (7.10). This yields the following equation:

$$\min_{\hat{\mathbf{f}}} E\left[Tr(\mathbf{ee}^t)\right] = E\left[Tr\left\{(\mathbf{f} - \mathbf{Tg})(\mathbf{f} - \mathbf{Tg})^t\right\}\right]$$

$$= E\left[Tr\left\{\mathbf{ff}^t - \mathbf{T}(\mathbf{Dff}^t + \mathbf{vf}^t) - (\mathbf{ff}^t\mathbf{D}^t + \mathbf{fv}^t)\mathbf{T}^t\right.\right. \tag{7.13}$$

$$\left.\left. + \mathbf{T}(\mathbf{Dff}^t\mathbf{D}^t + \mathbf{vf}^t\mathbf{D}^t + \mathbf{Dfv}^t + \mathbf{vv}^t)\mathbf{T}^t\right\}\right]$$

We have $Tr(\mathbf{B}) = Tr(\mathbf{B}^t)$. Since the trace is linear, it can be interchanged with the expectation operator. Equation (7.13) can be simplified using some assumptions. The noise is assumed to be independent of the required HR image. This assumption leads to:

$$E[\mathbf{fv}^t] = E[\mathbf{v}^t\mathbf{f}] = [\mathbf{0}] \tag{7.14}$$

The auto-correlation matrices for the image and noise can be defined as:

$$E[\mathbf{ff}^t] = \mathbf{R}_f \tag{7.15}$$

and

$$E[\mathbf{vv}^t] = \mathbf{R}_v \tag{7.16}$$

The matrix $\mathbf{R_v}$ is a diagonal matrix whose main diagonal elements are equal to the noise variance of the noisy LR image. Substituting from Equations (7.14), (7.15), and (7.16) in Equation (7.13) yields:

$$\min_{\mathbf{f}} E\left[Tr(\mathbf{ee}^t)\right] = Tr\left\{\mathbf{R_f} - 2\mathbf{TDR_f} + \mathbf{TDR_fD^tT^t} + \mathbf{TR_vT^t}\right\} \qquad (7.17)$$

By differentiating Equation (7.17) with respect to $\mathbf{T}$ and setting the result equal to zero, the operator $\mathbf{T}$ can be obtained as:

$$\mathbf{T} = \mathbf{R_fD^t}(\mathbf{DR_fD^t} + \mathbf{R_v})^{-1} \qquad (7.18)$$

Thus, the LMMSE estimate of the HR image will be given by [81,82]:

$$\hat{\mathbf{f}} = \mathbf{R_fD^t}\left(\mathbf{DR_fD^t} + \mathbf{R_v}\right)^{-1}\mathbf{g} \qquad (7.19)$$

In the implementation of the LMMSE model, three major problems are encountered [81,82]. The first is the estimation of the auto-correlation matrix of the HR image, which is not available prior to applying the model. The second problem is the noise variance estimation of the noisy LR image. The third problem is the large-dimension matrix inversion required in Equation (7.19). To solve the problem of estimating the auto-correlation of the HR image, the matrix $\mathbf{R_f}$ in Equation (7.19) can be written in the form [81,82]:

$$\mathbf{R_f} = \begin{bmatrix} \mathbf{R}_{0,0} & \mathbf{R}_{0,1} & \cdots & \mathbf{R}_{0,N-1} \\ \mathbf{R}_{1,0} & \mathbf{R}_{1,1} & \cdots & \vdots \\ \vdots & \ddots & \ddots & \vdots \\ \mathbf{R}_{N-1,0} & \cdots & \cdots & \mathbf{R}_{N-1,N-1} \end{bmatrix} \qquad (7.20)$$

where

$$\mathbf{R}_{i,j} = E\left[\mathbf{f}_i\mathbf{f}_j^t\right] \qquad (7.21)$$

$\mathbf{f}_i$ and $\mathbf{f}_j$ are the ith and j^{th} column partitions of the lexicographically-ordered vector $\mathbf{f}$. Often, pixels in an image possess no predictable correlation beyond a correlation distance d. If we assume that $d = 0$, then the matrix $\mathbf{R_f}$ can be approximated by a diagonal matrix in the form [81,82]:

$$\mathbf{R_f} = \begin{bmatrix} \mathbf{R}_{0,0} & \mathbf{0} & \cdots & \mathbf{0} \\ \mathbf{0} & \mathbf{R}_{1,1} & \cdots & \vdots \\ \vdots & \ddots & \ddots & \mathbf{0} \\ \mathbf{0} & \cdots & \mathbf{0} & \mathbf{R}_{N-1,N-1} \end{bmatrix} \qquad (7.22)$$

If the samples of each column are assumed uncorrelated except for each pixel with itself, each matrix $\mathbf{R}_{ii}$ can be approximated by a diagonal matrix for $i = 0, 1, \ldots$, $N - 1$ as follows [81,82]:

$$\mathbf{R}_{ii} = \begin{bmatrix} R_f(i,0) & 0 & \cdots & & 0 \\ 0 & R_f(i,1) & \ddots & & \vdots \\ \vdots & & \ddots & \ddots & 0 \\ 0 & & \cdots & 0 & R_f(i,N-1) \end{bmatrix} \qquad (7.23)$$

The main diagonal elements of the matrix $\mathbf{R}_{ii}$ can be approximated from a polynomial interpolation of the available LR image. For an image $f'(n_1, n_2)$, the auto-correlation at the spatial location (n_1, n_2) can be estimated from the following relation [81,82]:

$$R_f(n_1, n_2) \cong \frac{1}{w^2} \sum_{K=1}^{w} \sum_{l=1}^{w} f'(K,l) f'(n_1 + k, n_2 + l) \qquad (7.24)$$

where $R_f(n_1, n_2)$ is the auto-correlation at spatial position (n_1, n_2). The image $f'(n_1, n_2)$ may be taken as the bilinear, Keys', cubic spline, or cubic O-MOMS interpolated image. Thus, the matrix $\mathbf{R}_f$ can be approximated by a diagonal sparse matrix.

The second problem is the noise variance estimation. This problem can be solved by estimating the noise variance from the available LR image. The noise variance of the image is taken as the variance of a flat area in that image.

The third problem is the matrix inversion process required for estimating the HR image. This matrix inversion is of order $M^2 \times M^2$ for an $M \times M$ LR image. The problem of inverting this large-dimension matrix is solved depending on the approximation of $\mathbf{R}_f$ as a diagonal matrix yielding a sparse matrix inversion process in Equation (7.19).

7.4 Maximum Entropy Image Interpolation

A mathematical model for image interpolation has been derived based on the maximization of the *a priori* entropy of the HR image. If the samples of the required HR image are assumed to have unit energy, they can be treated as if they are probabilities, possibly of so many photons that are present at the ith sample of the required HR image [82]. The required HR image is assumed to be treated as light quanta associated with each pixel value. Thus, the entropy of the required HR image is defined as follows [82]:

$$H_e = -\sum_{i=1}^{N^2} f_i \log_2(f_i) \qquad (7.25)$$

where f_i is the sampled signal. This equation can be written in the vector form as follows:

$$H_e = -\mathbf{f}^t \log_2(\mathbf{f}) \tag{7.26}$$

For image interpolation, to maximize the entropy subject to the constraint that $\|\mathbf{g} - \mathbf{Df}\|^2 = \|\mathbf{v}\|^2$, the following cost function must be minimized [82]:

$$\psi(\mathbf{f}) = \mathbf{f}^t \log_2(\mathbf{f}) - \lambda \Big[\|\mathbf{g} - \mathbf{Df}\|^2 - \|\mathbf{v}\|^2 \Big] \tag{7.27}$$

where λ is a Lagrangian multiplier. Differentiating both sides of the above equation with respect to $\mathbf{f}$ and equating the result to zero:

$$\frac{\partial \psi(\mathbf{f})}{\partial \mathbf{f}} = 0 = \frac{1}{In(2)} \Big\{ 1 + In(\hat{\mathbf{f}}) \Big\} + \lambda \Big[2\mathbf{D}^t (\mathbf{g} - \mathbf{D}\hat{\mathbf{f}}) \Big] \tag{7.28}$$

Solving for the estimated HR image:

$$In(\hat{\mathbf{f}}) = -1 - \lambda In(2) \Big[2\mathbf{D}^t \Big(\mathbf{g} - \mathbf{D}\hat{\mathbf{f}} \Big) \Big] \tag{7.29}$$

Thus:

$$\hat{\mathbf{f}} = \exp \Big[-1 - \lambda In(2) \Big| 2\mathbf{D}^t \Big(\mathbf{g} - \mathbf{D}\hat{\mathbf{f}} \Big) \Big| \Big] \tag{7.30}$$

Expanding the above equation using Taylor expansion and neglecting all but the first two terms, since $\mathbf{g} - \mathbf{D}\hat{\mathbf{f}}$ must be a small quantity, leads to the following form:

$$\hat{\mathbf{f}} \cong -\lambda In(2) \Big[2\mathbf{D}^t \Big(\mathbf{g} - \mathbf{D}\hat{\mathbf{f}} \Big) \Big] \tag{7.31}$$

Solving for $\hat{\mathbf{f}}$ leads to [82]:

$$\hat{\mathbf{f}} \cong \Big(\mathbf{D}^t \mathbf{D} + \eta \mathbf{I} \Big)^{-1} \mathbf{D}^t \mathbf{g} \tag{7.32}$$

where $\eta = -1/(2\lambda In(2))$.

We can use a direct inversion solution for Equation (7.32). This solution is based on the direct inversion of the term $(\mathbf{D}^t\mathbf{D} + \eta\mathbf{I})$. This matrix inversion can be performed easily depending on the special nature of this sparse diagonal matrix. The matrix $\mathbf{D}^t\mathbf{D}$ is a diagonal matrix. This is easily verified by noticing that the operation $\mathbf{D}^t\mathbf{D}$ stands for decimation followed by interpolation. Thus, if $\mathbf{D}$ decimates by a factor r, applying $\mathbf{D}^t\mathbf{D}$ causes all positions $(1 + n_1 r, 1 + n_2 r)$

for integer (n_1, n_2) to stay unchanged, whereas the remaining pixels are replaced by zeros.

Thus, $\mathbf{D'D}$ stands for a masking operation represented by a diagonal matrix. The effect of the term $\eta\mathbf{I}$ is to remove the ill-posedness nature of the inverse problem of the term $\mathbf{D'D}$ by redistributing its Eigenvalues to avoid singularity. It is clear that the term $(\mathbf{D'D} + \eta\mathbf{I})$ represents a sparse diagonal matrix that can be easily inverted.

7.5 Regularized Image Interpolation

Regularization theory, which was basically introduced by Tikhonov and Miller, provides a formal basis for the development of regularized solutions for ill-posed problems [84–86]. The stabilizing functional approach is one of the basic methodologies for the development of regularized solutions. According to this approach, an ill-posed problem can be formulated as the constrained minimization of a certain function called the stabilizing functional. The specific constraints imposed by the stabilizing functional approach on the solution depend on the form and the properties of the stabilizing functional used. From the nature of the problem, these constraints are necessarily related to the *a priori* information regarding the expected regularized solution [84–86].

According to the regularization approach, the solution of Equation (7.1) is obtained by the minimization of the cost function [82,87]:

$$\psi\left(\hat{\mathbf{f}}\right) = \left\| \mathbf{g} - \mathbf{D}\hat{\mathbf{f}} \right\|^2 + \lambda \left\| \mathbf{Q}\hat{\mathbf{f}} \right\|^2 \tag{7.33}$$

where $\mathbf{Q}$ is the 2-D regularization operator and λ is the regularization parameter. This minimization is accomplished by taking the derivative of the cost function yielding:

$$\frac{\partial \psi\left(\hat{\mathbf{f}}\right)}{\partial \hat{\mathbf{f}}} = 0 = 2\mathbf{D}'\left(\mathbf{g} - \mathbf{D}\hat{\mathbf{f}}\right) - 2\lambda \mathbf{Q}'\mathbf{Q}\hat{\mathbf{f}} \tag{7.34}$$

Solving for that $\hat{\mathbf{f}}$ that provides the minimum of the cost function yields [82,87]:

$$\hat{\mathbf{f}} = \left(\mathbf{D}'\mathbf{D} + \lambda \mathbf{Q}'\mathbf{Q}\right)^{-1} \mathbf{D}'\mathbf{g} \tag{7.35}$$

The rule of the regularization operator $\mathbf{Q}$ is to move the small Eigenvalues of $\mathbf{D}$ away from zero, while leaving the large Eigenvalues unchanged. It also incorporates

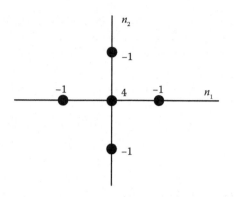

Figure 7.1   Two-dimensional Laplacian operator.

prior knowledge about the required degree of smoothness of **f** into the interpolation process.

The regularization operator **Q** is a finite difference matrix chosen to minimize the second or higher order difference energy of the estimated image. The 2-D Laplacian illustrated in Figure 7.1 is preferred for minimizing second order difference energy and it is the most popular regularization operator. It is the operator used in this chapter. The regularization parameter λ controls the trade-off between the fidelity to the data and the smoothness of the solution.

Regularization theory was previously applied to image restoration in an iterative manner or a frequency domain implementation. The regularized solutions have been extended to the problem of image interpolation. Unfortunately, the frequency domain solution is impossible. The possible solution is the iterative one that can be carried out as follows [82,87]:

$$\mathbf{f}_{i+1} = \mathbf{f}_i + \eta_0 \left\{ \mathbf{D}^t \mathbf{g} - \left(\mathbf{D}^t \mathbf{D} + \lambda \mathbf{Q}^t \mathbf{Q} \right) \mathbf{f}_i \right\} \tag{7.36}$$

where $\mathbf{f}_i$ is the obtained HR image at iteration i and η_0 is a convergence parameter. This method is a good solution that avoids the large computational cost involved in the matrix inversion process in Equation (7.35). The drawback of this method is the large number of iterations required to achieve a good HR image.

Another solution to the regularized image interpolation problem has been implemented by the segmentation of the LR image into overlapping segments and the separate interpolation of each segment using Equation (7.35) as an inversion process. It is clear that if a global regularization parameter is used, a single-matrix inversion process for a matrix of moderate dimensions is required because the term

$(\mathbf{D}^t\mathbf{D} + \lambda\mathbf{Q}^t\mathbf{Q})^{-1}$ is independent of the image to be interpolated. As a result, this solution is efficient from the computational cost perspective. The interpolation formula can be written in the following form [82,87]:

$$\hat{\mathbf{f}}_{i,j} = \left(\mathbf{D}^t\mathbf{D} + \lambda\mathbf{Q}^t\mathbf{Q}\right)^{-1}\mathbf{D}^t\mathbf{g}_{i,j} \qquad (7.37)$$

where $\mathbf{g}_{i,j}$ and $\hat{\mathbf{f}}_{i,j}$, are the $M^2 \times 1$ and $N^2 \times 1$ lexicographically ordered LR and the estimated HR blocks at position (i,j), respectively.

7.6 Simulation Examples

All the algorithms for image interpolation as an inverse problem have been tested on noisy LR images with different signal-to-noise ratios (SNRs). In the first experiment, the LR 128 × 128 woman image with SNR = 25 dB in Figure 7.2 has been used to test both the quality and computational cost of the interpolation algorithms as compared to traditional interpolation techniques.

The result of the LMMSE image interpolation algorithm with auto-correlation matrix estimated from the original HR image is given in Figure 7.3. The results of the LMMSE image interpolation algorithm with auto-correlation matrices estimated from polynomial image interpolation techniques are given in Figures 7.4 through 7.7. It is clear that the LMMSE algorithm gives good interpolation results with auto-correlation matrices estimated from different polynomial interpolation techniques.

Figure 7.2 Woman image. (a) Original image (256 × 256). (b) LR image, (128 × 128). SNR = 25 dB.

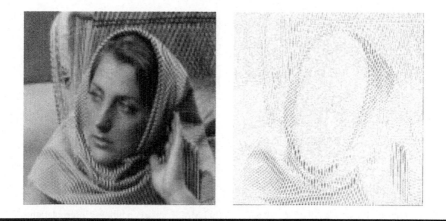

Figure 7.3   LMMSE interpolation using original image. (a) LMMSE interpolation, PSNR = 22.52 dB, c_e = 0.71, CPU = 12 sec. (b) Error image, MSE = 364.

The LMMSE interpolation algorithm is insensitive to the method of auto-correlation estimation. This algorithm requires the inversion of a matrix of dimensions 16384 × 16384, which is sparse in nature. The computational time required is the total time of polynomial interpolation, auto-correlation estimation, and matrix inversion.

The maximum entropy image interpolation algorithm has been tested on the same LR image with η = 0.001. This algorithm requires the inversion of a matrix of dimension 16384 × 16384, which is sparse in nature. The result is given in Figure 7.8. The maximum entropy solution requires less computational time as compared to the LMMSE algorithm and gives better PSNR values.

The adaptive least-squares image interpolation algorithm has also been tested on the same LR image. In the implementation of this algorithm, the LR image was segmented into overlapping blocks of 12 × 12 pixels each. Each block was interpolated separately to the size of 24 × 24 pixels, and 8 pixels were removed from the four sides of each block to yield small blocks of size 8 × 8 to avoid edge effects. By the process of segmentation, the adaptive least-squares technique requires the computation of a weight matrix of size 576 × 144 for each block, which is a moderate size. This allows the implementation of the least-squares algorithm for interpolating an image of any size, since the computational time will be linearly proportional to the number of blocks in the image (Figure 7.9). It is clear that the results obtained using this adaptive algorithm are very close to those obtained using the maximum entropy algorithm, but the computational time here is relatively large.

Both the iterative and inverse regularized image interpolation algorithms were tested on the same LR image with a global regularization parameter λ = 0.001. The results are given in Figure 7.10. In the inverse regularized algorithm, the LR image has been segmented using the same procedure followed in the least-squares

(a) Bilinear, PSNR = 21.19 dB, c_e = 0.53.

(b) LMMSE based on bilinear interpolation, PSNR = 22.12 dB, c_e = 0.7. CPU = 13 s.

(c) Cubic spline, PSNR = 21.02 dB, c_e = 0.5.

(d) LMMSE based on cubic spline interpolation, PSNR = 22.13 dB, c_e = 0.7. CPU = 15.3 s.

Figure 7.4   LMMSE interpolation results. (a) Bilinear, PSNR = 21.19 dB, c_e = 0.53. (b) LMMSE based on bilinear interpolation, PSNR = 22.12 dB, c_e = 0.7. CPU = 13 sec. (c) Cubic spline, PSNR = 21.02 dB, c_e = 0.5. (d) LMMSE based on cubic spline interpolation, PSNR = 22.13 dB, c_e = 0.7. CPU = 15.3 sec.

algorithm. By the process of segmentation and the usage of a global regularization parameter, this technique requires a single matrix inversion of size 576 × 576, which is a moderate size. The segmentation process allows the implementation of the inverse regularized interpolation algorithm for interpolating an image of any size, since the computational time will be linearly proportional to the number of blocks in the image. Results obtained using regularized image interpolation are good, but the computational cost required is greater than the costs of the LMMSE and maximum entropy algorithms.

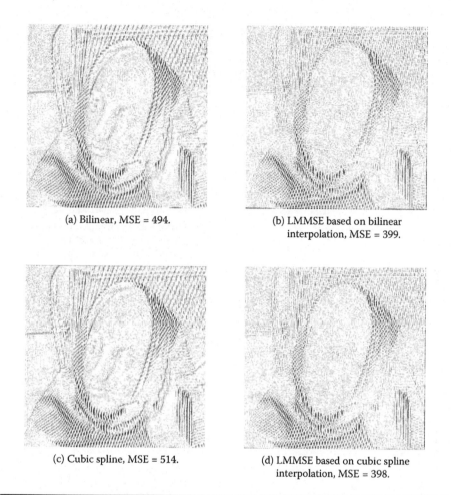

(a) Bilinear, MSE = 494.

(b) LMMSE based on bilinear interpolation, MSE = 399.

(c) Cubic spline, MSE = 514.

(d) LMMSE based on cubic spline interpolation, MSE = 398.

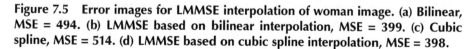

Figure 7.5   Error images for LMMSE interpolation of woman image. (a) Bilinear, MSE = 494. (b) LMMSE based on bilinear interpolation, MSE = 399. (c) Cubic spline, MSE = 514. (d) LMMSE based on cubic spline interpolation, MSE = 398.

All the image interpolation algorithms were also tested on the LR 128 × 128 test pattern image with SNR = 25 dB (Figure 7.11) and the results are given in Figures 7.12 through 7.19. In this experiment, the inverse regularized image interpolation algorithm gives the best interpolation results.

The LMMSE image interpolation algorithm has been tested for different estimates of noise variance in the cases of the test pattern and woman images. The results are given in Figure 7.20. It is clear from the figure that this algorithm is robust to small errors in estimating the noise variance. It is also better to use a lower estimate of the noise variance than to use a higher estimate.

(a) Cubic O-MOMS, PSNR = 21.07 dB, c_e = 0.5.

(b) LMMSE based on cubic O-MOMS interpolation, PSNR = 22.13 dB, c_e = 0.7. CPU =15.3 s.

(c) Keys', PSNR = 21.23 dB, c_e = 0.51.

(d) LMMSE based on Keys' interpolation, PSNR = 22.07 dB, c_e = 0.7, CPU = 13.5 s.

Figure 7.6   LMMSE interpolation results. (a) Cubic O-MOMS, PSNR = 21.07 dB, c_e = 0.5. (b) LMMSE based on cubic O-MOMS interpolation, PSNR = 22.13 dB, c_e = 0.7, CPU = 15.3 sec. (c) Keys', PSNR = 21.23 dB, c_e = 0.51. (d) LMMSE based on Keys' interpolation, PSNR = 22.07 dB, c_e = 0.7, CPU = 13.5 sec.

The effect of the parameter η on the maximum entropy image interpolation algorithm was studied on both the woman and test pattern images. The results are given in Figure 7.21. It is clear that this algorithm is insensitive to the choice of η in the range of 10^{-5} to 10^{-2} for both images.

The effect of the choice of the global regularization parameter λ on inverse regularized image interpolation algorithm was studied on both the test pattern and woman images, and the results are given in Figure 7.22. The effect of λ on the MSE

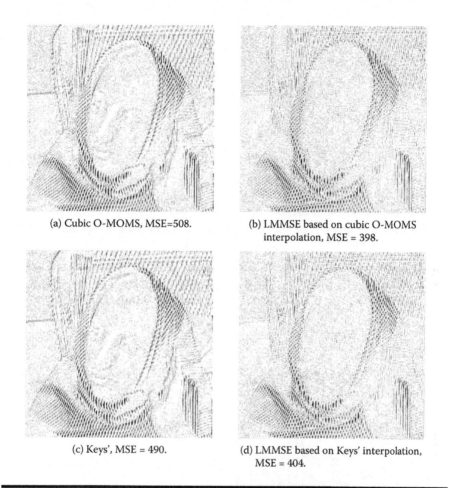

(a) Cubic O-MOMS, MSE=508.

(b) LMMSE based on cubic O-MOMS interpolation, MSE = 398.

(c) Keys', MSE = 490.

(d) LMMSE based on Keys' interpolation, MSE = 404.

Figure 7.7   Error images for LMMSE interpolation of woman image. (a) Cubic O-MOMS, MSE = 508. (b) LMMSE based on cubic O-MOMS interpolation, MSE = 398. (c) Keys', MSE = 490. (d) LMMSE based on Keys' interpolation, MSE = 404.

is small for λ in the range of 10^{-5} to 1 for both images. The effect of λ has also been studied for the iterative regularized interpolation algorithm. The results are shown in Figure 7.23. The effect of λ on the MSE is small for λ in the range of 10^{-5} to 10^{-2} for both images.

Figures 7.24 and 7.25 introduce comparisons of the different interpolation algorithms for both the woman and test pattern images, respectively. From the obtained results, it is clear that the maximum entropy algorithm gives the best interpolation results for the woman image that has so many details. On the other hand,

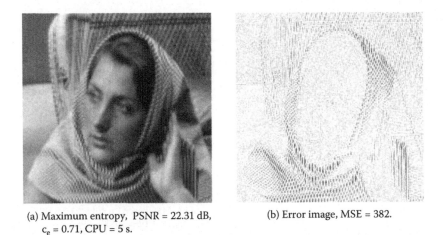

(a) Maximum entropy, PSNR = 22.31 dB,   (b) Error image, MSE = 382.
c_e = 0.71, CPU = 5 s.

Figure 7.8   Maximum entropy interpolation of woman image. (a) Maximum entropy, PSNR = 22.31 dB, c_e = 0.71, CPU = 5 sec. (b) Error image, MSE = 382.

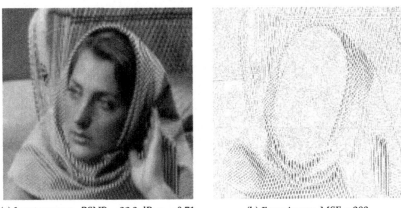

(a) Least squares, PSNR = 22.3 dB, c_e = 0.71,   (b) Error image, MSE = 383.
CPU = 449 s, I_{av} = 11.62.

Figure 7.9   Least-squares interpolation of woman image. (a) Least-squares, PSNR = 22.3 dB, c_e = 0.71, CPU = 449 sec, I_{av} = 11.62. (b) Error image, MSE = 383.

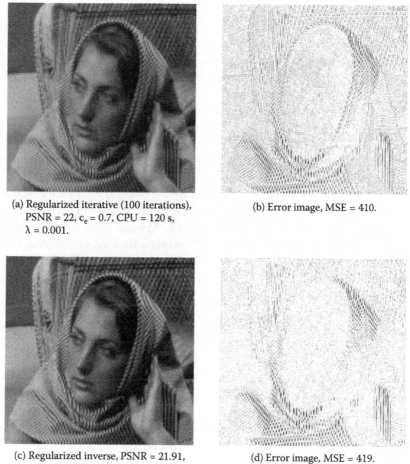

(a) Regularized iterative (100 iterations), PSNR = 22, c_e = 0.7, CPU = 120 s, λ = 0.001.

(b) Error image, MSE = 410.

(c) Regularized inverse, PSNR = 21.91, c_e = 0.63, CPU = 62 s, λ=0.001.

(d) Error image, MSE = 419.

Figure 7.10   Regularized interpolation of woman image. (a) Regularized iterative (100 iterations), PSNR = 22, c_e = 0.7, CPU = 120 sec, λ = 0.001. (b) Error image, MSE = 410. (c) Regularized inverse, PSNR = 21.91, c_e = 0.63, CPU = 62 sec, λ = 0.001. (d) Error image, MSE = 419.

the inverse regularized algorithm is the best for the test pattern image that has many edges with different orientations.

Other experiments on the image interpolation algorithms were performed on different noise-free and noisy images, and the results are tabulated in Table 7.1 and Table 7.2, respectively. From the results in the tables, it is clear that the algorithms of image interpolation as inverse problems give better results than the polynomial interpolation algorithms from the peak signal-to-noise ratio (PSNR) and edge preservation perspectives. Also, the inverse regularized image interpolation algorithm is the best from the PSNR perspective for most images containing many edges. For interpolating images with a large number of objects and details such as a building image, the maximum entropy interpolation algorithm is the most feasible.

7.7 Interpolation of Infrared Images

Ashiba et al. used the least-squares approach for the interpolation of infrared images [80]. First, they applied the model of down-sampling and filtering on the original infrared images to yield LR images down-sampled by a factor of two in both directions. After that, they investigated the cubic O-MOMS image interpolation algorithm, the warped-distance cubic O-MOMS algorithm, and the adaptive least-squares algorithm on the obtained LR images with a signal-to-noise ratio of 25 dB. The obtained results are given in Figures 7.26 and 7.27. The values of the PSNR and average number of iterations per block for the adaptive least-squares algorithm are given in the figures.

In the implementation of the least-squares algorithm, the LR image was segmented into overlapping blocks of 12×12 pixels each. Each block was interpolated separately to the size of 24×24 pixels and 8 pixels were removed from the four sides of each block to yield a small block of size 8×8 to avoid edge effects.

By the process of segmentation, the adaptive least-squares algorithm requires the computation of a weight matrix of size 576×144 for each block, which is a moderate size. The convergence parameter μ was set to 0.01 at the beginning of each block interpolation process. If the error decreased with increasing the number of iterations, μ is kept fixed. Otherwise, the value of μ is replaced by 0.01 to allow for the decrease of the estimation error.

The segmentation and overlapping of blocks allow the implementation of the adaptive least-squares algorithm for interpolating images of arbitrary sizes, since the computation time will be linearly proportional to the number of blocks in the image. It is clear that the result from this adaptive least-squares algorithm are better than those obtained using the cubic O-MOMS algorithm from the PSNR view. Furthermore, the implementation of the adaptive least-squares algorithm requires a small number of iterations per block and hence a short time.

The effects of errors in the filtering and down-sampling matrix $\mathbf{D}$ on the interpolation process in the adaptive least-squares algorithm was also studied for the plane and truck images. The errors in the matrix $\mathbf{D}$ are modeled by Gaussian noise added to its entries. The variation of the PSNR of the obtained HR image with the variance of this error is illustrated in Figure 7.28 for both the plane and truck images. This figure shows that small errors in the matrix $\mathbf{D}$ can be tolerated with small PSNR losses, while large errors degrade the performance significantly.

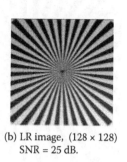

(b) LR image, (128 × 128)
SNR = 25 dB.

(a) Original image (256 × 256).

Figure 7.11 Test pattern image. (a) Original image (256 × 256). (b) LR image, (128 × 128). SNR = 25 dB.

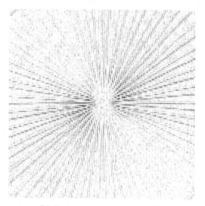

(a) LMMSE interpolation. PSNR = 26.15 dB
c_e = 0.81, CPU = 12 s.

(b) Error image, MSE = 158.

Figure 7.12 LMMSE interpolation using original image. (a) LMMSE interpolation. PSNR = 26.15 dB, c_e = 0.81, CPU = 12 sec. (b) Error image, MSE = 158.

(a) Bilinear, PSNR = 22.78 dB, c_e = 0.47.

(b) LMMSE based on bilinear interpolation, PSNR = 24.36 dB, c_e = 0.74. CPU = 13 s.

(c) Cubic spline, PSNR = 23.19 dB, c_e = 0.46.

(d) LMMSE based on cubic spline interpolation, PSNR = 24.38 dB, c_e = 0.74. CPU = 15.3 s.

Figure 7.13   LMMSE interpolation results. (a) Bilinear, PSNR = 22.78 dB, c_e = 0.47. (b) LMMSE based on bilinear interpolation, PSNR = 24.36 dB, c_e = 0.74, CPU = 13 sec. (c) Cubic spline, PSNR = 23.19 dB, c_e = 0.46. (d) LMMSE based on cubic spline interpolation, PSNR = 24.38 dB, c_e = 0.74. CPU = 15.3 sec.

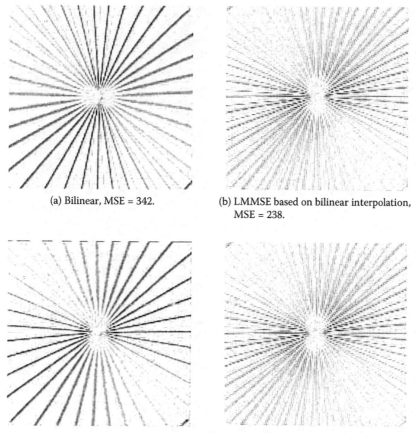

(a) Bilinear, MSE = 342.

(b) LMMSE based on bilinear interpolation, MSE = 238.

(c) Cubic spline. MSE = 312

(b) LMMSE based on cubic spline interpolation, MSE = 274.

Figure 7.14 Error images for LMMSE interpolation of test pattern image. (a) Bilinear, MSE = 342. (b) LMMSE based on bilinear interpolation, MSE = 238. (c) Cubic spline. MSE = 312. (d) LMMSE based on cubic spline interpolation, MSE = 274.

(a) Cubic O-MOMS, PSNR = 23.19 dB,
c_e = 0.47.

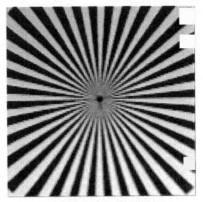

(b) LMMSE based on cubic O-MOMS
interpolation, PSNR = 24.37 dB,
c_e = 0.74, CPU = 15.3 s.

(c) Keys', PSNR = 23.12 dB, c_e = 0.46.

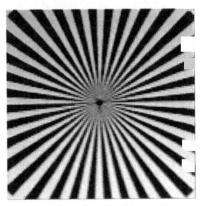

(d) LMMSE based on Keys' interpolation,
PSNR = 24.3 dB, c_e = 0.74. CPU = 13.5 s.

**Figure 7.15   LMMSE interpolation results. (a) Cubic O-MOMS, PSNR = 23.19 dB,
c_e = 0.47. (b) LMMSE based on cubic O-MOMS interpolation, PSNR = 24.37 dB,
c_e = 0.74, CPU = 15.3 sec. (c) Keys', PSNR = 23.12 dB, c_e = 0.46. (d) LMMSE based
on Keys' interpolation, PSNR = 24.3 dB, c_e = 0.74. CPU = 13.5 sec.**

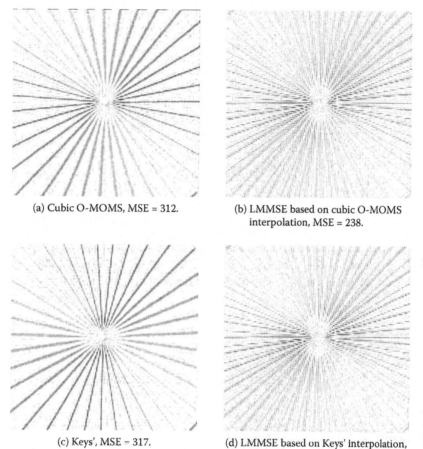

(a) Cubic O-MOMS, MSE = 312.

(b) LMMSE based on cubic O-MOMS interpolation, MSE = 238.

(c) Keys', MSE = 317.

(d) LMMSE based on Keys' interpolation, MSE = 242.

Figure 7.16   Error images for LMMSE interpolation of test pattern image. (a) Cubic O-MOMS, MSE = 312. (b) LMMSE based on cubic O-MOMS interpolation, MSE = 238. (c) Keys', MSE = 317. (d) LMMSE based on Keys' interpolation, MSE = 242.

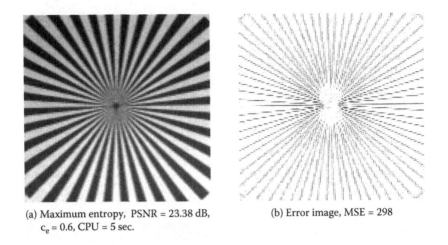

(a) Maximum entropy, PSNR = 23.38 dB,
c_e = 0.6, CPU = 5 sec.

(b) Error image, MSE = 298

Figure 7.17   Maximum entropy of test pattern image. (a) Maximum entropy, PSNR = 23.38 dB, c_e = 0.6, CPU = 5 sec. (b) Error image, MSE = 298.

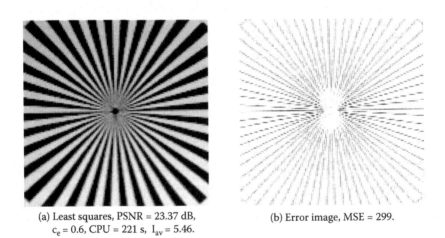

(a) Least squares, PSNR = 23.37 dB,
c_e = 0.6, CPU = 221 s, I_av = 5.46.

(b) Error image, MSE = 299.

Figure 7.18   Least-squares interpolation of test pattern image. (a) Least-squares, PSNR = 23.37 dB, c_e = 0.6, CPU = 221 s, I_{av} = 5.46. (b) Error image, MSE = 299.

(a) Regularized iterative (100 iterations), PSNR = 23.03, c_e = 0.6, CPU =120 s, λ = 0.001.

(b) Error image, MSE = 324.

(c) Regularized inverse, PSNR = 29.81, c_e = 0.89, CPU = 62 s, λ = 0.001.

(d) Error image, MSE = 68.

Figure 7.19 **Regularized interpolation of test pattern image. (a) Regularized iterative (100 iterations), PSNR = 23.03, c_e = 0.6, CPU = 120 sec, λ = 0.001. (b) Error image, MSE = 324. (c) Regularized inverse, PSNR = 29.81, c_e = 0.89, CPU = 62 sec, λ = 0.001. (d) Error image, MSE = 68.**

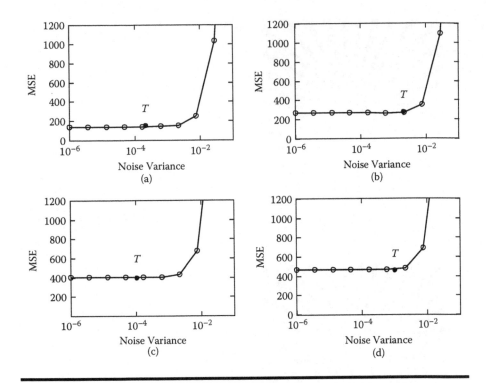

Figure 7.20 **Effect of noise variance estimation on LMMSE interpolation algorithm with auto-correlation estimated from cubic O-MOMS interpolation.** **(a) Test pattern image, SNR = 25 dB. (b) Test pattern image, SNR = 15 dB. (c) Woman image, SNR = 25 dB. (d) Woman image, SNR = 15 dB. T = point of true noise variance estimate.**

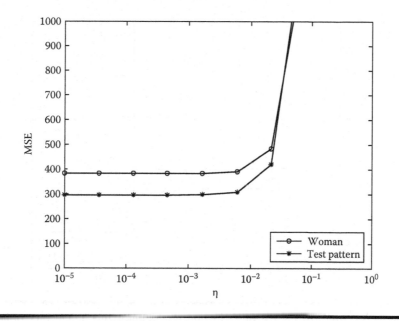

Figure 7.21 Effect of choice of η on maximum entropy image interpolation algorithm.

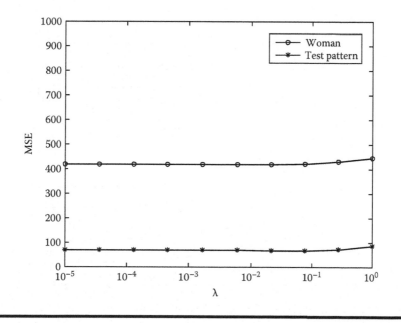

Figure 7.22 Effect of choice of λ on inverse regularized image interpolation algorithm.

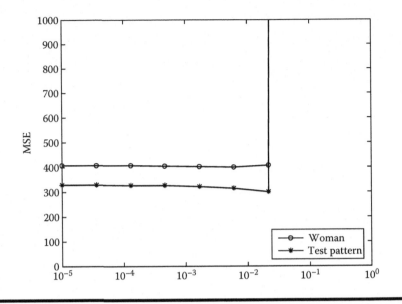

Figure 7.23 **Effect of choice of λ on iterative regularized image interpolation algorithm.**

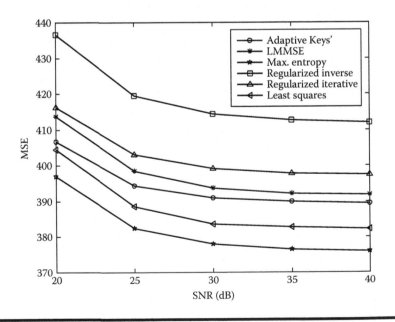

Figure 7.24 **Comparison of different interpolation algorithms for woman image.**

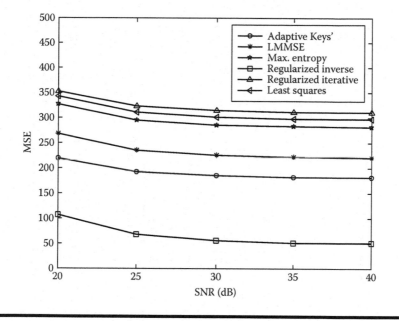

Figure 7.25 **Comparison of different interpolation algorithms for test pattern image.**

Table 7.1 Comparison of Inverse Interpolation Techniques for Noise-Free Images

Image	Least Squares	LMMSE	Maximum Entropy	Regularized Iterative	Regularized Inverse
Cameraman (128 × 128)	PSNR = 25.34 $c_e = 0.73$, $I_{av} = 21$ CPU = 810 sec	PSNR = 25 $c_e = 0.77$ CPU = 13 sec	PSNR = 25.36 $c_e = 0.73$ CPU = 5 sec	PSNR = 25.2 $c_e = 0.71$ CPU = 120 sec	PSNR = 26.83 $c_e = 0.84$ CPU = 62sec
Lenna (64 × 64)	PSNR = 23.52 $c_e = 0.65$ $I_{av} = 5.74$ CPU = 59 sec	PSNR = 23.6 $c_e = 0.74$ CPU = 3.3 sec	PSNR = 23.64 $c_e = 0.66$ CPU = 0.73 sec	PSNR = 23.25 $c_e = 0.66$ CPU = 8.6 sec	PSNR = 24.79 $c_e = 0.78$ CPU = 17sec
Mandrill (128 × 128)	PSNR = 19.15 $c_e = 0.65$ $I_{av} = 24.04$ CPU = 918 sec	PSNR = 19.07 $c_e = 0.66$ CPU = 13 sec	PSNR = 19.16 $c_e = 0.65$ CPU = 5 sec	PSNR = 19.1 $c_e = 0.64$ CPU = 120 sec	PSNR = 19.37 $c_e = 0.68$ CPU = 62sec
Building (64 × 64)	PSNR = 18.78 $c_e = 0.51$ $I_{av} = 13.53$ CPU = 132 sec	PSNR = 18.74 $c_e = 0.63$ CPU = 3.3 sec	PSNR = 18.85 $c_e = 0.51$ CPU = 0.73 sec	PSNR = 18.72 $c_e = 0.51$ CPU = 8.6 sec	PSNR = 18.29 $c_e = 0.47$ CPU = 17sec
Plane (64 × 64)	PSNR = 25.88 $c_e = 0.78$ $I_{av} = 35.3$ CPU = 334 sec	PSNR = 26.11 $c_e = 0.83$ CPU = 3.3 sec	PSNR = 25.92 $c_e = 0.78$ CPU = 0.73 sec	PSNR = 25.15 $c_e = 0.78$ CPU = 8.6 sec	PSNR = 27.61 $c_e = 0.88$ CPU = 17sec

Table 7.2 Comparison of Inverse Interpolation Techniques for Noisy Images (SNR = 20 dB)

Image	Least Squares	LMMSE	Maximum Entropy	Regularized Iterative	Regularized Inverse
Cameraman (128 × 128)	PSNR = 24.59 $c_e = 0.74$ $I_{av} = 19.56$ CPU = 739 sec	PSNR = 24.25 $c_e = 0.77$ CPU = 13 sec	PSNR = 24.58 $c_e = 0.75$ CPU = 5 sec	PSNR = 24.3 $c_e = 0.74$ CPU = 120 sec	PSNR = 25.58 $c_e = 0.83$ CPU = 62sec
Lenna (64 × 64)	PSNR = 23.24 $c_e = 0.66$ $I_{av} = 5.74$ CPU = 59 sec	PSNR = 23.29 $c_e = 0.73$ CPU = 3.3 sec	PSNR = 23.33 $c_e = 0.65$ CPU = 0.73 sec	PSNR = 23 $c_e = 0.66$ CPU = 8.6 sec	PSNR = 24.32 $c_e = 0.78$ CPU = 17sec
Mandrill (128 × 128)	PSNR = 18.95 $c_e = 0.65$ $I_{av} = 23.04$ CPU = 854 sec	PSNR = 18.82 $c_e = 0.65$ CPU = 13 sec	PSNR = 18.96 $c_e = 0.64$ CPU = 5 sec	PSNR = 18.8 $c_e = 0.63$ CPU = 120 sec	PSNR = 19.12 $c_e = 0.66$ CPU = 62sec
Building (64 × 64)	PSNR = 18.96 $c_e = 0.51$ $I_{av} = 13.53$ CPU = 136 sec	PSNR = 18.64 $c_e = 0.63$ CPU = 3.3 sec	PSNR = 18.75 $c_e = 0.51$ CPU = 0.73 sec	PSNR = 18.64 $c_e = 0.5$ CPU = 8.6 sec	PSNR = 18.19 $c_e = 0.47$ CPU = 17sec
Plane (64 × 64)	PSNR = 25.03 $c_e = 0.77$ $I_{av} = 33.09$ CPU = 314 sec	PSNR = 25.19 $c_e = 0.82$ CPU = 3.3 sec	PSNR = 25.07 $c_e = 0.77$ CPU = 0.73 sec	PSNR = 24.47 $c_e = 0.78$ CPU = 8.6 sec	PSNR = 26.18 $c_e = 0.87$ CPU = 17sec

(a) LR image. SNR = 25 dB.

(b) Cubic O-MOMS interpolation
PSNR = 23.92 dB.

(c) Warped-distance cubic O-MOMS
interpolation, PSNR = 24.1 dB.

(d) Adaptive least squares interpolation,
PSNR = 25.78 dB, I_{av} = 8.22 iterations,
CPU time = 4.6 second.

Figure 7.26   Interpolation of plane image with SNR = 25 dB. (a) LR image, SNR = 25 dB. (b) Cubic O-MOMS interpolation, PSNR = 23.92 dB. (c) Warped-distance cubic O-MOMS interpolation, PSNR = 24.1 dB. (d) Adaptive least-squares interpolation, PSNR = 25.78 dB, I_{av} = 8.22 iterations, CPU = 4.6 sec.

(a) LR image.

(b) Cubic O-MOMS interpolation, PSNR = 20.7 dB

(c) Warped-distance cubic O-MOMS interpolation, PSNR = 20 dB.

(d) Adaptive LS interpolation, PSNR = 21.75 dB, I_{av} = 19.66 iterations, CPU time = 7.2 second.

Figure 7.27 Interpolation of truck image with SNR = 25 dB. (a) LR image. (b) Cubic O-MOMS interpolation, PSNR = 20.7 dB. (c) Warped-distance cubic O-MOMS interpolation, PSNR = 20 dB. (d) Adaptive least-squares interpolation, PSNR = 21.75 dB. I_{av} = 19.66 iterations. CPU = 7.2 sec.

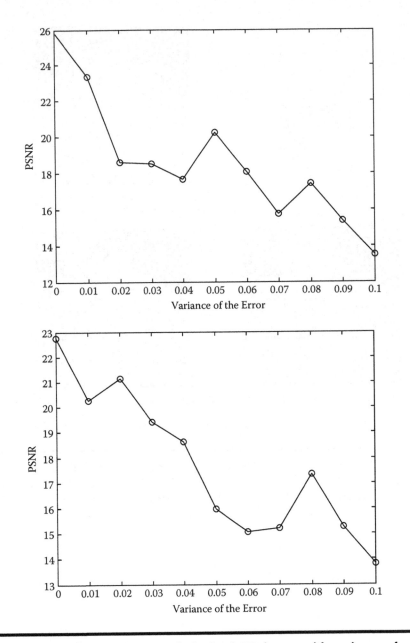

Figure 7.28 **Variation of PSNR of obtained HR image with variance of error in down-sampling operator used in interpolation process. (Top) Plane image. (Bottom) Truck image.**

Chapter 8

Image Registration

8.1 Introduction

Image registration is the process of overlaying two or more images of the same scene taken at different times, from different viewpoints and/or by different sensors. By image registration, the reference and the sensed images are aligned [88–90]. Image registration is an important step in all image processing fields that require the integration of information from multiple observations into a single image.

Image registration is a fundamental inverse problem in imaging. It represents a critical pre-processing step to many modern image processing tasks such as multichannel image restoration, image super-resolution (SR), and video compression. Image registration is a specific case of the more general problem of motion estimation in an image sequence, wherein the observed data follows the following form [90]:

$$f_s(n_1,n_2,t) = f_0(n_1 - v_1(n_1,n_2,t), n_2 - v_2(n_1,n_2,t), t) + \varepsilon(n_1,n_2,t) \tag{8.1}$$

where $f_0(n_1,n_2,t)$ and $f_s(n_1,n_2,t)$ are the reference and the shifted image functions at time t with spatial coordinates (n_1,n_2) and $\varepsilon(n_1,n_2,t)$ is an additive white Gaussian noise (AWGN) with variance σ^2.

The vector $\vec{v}(n_1,n_2,t) = [v_1(n_1,n_2,t), v_2(n_1,n_2,t)]^t$ is an unknown vector field characterizing the evolution of the image sequence in time. The estimation of this vector is a nonlinear problem, and this is the heart of the image registration process. If $v_1(n_1,n_2,t)$ and $v_2(n_1,n_2,t)$ are constants for the whole image, the registration problem will be similar to the problem of time delay estimation (TDE) or motion estimation treated in many signal processing applications [90,91].

During recent decades, several image acquisition devices have been developed and their development invoked research on automatic image registration. A large number

195

of researchers have worked in the field of image registration and several registration methods have been proposed [88–90]. In this chapter, we do not get into details of these methods. Our objective is to summarize the main approaches and their interesting features because the reconstruction of a high-resolution (HR) image from multiple degraded observations is an application field that benefits from image registration.

8.2 Applications of Image Registration

In image registration, the relative shifts between low-resolution (LR) images compared to a reference LR image are estimated with fractional pixel accuracy. Obviously, accurate sub-pixel motion estimation is a very important factor in the success of an SR image reconstruction algorithm. Although there is a broad spectrum of methodologies for image registration, it is difficult to classify and compare techniques since each technique is often designed for a specific application and not necessarily for types of problems or data. However, most registration techniques involve searching over the space of transformations of a certain type (e.g., affine, polynomial, or elastic) to find the optimal transformation for a particular problem. Generally, applications of image registration can be divided into four main groups according to the manner of the image acquisition [88].

8.2.1 Different Viewpoints (Multi-View Analysis)

Images of the same scene are acquired from different viewpoints. The aim is to gain a larger two-dimensional (2-D) view or a three-dimensional (3-D) representation of the scanned scene. Examples of such applications are mosaicking or SR reconstruction of remotely sensed images and shape recovery in computer vision.

8.2.2 Different Times (Multi-Temporal Analysis)

Images of the same scene are acquired at different times, often on a regular basis, and possibly under different conditions. The aim is to find and evaluate changes in the scene that appear between the consecutive image acquisitions. Examples of such applications are SR reconstruction of remotely sensed images, automatic change detection, and monitoring tumors in medical images.

8.2.3 Different Sensors (Multi-Modal Analysis)

Images of the same scene are acquired by different sensors. The aim is to integrate the information obtained from different source streams to gain a more complex and detailed scene representation. Examples of such applications are fusion of remotely sensed images, enhancement of spectral resolution of images, and processing of magnetic resonance (MR) images.

8.2.4 Scene-to-Model Registration

Images of a scene and a model of the same scene are registered. The model can be a computer representation of the scene. Examples of such applications are registration of satellite images into maps, comparison of a patient's image with a digital anatomical atlas, specimen classification, target template matching with real time images, and automatic quality inspection.

8.3 Steps of Image Registration

Most image registration methods consist of the following four steps [88]:

1. Feature detection. Salient and distinctive objects (edges, closed regions, contours, and corners) are automatically detected. These features are used for further processing and represented by their point representatives (centers of gravity, line endings, and distinctive points) that are called control points (CPs).
2. Feature matching. The correspondence between the features detected in the sensed image and those detected in the reference image is established.
3. Transform model estimation. The type and parameters of the mapping functions aligning the sensed image with the reference image are estimated. These parameters are computed by means of the established feature correspondence.
4. Image resampling and transformation. The sensed image is transformed by means of the mapping functions. Image values in non-integer coordinates are computed by an appropriate interpolation technique.

The implementation of each step has several problems. In the feature detection step, it is required to decide which kinds of features are appropriate for the given task. The features should be distinctive objects that are spread over the images and are easily detectable. The detection method should have good localization accuracy and should not be sensitive to the assumed image degradation. In the feature matching step, problems caused by incorrect feature detection or by image degradation arise. The feature descriptors should be invariant to the image degradations. They should discriminate between different features. They should also be stable against unexpected feature variations [88].

The *a priori* known information about the acquisition process and the expected image degradations determine the type of the matching functions that should be used. Where no prior information is available, the mathematical model should be flexible enough to deal with different forms of degradations. The accuracy of the feature detection method, the reliability of feature correspondence estimation, and the acceptable estimation error should also be considered in the choice of the mapping functions. A decision must determine which differences between the images should be removed and which differences should be maintained [88].

Another important topic is the choice of the appropriate type of resampling technique, which depends on the trade-off between the required accuracy of interpolation and the computational cost.

There are several criteria that can be used to categorize registration methods. The most frequently used ones are the application area, the dimensionality of the data, the type and complexity of the assumed image deformations, the computation cost, and the essential ideas of the registration algorithm [88]. In the following sections, we are going to discuss the implementation of each registration step.

8.3.1 Feature Detection Step

In this step, salient features in images are extracted and significant regions are understood as features. The compatibility of the feature sets in the sensed and the reference images is assured by the invariance and the accuracy of the feature detector. The number of common elements of the detected sets of features should be sufficiently high, regardless of the change of image geometry, radiometric conditions, presence of additive noise, and changes in the scanned scene [88]. This step does not work directly with image intensity values. Thus, it is suitable, when illumination changes are expected or multi-sensor analysis is demanded.

This step may be omitted in some registration applications such as medical image registration. Medical images are not so rich in such details. On the other hand, this step is very important for images containing large numbers of distinctive features like natural scene images, satellite images, commercial images, and high-definition television (HDTV) images.

8.3.2 Feature Matching Step

In this step, the detected features in the reference and sensed images can be matched by means of image intensity values in their close neighborhoods. There are two major categories: area-based methods and feature-based methods.

8.3.2.1 Area-Based Methods

Area-based methods are preferably applied when the available images do not have distinctive details. Area-based methods are able to make the registration of images in the case of a small shift or rotation between the images [88]. Another restriction on area-based methods is that the images must be correlated or statistically dependent.

Area-based methods are sometimes called correlation-like methods or template matching. They merge the feature detection step with the matching step. These methods deal with the image without attempting to detect salient objects. In these methods, windows of predefined size are used for correspondence estimation during the second registration step.

The limitations of these methods arise if the images are deformed by complex transformations. A deformation makes the window unable to cover the same parts of the scene in the reference and sensed images. Another drawback of these methods is that the choice of windows without salient features may lead to incorrect matching with other windows in both the sensed and reference images.

8.3.2.1.1 Correlation-Like Methods

The correlation coefficient (CC) is computed for window pairs from the sensed and reference images. This CC maximum value is searched and used as a measure of similarity between windows. The equation of this CC is given by [88,92]:

$$CC(n_1, n_2) = \frac{\sum\limits_{W} (W - E(W))(I(n_1, n_2) - E(I(n_1, n_2)))}{\sqrt{\sum\limits_{W} (W - E(W))^2} \sqrt{\sum\limits_{I(n_1, n_2)} (I(n_1, n_2) - E(I(n_1, n_2)))^2}} \tag{8.2}$$

where W is a window of the reference image and $I(n_1, n_2)$ is a window of the sensed image at position (n_1, n_2). E denotes the mathematical expectation.

The main drawbacks of the correlation-like methods are the flatness of the similarity measure maxima due to the self-similarity of the images and the high computational cost. The maximum can be sharpened by pre-processing or by using edge or vector correlation. Despite these limitations, the correlation-like registration methods are still in use due to their easy hardware implementation.

8.3.2.1.2 Fourier Methods

If the obtained images were acquired under varying conditions or they are corrupted by frequency-dependent noise, the Fourier methods are preferred. Fourier methods achieve acceleration of computational speed. The cross-power spectrum of the sensed and reference images is estimated. A search is performed for the location of the peak in its inverse. The normalized cross-power spectrum between the reference image f_0 and the sensed image f_s is given by [88]:

$$P_{0s}(u, v) = \frac{\Im(f_0)\Im(f_s)^*}{\left|\Im(f_0)\Im(f_s)^*\right|} = e^{2\pi i (uv_1 + vv_2)} \tag{8.3}$$

where (u, v) are the frequency coordinates and (v_1, v_2) are the translation shifts in both directions. These methods show robustness against the correlated and frequency-dependent noise and non-uniform time-varying illumination disturbances. They are used for registering images with translation shifts. The computation time savings using this method are significant if the images to be registered are large.

8.3.2.1.3 Mutual Information Methods

Mutual information (MI) methods originated from the information theory. The MI is a measure of the statistical dependency between two available data sets: the sensed image and the reference image. It is particularly suitable for registration of images from different modalities. MI between two random variables X and Y is given by [88]:

$$MI(X,Y) = H_e(Y) - H_e(Y|X) = H_e(X) + H_e(Y) - H_e(X,Y) \qquad (8.4)$$

where $H_e(X) = -p_d(X)\big(\log\big(p_d(X)\big)\big)$ represents the entropy of the random variable X and $p_d(X)$ is the probability density of that random variable.

These methods are based on the maximization of the MI and work with the entire image intensities directly.

8.3.2.2 Feature-Based Methods

Feature-based matching methods are typically applied when the local structural information is more significant than the information carried by the image intensities. These methods allow the registration of images of completely different natures and can handle complex between-image distortions [88]. When two sets of features in the reference and sensed images are detected, the aim is to find the pair-wise correspondence between them using their spatial relations or various descriptors of features. The descriptors of features must be invariant to all assumed differences between images to be registered.

8.3.2.2.1 Methods Using Spatial Relations

These methods are used if the detected features are ambiguous or distorted. Clustering techniques are the most popular among spatial relations methods. These techniques try to match points connected by abstract edges or line segments. The assumed geometrical model is the similarity transform. For each pair of CPs from both the reference and sensed images, the parameters of the transformation that maps the points on each other are computed and represented as a point in the space of transform parameters. The parameters of the transformations that closely map the highest number of features tend to form a cluster, while mismatches fill the parameter space randomly. The cluster is detected, and its centroid is assumed to represent the most probable vector of matching parameters. Mapping function parameters are thus found simultaneously with the feature correspondence [88].

8.3.2.2.2 Methods Using Invariant Descriptors

The correspondence of features can be estimated using their description, which is preferred to be invariant to the expected image deformation. The description should achieve several conditions. The most important ones are invariance (the descriptions of the corresponding features from the reference and sensed image must be the same), uniqueness (two different features should have different descriptions), stability (the description of a feature that is slightly deformed in an unknown manner should be close to the description of the original feature), and independence (if the feature description is a vector, its elements should be functionally independent) [88].

Features from the sensed and reference images with the most similar invariant descriptions are paired as the corresponding ones. The choice of the type of the invariant description depends on the feature characteristics and the assumed geometric deformation of the images. While searching for the best matching feature pairs in the space of feature descriptors, the minimum distance rule with thresholding is usually applied.

8.3.2.2.3 Relaxation Methods

The majority of registration methods are based on the relaxation approach. This approach is one of the solutions to the consistent labeling problem (CLP). This means the labeling of each feature from the sensed image with the label of a feature from the reference image. The process of recalculating the pair of figures of merit or pair of labels in the reference and sensed images, considering the match quality of the feature pairs and their neighbors, is iteratively repeated until a stable situation is reached. This method can handle shifted images and it tolerates local image distortions [88].

8.3.2.2.4 Pyramids and Wavelets

In these methods, both the reference and sensed images are used at a coarser resolution, and then at different lower resolutions, forming pyramids of matching. The wavelet packet transform is a useful tool in these methods. In general, this coarse-to-fine hierarchical strategy applies the usual registration methods, but starts with the reference and sensed images on a coarse resolution. Then, they gradually improve the estimates of the correspondence or the mapping function parameters while going toward finer resolutions.

At every level, these methods considerably decrease search space and thus save computational costs. Another important advantage is the fact that the registration with respect to large-scale features is achieved first, and then small corrections are made for finer details. On the other hand, this strategy fails if a false match is identified on a coarser level. To overcome this, a backtracking or consistency check should be incorporated into the algorithm [88].

8.3.3 Transform Model Estimation

After the feature correspondence between the sensed image and the reference image has been established, a mapping function is chosen. The role of this mapping function is to transform the sensed image to overlay it over the reference one. Whenever the mapping process is performed on the reference image, the CPs from both the transformed and the sensed images should be as close as possible. The type of the chosen mapping function should correspond to the assumed geometric deformation of the sensed image to the method of image acquisition and to the required accuracy of registration [88].

Models of mapping functions can be divided into two main categories: global and local mapping functions, according to the amount of image data they use as their support. Global models use all CPs for estimating one set of mapping function parameters valid for the entire image. On the other hand, local models treat the image as a composition of patches and the function parameters, depending on the location of their support in the image [88].

8.3.3.1 Global Mapping Models

These models such as rotation, translation, and scaling, often achieve shape-preserving mapping. One of the global mapping functions is defined as follows [88]:

$$m_1 = a(n_1 \cos(\varphi) + n_2 \sin(\varphi)) + v_1 \qquad (8.5)$$

$$m_2 = a(n_1 \sin(\varphi) + n_2 \cos(\varphi)) + v_2 \qquad (8.6)$$

where (n_1, n_2) and (m_1, m_2) are the original and transformed coordinates, respectively, and φ is the rotation angle. The variable a is a scalar and v_1 and v_2 are spatial shifts in the horizontal and vertical directions, respectively. Another global mapping model is the affine transform defined as follows [88]:

$$m_1 = a_0 + a_1 n_1 + a_2 n_2 \qquad (8.7)$$

$$m_2 = b_0 + b_1 n_1 + b_2 n_2 \qquad (8.8)$$

where a_0, a_1, a_2, b_0, b_1 and b_2 are constants. This model is used for multi-view registration when the distance of the scanning camera to the scene is large in comparison to the size of the scanned area. If this condition is not achieved, another model is used [1]:

$$m_1 = \frac{a_0 + a_1 n_1 + a_2 n_2}{1 + c_1 n_1 + c_2 n_2} \qquad (8.9)$$

$$m_2 = \frac{b_0 + b_1 n_1 + b_2 n_2}{1 + c_1 n_1 + c_2 n_2} \qquad (8.10)$$

where $a_0, a_1, a_2, b_0, b_1, b_2, c_1$ and c_2 are constants.

In general, the number of CPs used is usually higher than the minimum number required for the determination of the mapping function. The parameters of the mapping function are then computed by means of the least-squares fit, so that the polynomials minimize the sum of squared errors at the CPs. These mapping functions do not map the CPs onto their counterparts exactly. This approach is very effective and accurate for satellite images.

8.3.3.2 Local Mapping Models

In medical imaging, the images are deformed locally and therefore global mapping models cannot handle the registration of these images. In this case, local areas of the image should be registered with the available information about the existing local geometric distortion. There are several methods for local mapping registration such as the weighted least squares and the weighted mean methods [88].

8.3.4 Image Resampling and Transformation

The mapping functions constructed during the previous step are used to transform the sensed image to the coordinates of the reference image and thus register the images. The transformation can be realized in a forward or backward manner. Each pixel from the sensed image can be directly transformed using the estimated mapping function. This approach, called a forward method, is complicated to implement, as it can produce holes and/or overlaps in the transformed image due to the discretization and rounding. Hence, the backward approach is usually chosen [88].

The registered image data from the sensed image are determined using the coordinates of the target pixel (the same coordinate system as the reference image) and the inverse of the estimated mapping function. The image interpolation takes place in the sensed image after transformation on the regular grid.

8.4 Evaluation of Image Registration Accuracy

Estimation of the accuracy of a registration method is an essential part of the registration process and is required regardless of the images, the registration method, and the application area. The importance of this step is attributed to factors. First, errors can be dragged into the registration process at each of its stages. Second, it is hard to distinguish between registration inaccuracies and actual physical differences in image contents. Several error classes may occur during the registration process. These classes are [88]:

1. Localization error. This error arises from the displacement of the CP coordinates due to their inaccurate detection. It can be reduced by selecting an optimal feature detection algorithm for the given data.

2. Matching error. This error is measured by the number of false matches when establishing the correspondence between CP coordinates. A consistency check can be used to identify this type of error.

3. Alignment error. This error relates to the difference between the mapping model used for the registration and the actual between-image geometric distortion. This type of error can be evaluated in several ways such as a consistency check, the mean square error (MSE) at the control points (CPs) (control point error—CPE) and the test point error (TPE).

Chapter 9

Image Fusion

9.1 Introduction

Image fusion is a process by which information from different images of the same scene are incorporated into a single image. Hence, we can define image fusion as the representation of visual information contained in a number of input images into a single fused image [93–98]. The importance of image fusion lies in the fact that each observation image contains complementary information. When this complementary information is integrated with that of another observation, an image with the maximum amount of information is obtained.

Image fusion has been used in several application areas. In remote sensing and astronomy, multi-sensor fusion is used to achieve high spatial and spectral resolutions by combining images from two sensors—one with high spatial resolution and the other with high spectral resolution [95,99,100]. Numerous applications such as the fusion of computed tomography (CT), magnetic resonance (MR), and positron emission tomography (PET) have revolutionized medical imaging.

Plenty of applications that use multi-sensor fusion of visible and infrared images have appeared in military, security, and surveillance areas. In the case of multi-view fusion, a set of images of the same scene taken by the same sensor but from different viewpoints is fused to obtain an image with a higher resolution than the sensor normally provides or to recover a three-dimensional (3-D) representation of the scene. The multi-temporal approach has two objectives. Images of the same scene are acquired at different times to find and evaluate changes in the scene or obtain a less degraded image of the scene [101].

Medical imaging is used to detect changes of organs and tumors; remote sensing for monitoring changes on land. The list of applications mentioned above illustrates the diversity of problems we face when fusing images. It is impossible to design a universal method applicable to all image fusion tasks. Every method should take into account the fusion purpose and characteristics of individual sensors, the specific imaging conditions, imaging geometry, noise corruption, required accuracy, and application-dependent data properties.

9.2 Objectives of Image Fusion

The images to be fused always contain both complementary and redundant information. The integration of this redundant information reduces the uncertainty and increases the accuracy of the features. On the other hand, the complementary information provides more detailed information regarding features not available from a single source. The objectives of image fusion can be classified into the following categories [95,98]:

Image sharpening — One of the main objectives of image fusion is to increase the spatial resolution of the input image. For example, in satellite image fusion, the high-resolution (HR) panchromatic (PAN) images are fused to the low-resolution (LR) multi-spectral (MS) images. The outcome of this process is an MS fused image that has a high spatial resolution.

Feature enhancement — In some instances, features in an image may be distorted or blurry. By using detail information from other sensors and fusing this information into the image, one can enhance the features of the image. For instance, certain sensors such as infrared cameras do not require sources of illumination. These sensors collect and process the thermal energy radiated from an object to produce an image. However, the images produced by these cameras have much lower signal-to-noise ratios (SNRs) than the visual images. As a result, the fused image of infrared and visual inputs reveals more enhanced features.

Improved classification — One way of examining a scene is to analyze its different textures. In some instances, a single sensor does not provide enough information to discriminate between two textures. Therefore, using more sensors and fusing their spectral responses achieves a better texture classification.

Robotic vision — One of the problems in robotic vision is self localization by which a robot determines its location. Most single-vision sensors fail to provide reliable outcomes due to low SNRs and other sensory limitations. By fusing the information from more sensors, a robot can distinguish sharp corners and edges from walls.

Biomedical applications — Image fusion provides a diagnostic tool that facilitates the detection and classification of diseases from multiple modalities such as MR or CT scans. Another application of fusion in the medical field is improving the visual information for edge detection and enhancement of features for the detection of lesions.

Clutter removal — On many occasions in remote sensing, the actual scene of interest is covered by clouds or clutter. In some instances, even the shadows of the cloud influence the way the images are interpreted. Since infrared radiation is not blocked by clouds, fusion of visible and infrared images preserves covered features.

9.3 Implementation of Image Fusion

Multi-sensor image fusion can be divided into pixel, feature, and decision levels. In pixel level fusion, the fusion works directly on the pixels obtained at the sensor output. On the other hand, feature level fusion works on image features extracted from source images. Finally, decision level image fusion works on merging the interpretations of different images obtained after an image is understood [99–103].

Figure 9.1 illustrates a system that uses image fusion at all the three levels of processing [99–103]. This general structure could be used as a basis for any image processing system.

There are interesting applications for both single-sensor and multi-sensor image fusion. Figure 9.2 illustrates a single-sensor image fusion system. The sensor shown could be a visible-band sensor such as a digital camera. This sensor captures the real world as a sequence of images. The sequence is then fused in a single image and used either by a human operator or by a computer to perform a certain task. For example, in object detection, a human operator searches a scene to detect objects such as intruders in a security area [99–103].

This kind of system has some limitations due to the capabilities of the imaging sensor used. The conditions under which the system can operate, the dynamic range, and the resolution are all limited by the capabilities of the sensor. For example,

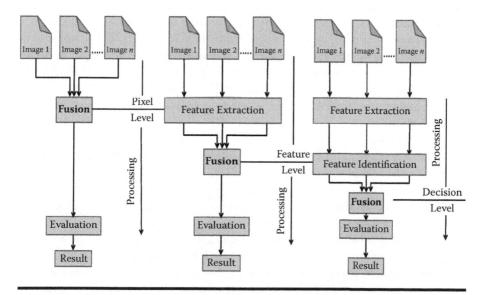

Figure 9.1   Information fusion at all three processing levels.

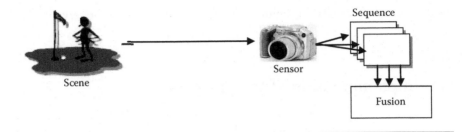

Figure 9.2 Single-sensor image fusion system.

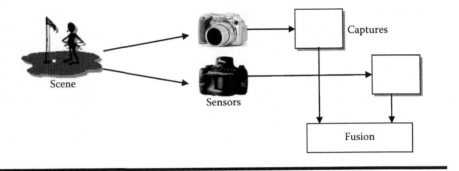

Figure 9.3 Multi-sensor image fusion system.

a visible-band sensor such as a digital camera is appropriate for a brightly illumi-
nated environment such as a daylight scene, but is not suitable for poorly illumi-
nated situations at night or under adverse conditions such as fog or rain [99–103].

Multi-sensor image fusion systems overcome the limitations of a single-sensor
vision system by combining the images from these sensors to form a composite
image. Figure 9.3 shows a multi-sensor image fusion system. For example, to better
identify the objects in remote sensing images, the MS images with high spectral
resolution and low spatial resolution and the PAN images with high spatial resolution and low spectral resolution, need to be fused. In medical applications, multiple
CT and MR images are fused to improve the information content for diagnosis
[99–103]. The benefits of multi-sensor image fusion include [99–103]:

Extended range of operation — Multiple sensors utilizing different operating
conditions can be deployed to extend the effective range of operation. For example,
different sensors can be used for day/night operation.

Reduced uncertainty — Combined information from multiple sensors can
reduce the uncertainty associated with a sensing or decision process.

Increased reliability — The fusion of multiple measurements can reduce noise
and therefore improve the reliability of the measured quantity.

Robust system performance — Redundancy in multiple measurements can
help in systems robustness. In case one or more sensors fail or the performance of a
particular sensor deteriorates, the system can depend on the other sensors.

Compact representation of information — Fusion leads to a compact representation of information present in all observation images.

9.4 Pixel Level Image Fusion

Pixel level (PL) image fusion is the simplest approach. It represents fusion of visual information of the same scene from any number of registered images obtained using different sensors. The goal of PL image fusion is to represent the visual information present in any number of input images in a single fused image without introducing distortion or loss of information.

In simpler terms, the main condition for successful fusion is that all visible information in the input images should be visible in the fused image. However, although theoretically possible, due to the redundant nature of multi-sensor information (for example, slightly different signatures of the same object in different sensor modalities), the complete representation of all visual information from a number of input images into a single one is seldom achieved in practice.

Thus, the practical goal of PL image fusion is modified to the fusion or preservation in the output fused image of the most important visual information in an input image set. The main requirement of the fusion process then is to identify the most significant features in the input images and transfer them without loss into the fused image. What defines important visual information is generally application dependent. In most applications and in image fusion for display purposes in particular, the definition is perceptually important information [99–103].

A simple diagram of a system using PL image fusion is shown in Figure 9.4. Only two imaging sensors survey the environment, producing two different representations of the same scene. The representations of the environment are, again, in the form of image signals that are corrupted by noise arising from atmospheric aberrations, sensor design, quantization, etc.

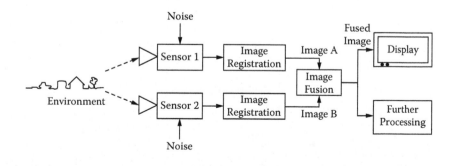

Figure 9.4   Basic structure of multi-sensor system using PL image fusion.

The image signals produced by the sensors are first registered. In Figure 9.4, the registered input images are fused and the resulting fused image can then be used directly for display purposes or passed on for further processing.

For PL image fusion, input images must be of the same scene. Furthermore, inputs are assumed to be spatially registered and of equal size. The simplest methods for PL image fusion are averaging and weighted averaging. These methods include only averaging of the images to be fused. Although simple, the method is not feasible in some applications due to contrast reduction side effects [99–103].

9.5 Principal Component Analysis Fusion

Principal component analysis (PCA) is a statistical technique that can be used for image fusion. The following steps are applied for the fusion of two images using the PCA [104–109]:

1. The two images are decomposed into their principal components (PCs).
2. The first PC of the first image is replaced by the first PC of the second image.
3. An inverse transformation is applied to reconstruct a new image.

PCA can be used for the fusion of PAN and MS images in satellite image fusion. It converts an MS image with correlated bands into a set of uncorrelated components. The first component resembles the PAN image. It is, therefore, replaced by the HR PAN image for the fusion. Before the replacement, the PAN image must be matched to the first component. The PAN image is fused into the LR MS bands by performing an inverse principal component transform (PCT). The block diagram of the PCA image fusion method for satellite images is shown in Figure 9.5. The algorithm that replaces the principal component of the MS image with the PAN image allows the spatial details of the PAN image to be incorporated into the MS image [104–109]. We can summarize in steps the PCA fusion method for satellite images as follows:

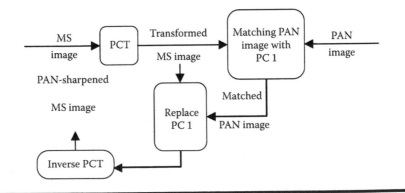

Figure 9.5 Satellite image fusion with PCA method.

1. The re-sampled bands of the MS image to the same resolution as the PAN image are transformed with the PCT.
2. The PAN image is histogram matched to the first PC. This is done to compensate for the spectral differences between the two images that occur due to differences in sensors or acquisition dates and angles.
3. The first PC of the MS image is replaced by the histogram-matched PAN image.
4. The new merged MS image is obtained by computing the inverse PCT.

9.6 Wavelet Fusion

Wavelet-based image fusion methods include both the discrete wavelet transform (DWT) and the discrete wavelet frame transform (DWFT) methods [104–109]. The main difference between these methods is that the DWT uses decimation and the DWFT does not use decimation. This makes the DWT method very sensitive to registration errors.

9.6.1 DWT Fusion

Figure 9.6 depicts a diagram of the fusion of two images using the DWT [104–109]. It can be defined considering the wavelet transform W of two registered input images $I_1(n_1, n_2)$ and $I_2(n_1, n_2)$ together with the fusion rule ϕ [104–109]. Several wavelet fusion rules can be used for selecting the wavelet coefficients.

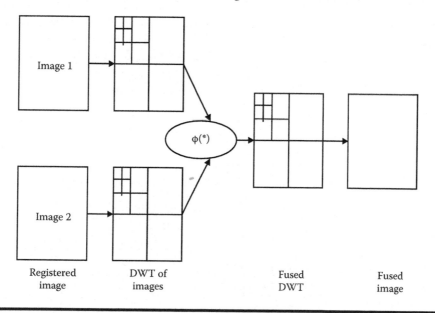

| Registered image | DWT of images | Fused DWT | Fused image |

Figure 9.6   DWT fusion.

Maximum frequency rule — It selects the coefficients with the highest absolute values. The high values indicate salient features like edges, and are thus incorporated into the fused image. This rule is applied at all resolutions under consideration.

Weighted average rule — It generates a coefficient via a weighted average of the coefficients of the two images. The weights are based on the correlation between the two images.

Standard deviation rule — It calculates an activity or energy metric associated with each coefficient. A decision map is created to indicate the source image from which the coefficient has to be selected.

Window-based verification rule — It creates a binary decision map to choose between each pair of coefficients using a majority filter.

The maximum frequency rule is the most frequently used. It selects the maximum coefficients from the wavelet transformed images [104–109]. Then the inverse wavelet transform W^{-1} is computed, and the fused image $I(n_1, n_2)$ is reconstructed [104–109]:

$$I(n_1, n_2) = W^{-1}(\phi(W(I_1(n_1, n_2)), W(I_2(n_1, n_2)))) \qquad (9.1)$$

DWT fusion can also be used with satellite images in which the PAN image is fused with one of the MS bands as shown in Figure 9.7.

9.6.2 DWFT Fusion

The purpose of the DWFT is to decompose the image into additive components, each of which is a sub-band of the image. This step isolates the different frequency components of the image into different planes without down-sampling as in the DWT. The *a trous* algorithm below is used for this purpose [110,111]. Given an image P, it is possible to construct the sequence of approximations:

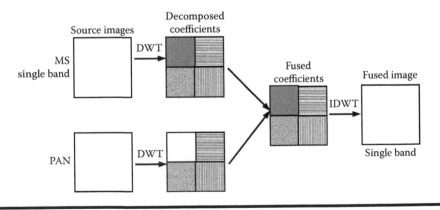

Figure 9.7   Block diagram of DWT fusion for satellite images.

$$f_1(P) = P_1, f_2(P_1) = P_2, f_3(P_2) = P_3, \ldots\ldots\ldots\ldots, f_n(P_{n-1}) = P_n \qquad (9.2)$$

where n is an integer which is preferred to be 3. To construct this sequence, successive convolutions with a certain low-pass kernel are performed. The functions f_1, f_2, f_3, and f_n mean convolutions with this kernel given by [110,111]:

$$H = \frac{1}{256} \begin{bmatrix} 1 & 4 & 6 & 4 & 1 \\ 4 & 16 & 24 & 16 & 4 \\ 6 & 24 & 36 & 24 & 6 \\ 4 & 16 & 24 & 16 & 4 \\ 1 & 4 & 6 & 4 & 1 \end{bmatrix} \qquad (9.3)$$

The wavelet detail planes are computed as the differences between two consecutive approximations P_{l-1} and P_l [110,111], i.e.,

$$\Delta_l = P_{l-1} - P_l, \quad (l = 1, 2, \ldots\ldots\ldots, n) \qquad (9.4)$$

Thus, the curvelet reconstruction formula is given by [110,111]:

$$P = \sum_{l=1}^{n} \Delta_l + P_n \qquad (9.5)$$

where the planes Δ_l contain high-frequency details and P_n is a low-frequency approximation component. The decomposition and reconstruction stages of the additive wavelet transform (AWT) are shown in Figures 9.8 and 9.9, respectively.

As shown in Figure 9.10, in the DWFT fusion method for satellite images, the source images are first decomposed using the DWFT. After that, wavelet coefficients from the lowest-frequency approximation component of each spectral band of the MS image and the high-frequency detail components from the PAN image are combined, and the fused image is then reconstructed by performing the inverse DWFT. The performance of the DWFT fusion method outperforms the DWT method [110,111].

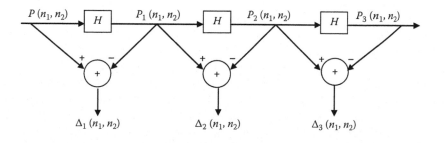

Figure 9.8   Decomposition in AWT.

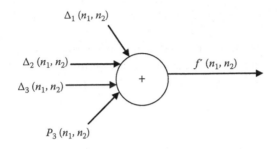

Figure 9.9   Reconstruction in AWT.

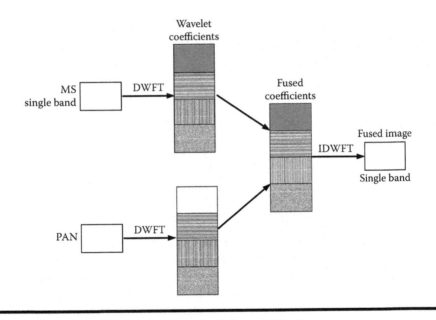

Figure 9.10   DWFT image fusion method for satellite images.

9.7 Curvelet Fusion

As mentioned in the previous section, the DWT is the most famous tool for image and signal analysis because of its advantageous property that helps localize point singularities in a signal or an image [104–107]. A major disadvantage of the DWT in image processing is that it gives a large number of coefficients in all scales corresponding to the edges of the image. This means that many coefficients are required to exactly reconstruct the edges of an image. This makes the DWT inefficient for handling long curved edges. Recent approaches like the ridgelet transform and the curvelet transform are more efficient in handling long linear and curvilinear singularities in an image [104–107].

In the curvelet transform, the AWT is used instead of the DWT to decompose the image into different sub-bands called the detail planes and the approximation plane, and each sub-band of the detail planes is then partitioned into small tiles. Finally, the ridgelet transform is applied on each tile [104–107]. In this way, the image edges can be represented efficiently by the ridgelet transform, because the image edges will now be almost like small straight lines. Thus, the curvelet transform is an extension of the ridgelet transform to detect curved edges effectively. The algorithm of the curvelet transform can be summarized in the following steps [104–107]:

1. The image is split into four sub-bands Δ_1, Δ_2, Δ_3 and P_3 using the AWT.
2. Tiling is performed on the sub-bands Δ_1, Δ_2, and Δ_3.
3. The discrete ridgelet transform is performed on each tile of the sub-bands Δ_1, Δ_2, and Δ_3.

Figure 9.11 is a diagram of the curvelet transform steps. A detailed description of these steps will be presented in the following sub-sections.

The curvelet transform can be used to fuse images with different modalities such as MR and CT images. The steps of the curvelet fusion approach to MR and CT images can be summarized as follows [104–107]:

1. The MR and the CT images are registered.
2. The curvelet transform steps are performed on both images.
3. The maximum frequency fusion rule is used for the fusion of the ridgelet transforms of the tiles in the sub-bands Δ_1, Δ_2, and Δ_3 of both images.
4. An inverse curvelet transform step is performed on P_3 of the MR image and the fused sub-bands Δ_{1f}, Δ_{2f}, and Δ_{3f}.
5. A post-processing high-pass filtering step can be used to sharpen the fusion result if blurring is due to the digital approximation of the ridgelet transform.
6. A wavelet denoising step can be used if the fusion is performed in a noisy environment.

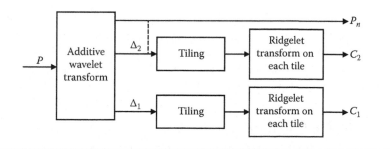

Figure 9.11   Diagram of discrete curvelet transform.

These steps are expected to merge the details in both images into a single image with many more details. The objective of the post-processing high-pass filtering step is to enhance edges in the fusion results. This step can be accomplished through the use of a high-pass filter mask H_F such as [104–107]:

$$H_F = \begin{bmatrix} 0 & -1 & 0 \\ -1 & 5 & -1 \\ 0 & -1 & 0 \end{bmatrix} \tag{9.6}$$

Also, in some fusion cases, white Gaussian noise may affect fusion result. Wavelet denoising can be used to reduce the noise effects. The denoising step can be implemented through a hard thresholding strategy on the detail coefficients in the wavelet domain as follows [104–107]:

$$W_H(x) = \begin{cases} x & if \ |x| > t \\ 0 & if \ |x| \le t \end{cases} \tag{9.7}$$

where x is the magnitude of the detail wavelet coefficients and t is the threshold.

9.7.1 Sub-Band Filtering

Sub-band filtering is similar to the AWT. The next step is finding a transformation capable of representing edges with different slopes and orientations. A possible solution is the ridgelet transform that may be interpreted as the one-dimensional (1-D) wavelet transform in the Radon domain. An inconvenience with the ridgelet transform is that it is not capable of representing curves. To overcome this drawback, the input image is partitioned into square blocks and the ridgelet transform is applied on each block. Assuming a piecewise linear model for the contour, each block will contain straight edges only that may be analyzed by the ridgelet transform.

9.7.2 Tiling

Tiling is the process by which the detail planes are divided into overlapping tiles. These tiles are small in dimensions to transform curved lines into small straight lines in the sub-bands Δ_1, Δ_2, and Δ_3 [104–107]. Tiling improves the ability of the curvelet transform to handle curved edges. Small tiles of dimensions 12×12 pixels with two pixels of overlapping from each side are suitable for this fusion purpose.

9.7.3 Ridgelet Transform

The motivation for this transform arose from the need to find a sparse representation of functions that have discontinuities along lines [104–107]. The ridgelet transform belongs to the family of discrete transforms employing basis functions. To facilitate its mathematical representation, it can be viewed as a wavelet analysis in the Radon domain. The Radon transform is a tool for shape detection. The ridgelet transform is, primarily, a tool for ridge detection or shape detection of the objects in an image. The two-dimensional (2-D) continuous ridgelet transform in $\mathbf{R}^2$ can be defined through the introduction of the following basis function [104–107]:

$$\psi_{a,b,\theta} = a^{-1/2} \psi \left(\frac{(x_1 \cos\theta + x_2 \sin\theta - b)}{a} \right) \tag{9.8}$$

for each $a > 0$, each $b \in \mathbf{R}$, and each $\theta \in [0, 2\pi]$. This function is constant along lines $x_1 \cos\theta + x_2 \sin\theta = const$. Thus, the ridgelet transform of an image $f(x_1, x_2)$ is represented by [104–107]:

$$R_f(a,b,\theta) = \int_{-\infty}^{\infty} \int_{-\infty}^{\infty} \psi_{a,b,\theta}(x_1,x_2) f(x_1,x_2) dx_1 dx_2 \tag{9.9}$$

This transform is invertible and the reconstruction formula is given by [104–107]:

$$f(x_1,x_2) = \int_0^{2\pi} \int_{-\infty}^{\infty} \int_0^{\infty} R_f(a,b,\theta) \psi_{a,b,\theta}(x_1,x_2) \frac{da}{a^3} db \frac{d\theta}{4\pi} \tag{9.10}$$

The Radon transform for an object f is the collection of line integrals indexed by $(\theta, t) \in (0, 2\pi) \times \mathbf{R}$ and is given by [104–107]:

$$Rf(\theta,t) = \int_{-\infty}^{\infty} \int_{-\infty}^{\infty} f(x_1,x_2) \delta(x_1 \cos\theta + x_2 \sin\theta - t) dx_1 dx_2 \tag{9.11}$$

Thus, the ridgelet transform can be represented in terms of the Radon transform as follows [104–107]:

$$R_f(a,b,\theta) = \int_{-\infty}^{\infty} Rf(\theta,t) a^{-1/2} \psi \left(\frac{(t-b)}{a} \right) dt \tag{9.12}$$

For practical applications, we require a discrete implementation of the ridgelet transform, which is a challenging problem because the Radon transform is polar in nature. An approximation for the Radon transform of digital data can be obtained using the fast Fourier transform (FFT) [104–107]. First, the 2-D FFT of the given image is computed. Then, the result in the frequency domain is used to evaluate the frequency values in a pseudo-polar style. This conversion from a Cartesian to a polar grid could be achieved by interpolation. By applying the 1-D inverse fast Fourier transform (IFFT) for each ray, the Radon projections are obtained [104–107].

For our implementation of the Cartesian-to-polar conversion, we used a pseudo-polar grid in which the pseudo-radial variable has level sets that are squares rather than circles. This grid has often been called the concentric squares grid in the signal processing literature. In the medical tomography literature, it is sometimes called the rectopolar grid. The geometry of the rectopolar grid is illustrated in Figure 9.12 [104–107]. Concentric circles of linearly growing radii in the polar grid are replaced by concentric squares of linearly growing sides. The rays are spread uniformly, not in angle but in slope.

We select $2n$ radial lines in the frequency plane obtained by connecting the origin to the vertices (k_1, k_2) lying on the boundary of the array (k_1, k_2), such that k_1 or $k_2 \in \left\{ -\frac{n}{2}, \frac{n}{2} \right\}$. The polar grid is the intersection between the set of radial lines and the set of Cartesian lines parallel to the axes. The cardinality of the rectopolar grid is equal to $2n^2$ as there are $2n$ radial lines and n sampled values on each of these lines. As a result, data structures associated with this grid will have a rectangular format [104–107].

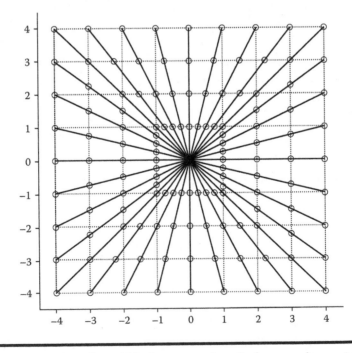

Figure 9.12 **Pseudo-polar grid in frequency domain for $n \times n$ image ($n = 8$).**

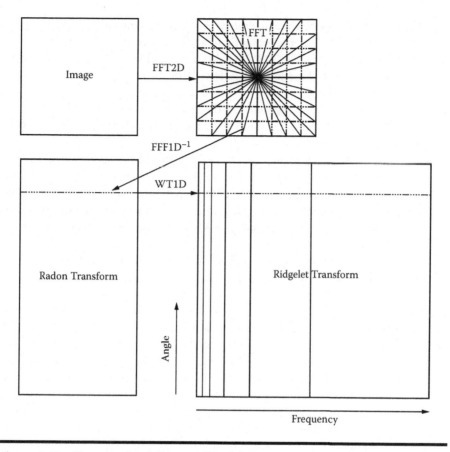

Figure 9.13 Flow graph of discrete ridgelet transform.

To complete the ridgelet transform, we must take a 1-D DWT along the radial variable in the Radon space [104–107]. Figure 9.13 shows the flow graph of the ridgelet transform.

9.8 IHS Fusion

This technique is used mainly for remote sensing image fusion. The widespread use of the intensity, hue, saturation (IHS) transform to merge remote sensing images is based on its ability to separate the spectral information of the red–green–blue (RGB) image into its two components (H) and (S), while isolating most of the spatial information in the intensity (I) component. Hue (H) is an attribute that describes a pure color. It is measured by an angle that represents the type of color. Saturation (S) gives a measure of the degree to which a pure color is diluted by white light. It represents how much whiteness is mixed in the color. Intensity is the most useful descriptor of monochromatic images. This quantity is definitely measurable and easily interpretable [108–111]. As shown in Figure 9.14, the usual steps to perform this fusion technique are as follows [108–111]:

1. The LR MS images are registered to the same size as the HR PAN image.
2. The R, G, and B bands of the MS image are transformed into their IHS components.
3. The HR PAN image is modified based on the MS image. This is usually performed by conventional histogram matching between the PAN image and the intensity component of the IHS representation after computing the histogram of both the PAN and this intensity component. The histogram of the intensity is used as a reference to which we match the histogram of the PAN image.
4. The modified PAN image is used as the new intensity component and an inverse IHS transformation is performed to obtain the RGB fused image.

The fused image provides the full details of the PAN image, but introduces some color distortion because of the low correlation between the PAN image and the intensity component. Because the intensity component is obtained as a linear combination of the different bands of the MS image, a high correlation between these bands and the PAN image is possible only if the bandwidth of the PAN image covers the entire range of bandwidths of all the MS original bands.

Different transformations have been developed to transfer a color image from the RGB space to the IHS space. The most common RGB-IHS conversion system is based on the following linear transformation [108–111]:

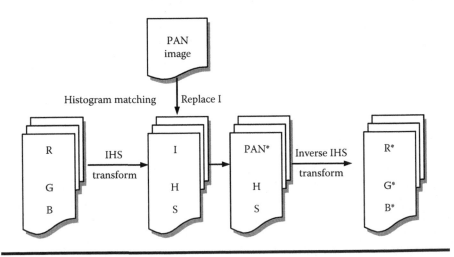

Figure 9.14 IHS image fusion process.

$$
\begin{pmatrix} I \\ V_1 \\ V_2 \end{pmatrix} = \begin{pmatrix} \dfrac{1}{3} & \dfrac{1}{3} & \dfrac{1}{3} \\ \dfrac{-\sqrt{2}}{6} & \dfrac{-\sqrt{2}}{6} & \dfrac{\sqrt{2}}{6} \\ \dfrac{1}{\sqrt{2}} & \dfrac{-1}{\sqrt{2}} & 0 \end{pmatrix} \begin{pmatrix} R \\ G \\ B \end{pmatrix}
\tag{9.13}
$$

$$
H = \tan^{-1}\left(\frac{V_2}{V_1}\right)
\tag{9.14}
$$

$$
S = \sqrt{V_1^2 + V_2^2}
\tag{9.15}
$$

where variables V_1 and V_2 can be considered as the x and y axes in the Cartesian coordinate system, while the intensity (I) indicates the z axis.

Fusion proceeds by replacing the intensity component (I) with the histogram matched PAN image (PAN^*). The fused image is then obtained by performing an inverse transformation from the IHS space back to the original RGB space as follows [108–111]:

$$
\begin{pmatrix} R^* \\ G^* \\ B^* \end{pmatrix} = \begin{pmatrix} 1 & \dfrac{-1}{\sqrt{2}} & \dfrac{1}{\sqrt{2}} \\ 1 & \dfrac{-1}{\sqrt{2}} & \dfrac{-1}{\sqrt{2}} \\ 1 & \sqrt{2} & 0 \end{pmatrix} \begin{pmatrix} PAN^* \\ V_1 \\ V_2 \end{pmatrix}
\tag{9.16}
$$

9.9 High-Pass Filter Fusion

In satellite image fusion, if the spectral bands are not perfectly spectrally overlapped with the PAN band, as occurs with the Spot4, Ikonos, and Quick-Bird images, the PCA method yields poor results in terms of the spectral fidelity of color representations [108,109]. To overcome this inconvenience, methods based on injecting high-frequency spatial details taken from the PAN image have been introduced and have demonstrated superior performance. Figure 9.15 shows the block diagram of the high-pass filter method in which the high-frequency spatial content of the PAN image is extracted using a high-pass filter and transferred to the re-sampled MS image. The mathematical model of this method for each band is given by:

$$
DN_{MS}^H = DN_{MS}^L + (DN_{PAN}^H - DN_{PAN}^L)
\tag{9.17}
$$

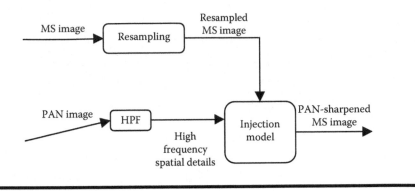

Figure 9.15 Image fusion with HPF method.

where DN means the digital numbers or pixel values, $DN_{PAN}^L = DN_{PAN}^H \otimes h$ is the smoothed version of the PAN image, and h is a low-pass filter such as the boxcar filter. DN_{PAN}^H is the PAN image with high-frequency spatial details, DN_{MS}^L is a certain band of the LR MS image, and DN_{MS}^H is its corresponding band of the PAN-sharpened MS image [108,109].

This method preserves a high percentage of the spectral characteristics of the MS image because the spectral information is associated with the low spatial frequencies of the MS image. The cut-off frequency of the filter has to be chosen in such a way that the included data does not influence the spectral information of the MS image [108,109].

9.10 Gram–Schmidt Fusion

In the Gram-Schmidt (GS) method, as described by its inventors [108,109], the spatial resolution of the MS image is enhanced by merging the HR PAN image with the low spatial resolution MS bands. The block diagram of this method is shown in Figure 9.16. The main steps of this method are as follows:

1. A low spatial resolution PAN image is simulated.
2. The Gram–Schmidt transformation (GST) is performed on the simulated low spatial resolution PAN image together with all the low spatial resolution spectral band images. The simulated low spatial resolution PAN image is employed as the first band in the GST.
3. The statistics of the high spatial resolution PAN image are adjusted to match the statistics of the first transform band that results from the GST to produce a modified high spatial resolution PAN image.
4. The modified high spatial resolution PAN image is substituted for the first transform band that results from the GST to produce a new set of transform bands. The inverse GST is performed on the new set of transform bands to produce the enhanced spatial resolution MS image.

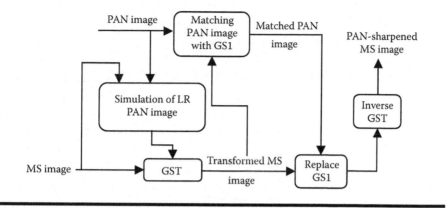

Figure 9.16   Image fusion with GS method.

In this method, the spectral characteristics of the low spatial resolution MS data are preserved in the high spatial resolution PAN-sharpened MS image.

9.11 Fusion of Satellite Images

The IHS fusion method usually integrates color and spatial features smoothly. It has the ability to merge the spatial details of the PAN image efficiently with the MS image. If the correlation between the intensity image and the PAN image is high, this method can also preserve the color information efficiently. Unfortunately, color distortion is significant due to the low correlation between the intensity image and the PAN image, especially when the natural color MS bands and the PAN image from Landsat-7, Ikonos, and Quick-Bird satellites are fused.

On the other hand, the DWFT image fusion method can preserve color information better than the other conventional fusion methods such as the DWT, IHS, and PCA methods because the HR spatial details from the PAN image are injected into all three LR MS bands. A hybrid fusion method can be used to make use of the sophisticated characteristics of both the IHS and the DWFT fusion methods to preserve both the spatial and spectral details. The steps of this method are shown in Figure 9.17 and can be summarized as follows [108,109]:

1. Registration. The PAN and the MS images are first registered.
2. IHS transform. The MS image is transformed into its IHS components.
3. Histogram matching. To improve the quality of the fused image, the histogram of the PAN image is matched to that of the intensity image.
4. DWFT. A single-level DWFT is performed on both the new PAN image and the intensity component of the IHS image.
5. Substitution. The approximation component of the intensity image is replaced by its average with the approximation component of the new PAN image and the detail components are left unchanged.

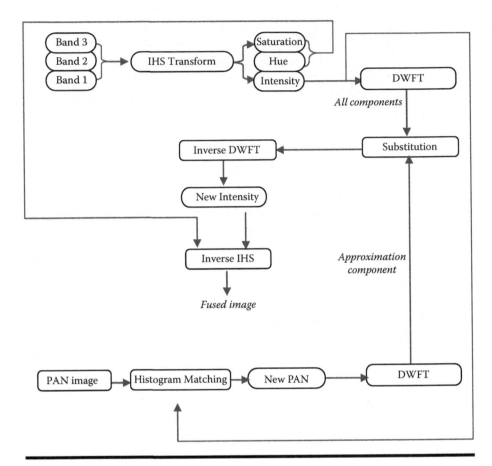

Figure 9.17 Hybrid fusion of satellite images.

6. Inverse DWFT. The new intensity image is obtained by applying the inverse DWFT on the approximation and detail components obtained from the previous step. The resulting new intensity image has a similar gray-scale distribution to that of the IHS intensity image, while it preserves the spatial details of the PAN image.

7. Inverse IHS. The final fused image is generated by transforming the new intensity image together with the hue and saturation components back into the RGB space.

Several experiments have been carried out to test the performance of the above-mentioned hybrid satellite image fusion method. Images from different satellites have been used in these experiments. The details of the images used in the simulation experiments are given in Table 9.1. Fusion experiments were carried out at different SNR values. The images used in these fusion experiments at different

Table 9.1   Details of Test Experiments

Satellite	Resolution	Image Size (Pixels)	Coverage Area
Landsat-5 (MS) Spot4 (PAN)	MS image: 30 m PAN image: 10m	256×256	London, UK
Landsat-7	MS image: 30 m PAN image: 15 m	256×256	Cairo, Egypt
Ikonos	MS image: 4 m PAN image: 1 m	256×256	Fredericton, NB, Canada

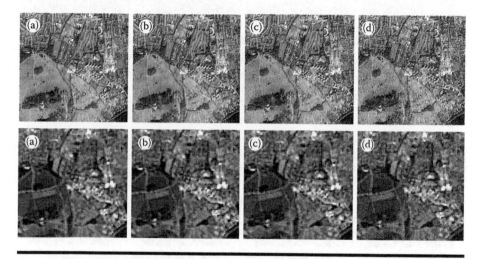

Figure 9.18   Landsat-5 MS image and Spot PAN image of an area of London. First row is for PAN images and second row is for MS images. (a) Original images. (b) Images at SNR = 25 dB. (c) Images at SNR = 15 dB. (d) Images at SNR = 5 dB.

SNRs are shown in Figures 9.18 through 9.20. Fusion results of all methods for experiments carried out at SNR = 25 dB are given in Figures 9.21 through 9.23, and the numerical evaluation metrics for these experiments are tabulated in Tables 9.2 through 9.6. From the obtained results, it is clear that both the integrated IHS and DWT method and the hybrid fusion method achieve the lowest discrepancy values, while the hybrid fusion method achieves higher correlation coefficients than the integrated IHS and DWT method. The discrepancy is defined as:

$$D_K = \frac{1}{M \times N} \sum_{i=1}^{M} \sum_{j=1}^{N} \left| F_{K,i,j} - L_{K,i,j} \right|; \quad K = R, G, B \qquad (9.18)$$

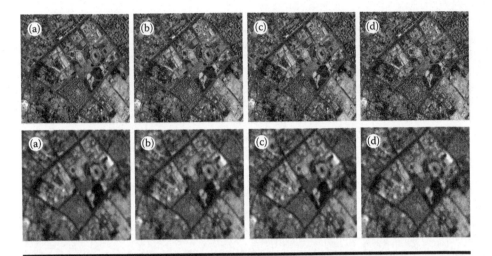

Figure 9.19 Landsat-7 PAN and MS images of an area of Cairo. First row is for PAN images and second row is for MS images. (a) Original images. (b) Images at SNR = 25 dB. (c) Images at SNR = 15 dB. (d) Images at SNR = 5 dB.

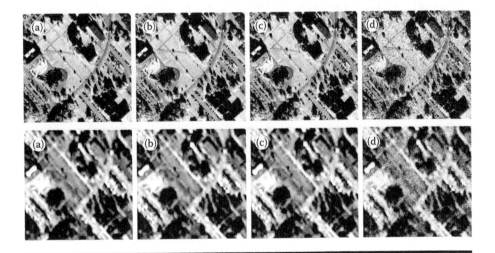

Figure 9.20 Ikonos PAN and MS images of an area of Canada. First row is for PAN images and second row is for MS images. (a) Original images. (b) Images at SNR = 25 dB. (c) Images at SNR = 15 dB. (d) Images at SNR = 5 dB.

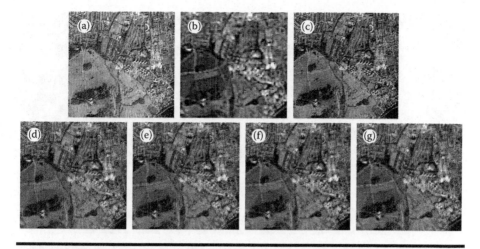

Figure 9.21 Fusion results of Spot PAN image and Landsat-5 MS image at SNR = 5 dB. (a) Spot PAN image. (b) Landsat-5 MS image. (c) IHS fusion result. (d) DWT fusion result. (e) DWFT fusion result. (f) Integrated IHS and DWT fusion result. (g) Hybrid fusion result.

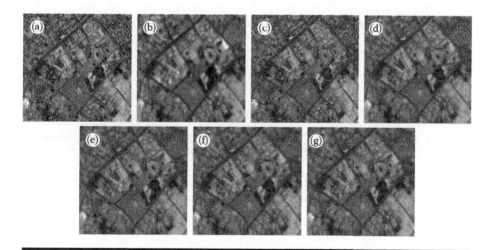

Figure 9.22 Fusion results of Landsat-7 images at SNR = 25 dB. (a) Spot PAN image. (b) Landsat-5 MS image. (c) IHS fusion result. (d) DWT fusion result. (e) DWFT fusion result. (f) Integrated IHS and DWT fusion result. (g) Hybrid fusion result.

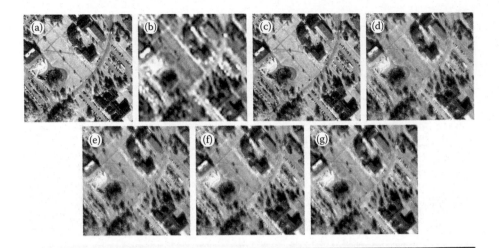

Figure 9.23   Fusion results of Ikonos images at SNR = 25 dB. (a) Spot PAN image. (b) Landsat-5 MS image. (c) IHS fusion result. (d) DWT fusion result. (e) DWFT fusion result. (f) Integrated IHS and DWT fusion result. (g) Hybrid fusion result.

Table 9.2   Evaluation Metrics for Fusion of Landsat-5 and Spot Images at SNR = 25 dB

Fusion Method	D_K			C_K		
	R	G	B	R	G	B
IHS	0.134	0.143	0.132	0.913	0.939	0.888
DWT	0.106	0.099	0.116	0.937	0.936	0.936
DWFT	0.105	0.098	0.115	0.972	0.971	0.971
Integrated IHS and DWT	0.090	0.096	0.089	0.864	0.886	0.836
Hybrid	0.089	0.095	0.088	0.893	0.917	0.864

where $F_{K,i,j}$ and $L_{K,i,j}$ are the pixel values at position (i,j) in the *Kth* band of the fused and original MS images, respectively. *M* and *N* are the dimensions of each band of the MS image. The spectral quality of the fused MS image is increased as D_K is decreased.

The spatial quality is measured using the correlation coefficient, C_K, between the high-frequency details of the *Kth* band of the fused image and the original PAN image, because the spatial information is mostly concentrated in high frequencies. The value of C_K can be estimated as follows:

Table 9.3 Evaluation Metrics for Fusion of Landsat-7 Images at SNR = 25 dB

Fusion Method	D_K			C_K		
	R	G	B	R	G	B
IHS	0.078	0.075	0.083	0.950	0.945	0.975
DWT	0.048	0.047	0.048	0.899	0.898	0.896
DWFT	0.048	0.046	0.048	0.954	0.953	0.952
Integrated IHS and DWT	0.045	0.044	0.048	0.865	0.858	0.888
Hybrid	0.045	0.043	0.048	0.915	0.906	0.942

Table 9.4 Evaluation Metrics for Fusion of Ikonos Images at SNR = 25 dB

Fusion Method	D_K			C_K		
	R	G	B	R	G	B
IHS	0.188	0.198	0.186	0.915	0.928	0.906
DWT	0.118	0.121	0.143	0.932	0.931	0.930
DWFT	0.117	0.120	0.142	0.966	0.966	0.965
Integrated IHS and DWT	0.106	0.112	0.106	0.879	0.889	0.876
Hybrid	0.106	0.111	0.105	0.908	0.919	0.904

Table 9.5 Spectral and Spatial Characteristics of Spot 4 Images

Band	Wavelength (µm)	Resolution (m)
1	0.50 to 0.59 (green)	20
2	0.61 to 0.68 (red)	20
3	0.78 to 0.89 (NIR)	20
4	1.58 to 1.75 (MIR)	20
PAN	0.61 to 0.68 (PAN)	10

$$C_K = \frac{\sum_{i=1}^{M}\sum_{j=1}^{N}(H_{k,i,j} - \bar{H})(P_{i,j} - \bar{P})}{\sqrt{\sum_{i=1}^{M}\sum_{j=1}^{N}(H_{k,i,j} - \bar{H})^2 \times \sum_{i=1}^{256}\sum_{j=1}^{256}(P_{i,j} - \bar{P})^2}} \qquad (9.19)$$

Table 9.6 Spectral and Spatial Characteristics of Quick-Bird Images

Band	Wavelength (µm)	Resolution (m)
1	0.45 to 0.52 (blue)	2.8
2	0.52 to 0.60 (green)	2.8
3	0.63 to 0.69 (red)	2.8
4	0.76 to 0.90 (NIR)	2.8
PAN	0.45 to 0.90 (PAN)	0.7

where H_k represents the high-frequency details image of F_k, $\bar{H}$ is the mean of H_k, $P_{i,j}$ is the pixel value at position (i,j) of the original PAN image, and $\bar{P}$ is the mean of the original PAN image. To extract the high-frequency details of any band, convolution with the following Laplacian filter is used.

$$\cdot L_F = \begin{bmatrix} -1 & -1 & -1 \\ -1 & 8 & -1 \\ -1 & -1 & -1 \end{bmatrix} \tag{9.20}$$

The spatial quality of the *Kth* band of the fused image is directly proportional to C_K.

Other experiments have been carried out to test the effect of noise on satellite image fusion. The variations of the discrepancy and correlation coefficient for each band of the fused MS image with the SNR in all simulation experiments with different fusion methods are shown in Figures 9.24 through 9.26. These figures reveal that the discrepancy is decreased with the increase in the SNR while the correlation coefficient is increased with the increase in the SNR for all bands of the fused images. The figures reveal also that the hybrid fusion method gives the lowest discrepancy values at all SNRs.

The feasibility of wavelet denoising in the satellite image fusion process was also studied and the results for all fusion methods with and without wavelet denoising are illustrated in Figures 9.27 through 9.41. These results show that wavelet denoising is feasible only for fusion at low SNR values. If the fusion is performed at high SNR values, denoising removes part of the high-frequency fused image details rather than noise, which decreases the correlation coefficients for all bands.

The equation of the hard thresholding is given by [73]:

$$f_{hard}(x) = \begin{cases} x & |x| \geq TH \\ 0 & |x| < TH \end{cases} \tag{9.21}$$

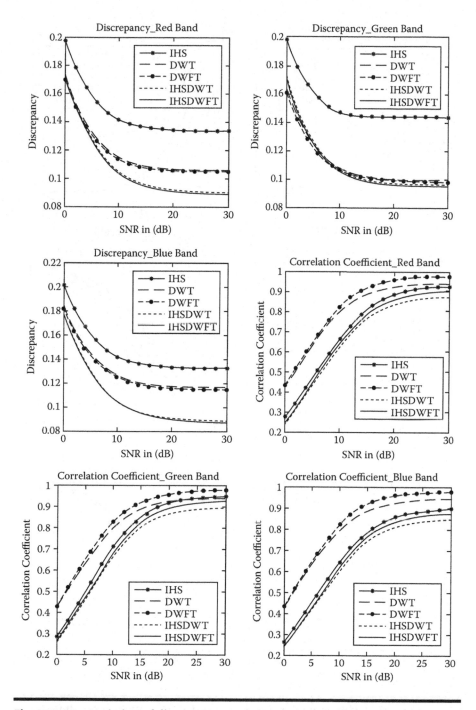

Figure 9.24 Variation of discrepancy and correlation coefficient for all bands of fused image with SNR for fusion of Landsat-5 and Spot images.

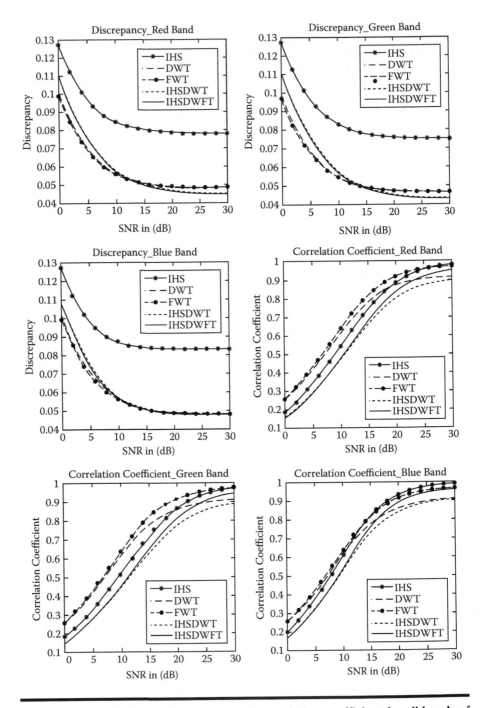

Figure 9.25 **Variation of discrepancy and correlation coefficient for all bands of fused image with SNR for fusion of Landsat-7 images.**

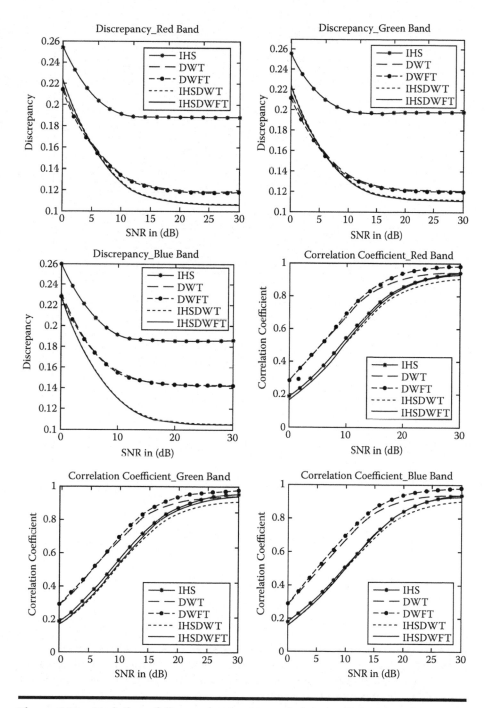

Figure 9.26 Variation of discrepancy and correlation coefficient for all bands of fused image with SNR for fusion of Ikonos images.

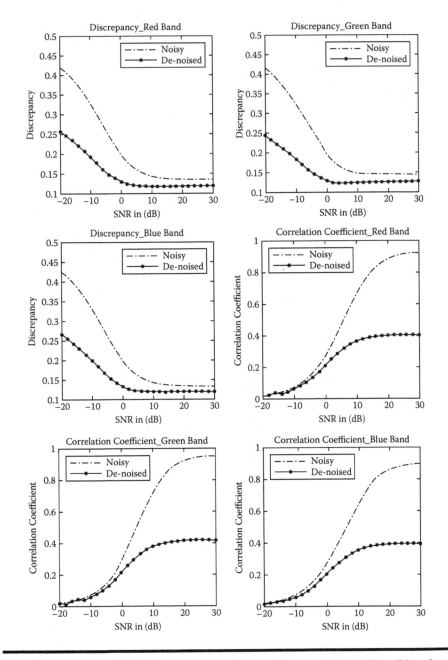

Figure 9.27 Variation of discrepancy and correlation coefficient for all bands of fused image with SNR for fusion of Landsat-5 and Spot images using IHS method with and without wavelet denoising.

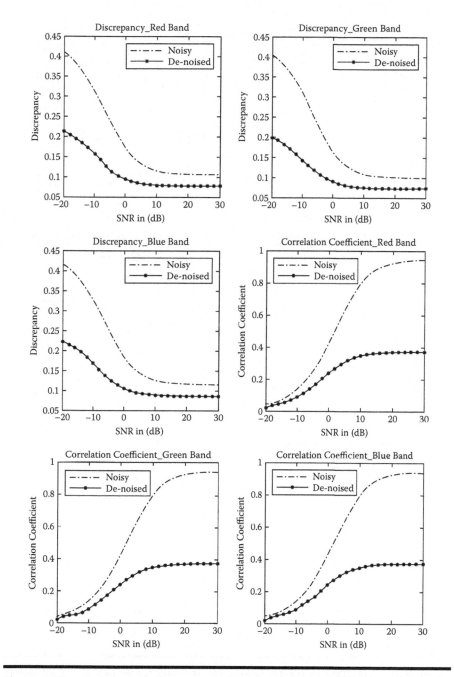

Figure 9.28 Variation of discrepancy and correlation coefficient for all bands of fused image with SNR for fusion of Landsat-5 and Spot images using DWT method with and without wavelet denoising.

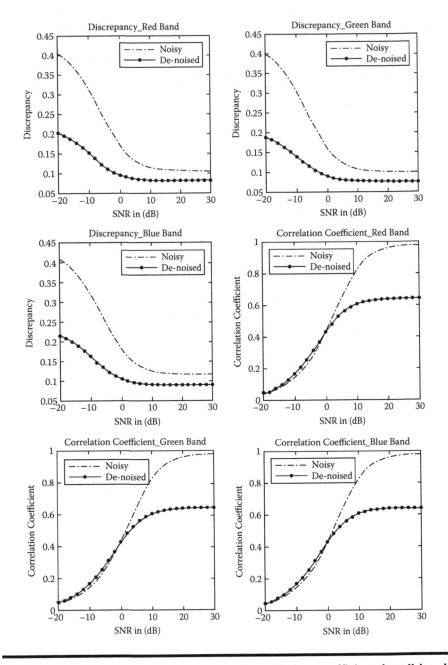

Figure 9.29 **Variation of discrepancy and correlation coefficient for all bands of fused image with SNR for fusion of Landsat-5 and Spot images using DWFT method with and without wavelet denoising.**

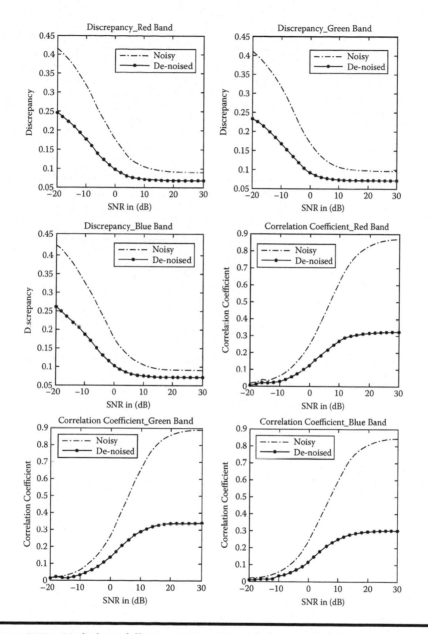

Figure 9.30 Variation of discrepancy and correlation coefficient for all bands of fused image with SNR for fusion of Landsat-5 and Spot images using integrated IHS and DWT method with and without wavelet denoising.

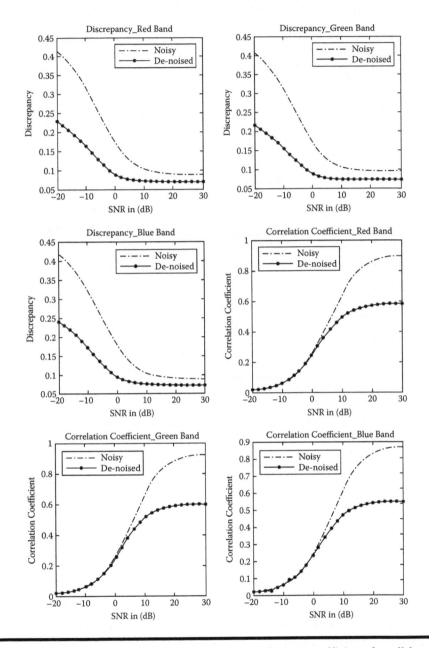

Figure 9.31 Variation of discrepancy and correlation coefficient for all bands of fused image with SNR for fusion of Landsat-5 and Spot images using proposed hybrid method with and without wavelet denoising.

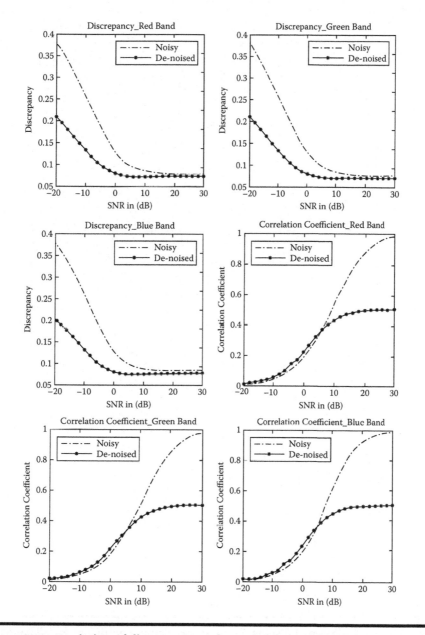

Figure 9.32 Variation of discrepancy and correlation coefficient for all bands of fused image with SNR for fusion of Landsat-7 images using IHS method with and without wavelet denoising.

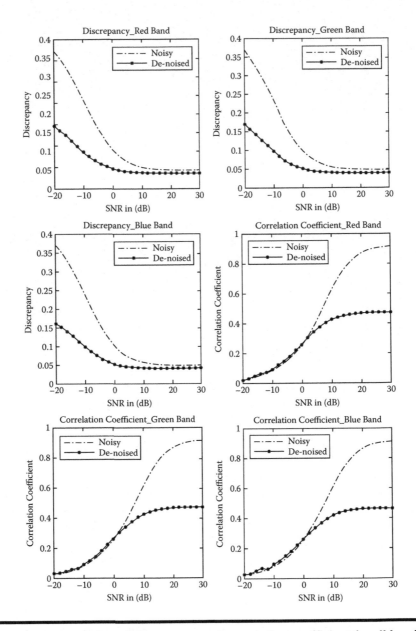

Figure 9.33 Variation of discrepancy and correlation coefficient for all bands of fused image with SNR for fusion of Landsat-7 images using DWT method with and without wavelet denoising.

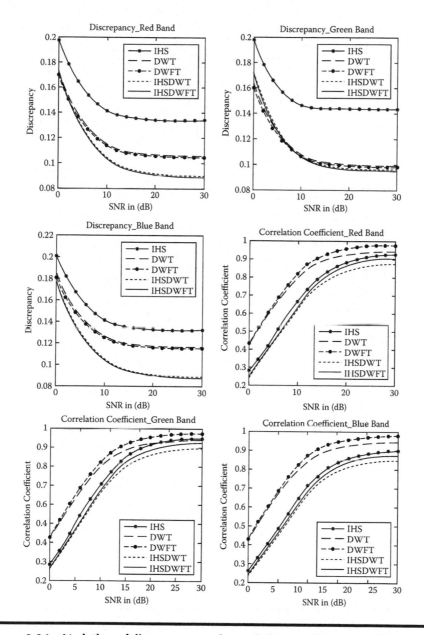

Figure 9.34   Variation of discrepancy and correlation coefficient for all bands of fused image with SNR for fusion of Landsat-7 images using DWFT method with and without wavelet denoising.

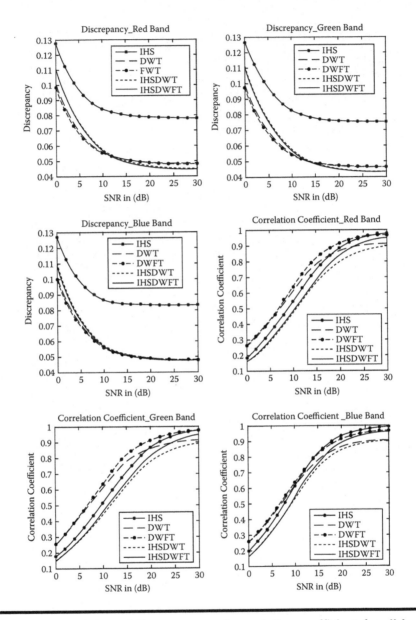

Figure 9.35 Variation of discrepancy and correlation coefficient for all bands of fused image with SNR for fusion of Landsat-7 images using integrated IHS and DWT method with and without wavelet denoising.

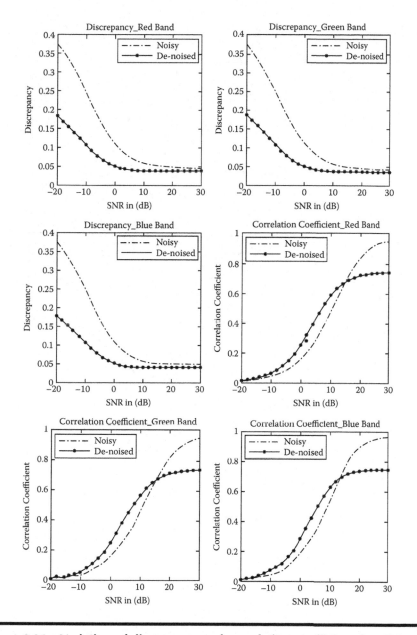

Figure 9.36 **Variation of discrepancy and correlation coefficient for all bands of fused image with SNR for fusion of Landsat-7 images using proposed hybrid method with and without wavelet denoising.**

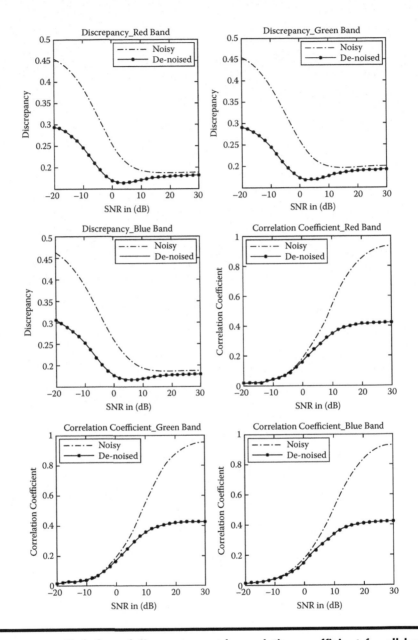

Figure 9.37   Variation of discrepancy and correlation coefficient for all bands of fused image with SNR for fusion of Ikonos images using IHS method with and without wavelet denoising.

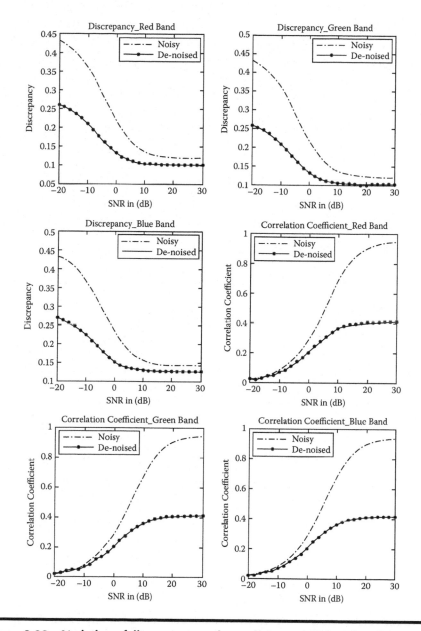

Figure 9.38 Variation of discrepancy and correlation coefficient for all bands of fused image with SNR for fusion of Ikonos images using DWT method with and without wavelet denoising.

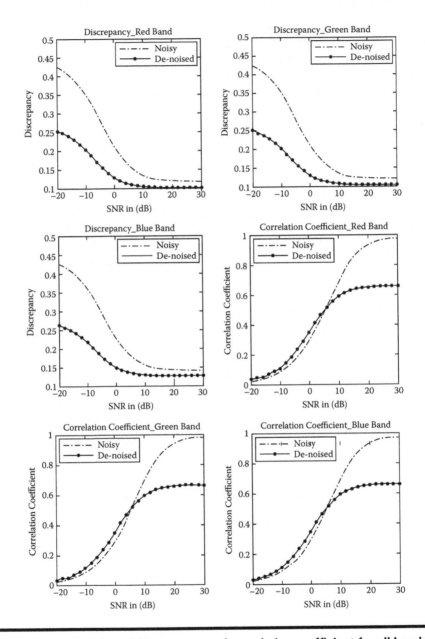

Figure 9.39 **Variation of discrepancy and correlation coefficient for all bands of fused image with SNR for fusion of Ikonos images using DWFT method with and without wavelet denoising.**

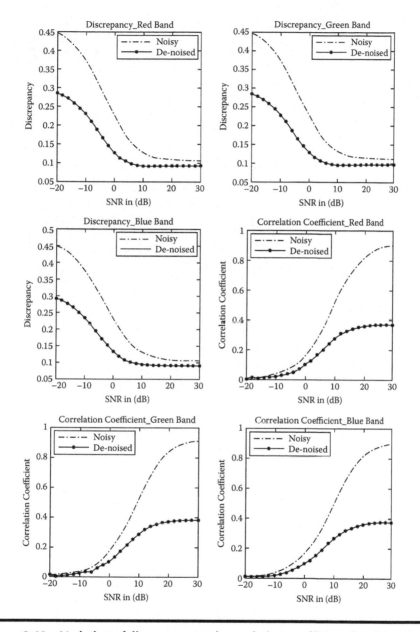

Figure 9.40 **Variation of discrepancy and correlation coefficient for all bands of fused image with SNR for fusion of Ikonos images using integrated IHS and DWT method with and without wavelet denoising.**

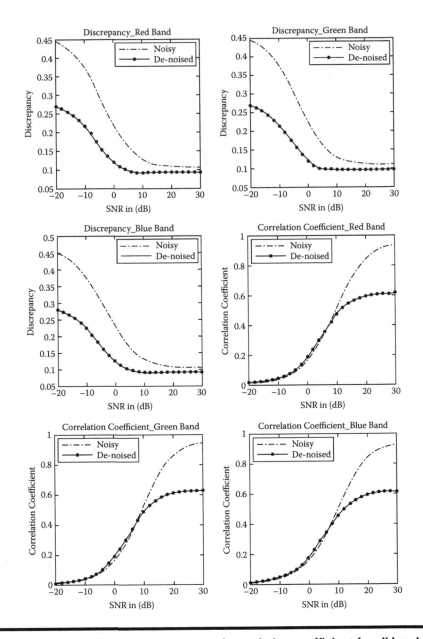

Figure 9.41   Variation of discrepancy and correlation coefficient for all bands of fused image with SNR for fusion of Ikonos images using hybrid method with and without wavelet denoising.

On the other hand, the equation of soft thresholding is given by [73]:

$$
f_{soft}(x) = \begin{cases} x & |x| \geq TH \\ 2x - TH & TH/2 \leq x < TH \\ TH + 2x & -TH < x \leq -TH/2 \\ 0 & |x| < TH/2 \end{cases}
\tag{9.22}
$$

TH denotes the threshold value and x represents the coefficients of the high-frequency components of the DWT.

Metwalli et al. presented another hybrid image fusion method to improve the process of injection of the spatial details extracted from the PAN image into the MS image as shown in Figure 9.42. The steps of this method are summarized as follows [109]:

1. The re-sampled bands of the MS image to the same resolution as the PAN image are transformed with the PCT.
2. The PAN image is smoothed with a Gaussian low-pass filter (GLPF).
3. The spatial details of the PAN image are extracted as the difference between the original PAN image and the smoothed one.
4. A linear combination between the extracted spatial details of the PAN image and the first principal component is performed using a gain factor, estimated as the ratio between the standard deviation of the first principal component PC_1 and the standard deviation of the PAN image. The gain factor is used to compensate for the difference in radiometry between the PAN image and PC_1.
5. The new first principal component and the other principal components are transformed with the inverse PCT to obtain the PAN-sharpened MS image.

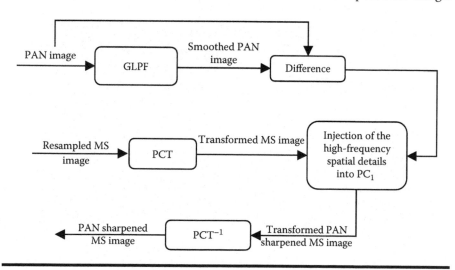

Figure 9.42   Hybrid image fusion.

The mathematical model for the injection of spatial details of the PAN image into the MS image is given by [109]:

$$DN^H_{PC_1} = DN^L_{PC_1} + \alpha * (DN^H_{PAN} - DN^L_{PAN}) \qquad (9.23)$$

where $DN^L_{PC_1}$ is the first principal component of the MS image, $DN^H_{PC_1}$ is the PAN-sharpened first principal component, and α is the gain parameter. The GLPF used to smooth the PAN image has the following kernel:

$$G(x, y) = \frac{1}{2\pi\sigma^2} e^{-\frac{x^2+y^2}{2\sigma^2}} \qquad (9.24)$$

The degree of smoothing is determined by the standard deviation of the kernel σ and the window size of the filter. The smoothing by the GLPF has no ringing effect as in the case of the Butterworth low-pass filter [109]. The difference between the original PAN image and the smoothed PAN image is taken as the high-frequency detail image that will be injected into the first principal component of the MS image.

Simulation experiments have been performed to test the hybrid fusion method and compare it with the traditional fusion methods. Two different types of data have been used in these simulation experiments as follows:

1. Spot4 data for a part of Cairo, Egypt with spectral and spatial properties as shown in Table (9.5). Figure 9.43 shows the MS and the PAN images of the Spot4 data. The ratio between the spatial resolution of the PAN image and the spatial resolution of the MS image is 1/2.

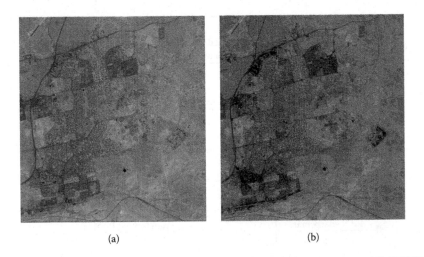

(a)   (b)

Figure 9.43   Spot4 images. (a) MS image. (b) PAN image.

2. Quick-Bird Data for a part of Sundarbans, India with spectral and spatial properties as shown in Table 9.6. Figure 9.44 shows the MS and the PAN images of Quick-Bird data. The ratio between the spatial resolution of the PAN image and the spatial resolution of the MS image is 1/4.

The test areas include ground conditions such as urban areas, vegetation, and water supplies in order to analyze the fusion methods for a variety of spectral and spatial contents.

The hybrid image fusion method was tested on the Spot4 data shown in Figure 9.43 and the fusion result is given in Figure 9.44. Similarly, it was tested on the Quick-Bird data shown in Figure 9.45, and the fusion result is given in Figure 9.46. Figure 9.47 shows the variation of the root mean square error (RMSE) of each band in the PAN-sharpened MS image with the σ and the window size of the GLPF for the fusion of the Spot4 images.

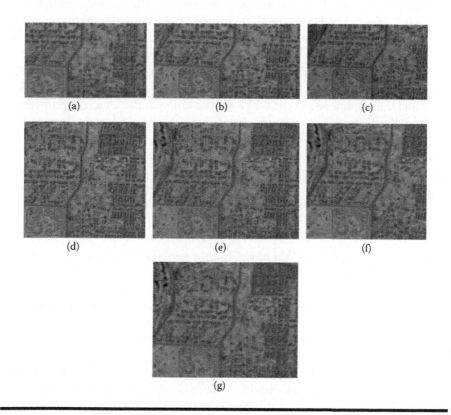

(a) (b) (c)

(d) (e) (f)

(g)

Figure 9.44 Fusion results for Spot4 images. (a) Original MS image resampled to PAN resolution. (b) Proposed method. (c) PCA method. (d) HPF method. (e) GS method. (f) DWT method. (g) SWT method.

Figure 9.45   Quick-Bird images. (a) MS image. (b) PAN image.

Figure 9.48 shows the variation of the correlation coefficient for high-frequency details (c_{rh}) of each band in the PAN-sharpened MS image with the σ and the window size of the GLPF for the fusion of the Spot4 images. Figure 9.49 shows the variation of the RMSE of each band in the PAN-sharpened MS image with the σ and the window size of the GLPF for the fusion of the Quick-Bird images. Figure 9.50 shows the variation of the c_{rh} of each band in the PAN-sharpened MS image with the σ and the window size of the GLPF for the fusion of the Quick-Bird images.

From these figures, we notice that increasing the filter window size beyond 5×5 for the Spot4 images and beyond 7×7 for the Quick-Bird images does not add more into the spatial quality of the PAN-sharpened MS images, but leads to more spectral distortion. As σ is increased, the spatial quality for all bands is increased until a certain limit is reached. Beyond this limit, increasing σ does not enhance the spatial quality of the PAN-sharpened MS image, but leads to more spectral distortion.

In general, it was concluded from the experiments that when the ratio between the spatial resolution of the PAN image and the spatial resolution of the MS image is 1/2, a GLPF with a window size of 5×5 and σ of about 1.5 produces a PAN-sharpened MS image with as minimum spectral distortion as possible and good spatial resolution. When the ratio between the spatial resolution of the PAN image and the spatial resolution of the MS image is 1/4, a GLPF with window size of 7×7 and σ of about 1.5 produces a sharpened MS image with good spatial and spectral properties.

Simulation experiments were conducted to compare between the performances of the hybrid fusion method, the PCA method, the high pass filtering (HPF) method, the GS method, the discrete wavelet transform (DWT) method, and the stationary wavelet transform (SWT) method. For the hybrid fusion method, windows of size 5×5 for the Spot4 data and size 7×7 for the Quick-Bird data were used. The value of σ used in the simulations was 1.5. For the DWT and SWT

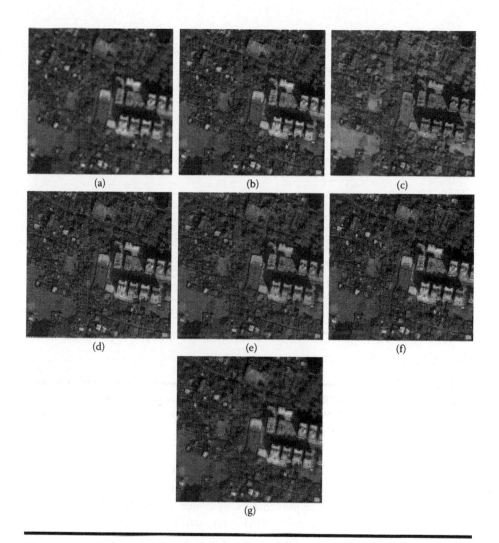

Figure 9.46 Fusion results for Quick-Bird images. (a) Original MS image resampled to PAN resolution. (b) Hybrid method. (c) PCA method. (d) HPF method. (e) GS method. (f) DWT method. (g) SWT method.

methods, biorthogonal filters with a single decomposition level for the Spot4 data and two decomposition levels for Quick-Bird data were used.

Figures 9.44 and 9.46 show the results of the different fusion methods for both Spot4 and the Quick-Bird data, respectively. Tables 9.3 and 9.4 show the RMSE between the corresponding bands of the original MS image and the PAN-sharpened MS image using the different fusion methods for the Spot4 and the Quick-Bird data, respectively. These results show that the hybrid fusion method gives the lowest RMSE values for the different bands.

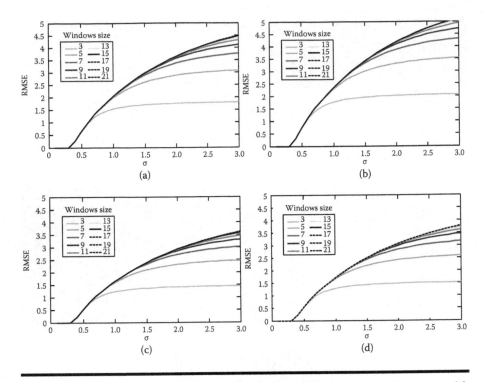

Figure 9.47 Variation of RMSE of each band in PAN-sharpened MS image with σ and window size of GLPF for fusion of Spot4 images. (a) Band 1. (b) Band 2. (c) Band 3. (d) Band 4.

Tables 9.5 and 9.6 show the values of c_{rh} for all bands of the obtained MS images with the different fusion methods. The results in these two tables give indications for the spatial improvements obtained from the different fusion methods. The HPF method showed the lowest spatial improvement. The hybrid fusion method gives spatial enhancement results comparable to the PCA, the GS, the DWT, and the SWT methods and has the best preservation of spectral characteristics.

It was concluded from the experiments that when the ratio between the spatial resolution of the PAN image and the spatial resolution of the MS image is 1/2, a GLPF with window size of 5×5 and σ of about 1.5 produces a sharpened MS image with the least spectral distortion and the best spatial quality. When the ratio between the spatial resolution of the PAN image and the spatial resolution of the MS image is 1/4, a GLPF with window size of 7×7 and σ of about 1.5 produces a sharpened MS image with the least spectral distortion and the best spatial quality. The experiments have shown that the hybrid fusion method has significantly reduced the spectral distortion compared to the PCA, the HPF, the GS, the DWT, and the SWT fusion methods. The spatial quality of the hybrid fusion method

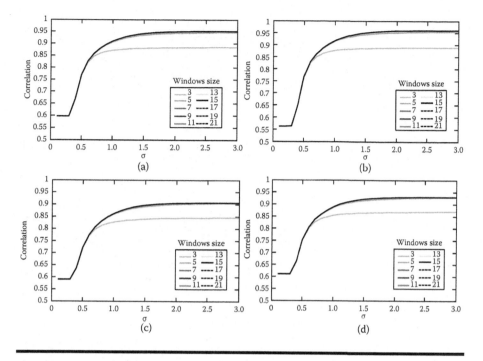

Figure 9.48 Variation of c_{rh} of each band in PAN-sharpened MS image with σ and window size of GLPF for fusion of Spot4 images. (a) Band 1. (b) Band 2. (c) Band 3. (d) Band 4.

is higher than the HPF fusion method and comparable to the PCA, the GS, the DWT, and the SWT fusion methods.

9.12 Fusion of MR and CT Images

Image fusion has been widely used for medical applications, especially for the fusion of MR and CT images. In most cases, image fusion is only a preparatory step to specific tasks such as human monitoring for patients, and thus the performance of the fusion algorithm must be measured in terms of the improvement achieved in the subsequent tasks.

Image quality metrics are classified into subjective and objective metrics. Subjective metrics depend on rating goodness scales. The subjective evaluation of an image is performed by viewers and each image is given a certain grade on the rating scale. The average rating of all viewers is taken as a measure of image quality. The subjective test emphatically concentrates on the fidelity of the image and at the same time considers image intelligibility. When performing a subjective test, viewers focus on the differences between the reconstructed image and an original image.

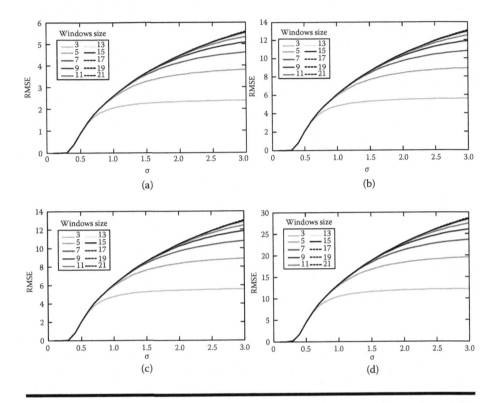

Figure 9.49 **Variation of RMSE of each band in PAN-sharpened MS image with σ and window size of GLPF for fusion of Quick-Bird images. (a) Band 1. (b) Band 2. (c) Band 3. (d) Band 4.**

Unfortunately, in image fusion experiments, the availability of a ground truth is not guaranteed [104–107] so subjective tests are not suitable for this case.

On the other hand, objective image quality metrics are numerical and based on certain criteria. One of the most popular objective image quality metrics in image processing is the peak signal-to-noise ratio (PSNR) that requires knowledge of the original image [104–107]. For image fusion applications, there is no reference or original image. To solve this problem for the fusion of MR and CT images, two PSNRs (the $PSNR_C$ and the $PSNR_M$) were defined. The $PSNR_C$ is the PSNR between the fusion result and the CT image, and the $PSNR_M$ is the PSNR between the fusion result and the MR image. The evaluation of the PSNR requires the estimation of the mean square error (MSE) between the reconstructed image (the fusion result in our case) and the original, which is either the CT image or the MR image. A high value of $PSNR_C$ and $PSNR_M$ is an indication of the closeness of the fusion result to the CT image and the MR image, respectively.

The PSNR metric alone is not enough for the evaluation of the quality of fusion results. A new metric, namely the similarity metric, has been presented to assist in

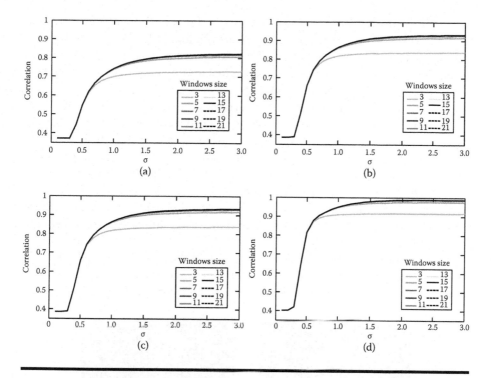

Figure 9.50 Variation of c_{rh} of each band in PAN-sharpened MS image with σ and window size of GLPF for fusion of Quick-Bird images. (a) Band 1. (b) Band 2. (c) Band 3. (d) Band 4.

evaluation with the PSNR. It is a metric for the correlation between edge pixels of the fusion result and the edge pixels of either the CT or the MR image. The objective of this metric is to measure how edges are preserved in the fusion result. As in the case with the PSNR, two similarity values can be calculated: the S_C and the S_M. The S_C is the similarity between the fusion result and the CT image, and the S_M is the similarity between the fusion result and the MR image. The similarity estimation steps are as follows [104–107]:

1. Edge detection of the fusion result using a suitable edge detector such as the Canny edge detector, which is one of the best.
2. Edge detection of the original image.
3. Estimation of the ratio between the similar edge pixels in the fusion result and those of either the CT image or the MR image.

$$S = N_s/N_t \qquad (9.25)$$

where N_s is the number of similar edge pixels between the fusion result and the original image, and N_t is the total number of edge pixels in the fusion result.

Two experiments are included to test the performance of the curvelet fusion of MR and CT images and compare it to the traditional PCA and DWT fusion techniques. In these experiments, four metrics were used to evaluate the fusion results in addition to the visual quality. These metrics include the two PSNR metrics and the two similarity metrics mentioned above.

The objective is to maximize all these metrics. The maximization of the PSNR metrics alone does not necessarily reveal the success of the fusion algorithm because they are maximized when the MSE is minimized. In the cases of fusions of very similar images, the MSE between the fusion result and the original image is small, which means a large PSNR without a guarantee for high quality fusion results.

This point reveals why the PSNR results are not used in the evaluation of pixel level fusion algorithms. To avoid this problem, the similarity metrics are used for performance evaluation of the fusion algorithms. If the similarity values are high, we can look at the PSNR values.

In the first experiment, the MR and CT scans of the brain shown in Figure 9.51 are used. The DWT components of both images used for the DWT fusion are shown in Figure 9.52. The AWT components of both images used to generate the

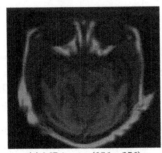 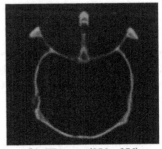

(a) MR image (256 × 256)	(b) CT image (256 × 256)

Figure 9.51   Original images for case 1.

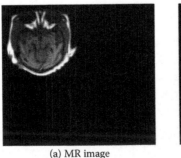 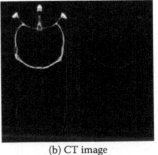

(a) MR image	(b) CT image

Figure 9.52   DWT of images of case 1.

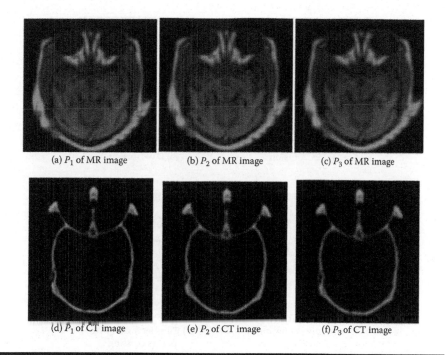

(a) P_1 of MR image   (b) P_2 of MR image   (c) P_3 of MR image

(d) P_1 of CT image   (e) P_2 of CT image   (f) P_3 of CT image

Figure 9.53   AWT approximation planes of MR and CT images for case 1 after successive convolutions with low-pass kernel.

detail planes required for curvelet fusion are shown in Figure 9.53. The fusion results of this experiment are shown in Figure 9.54. A high-pass filtering step was used to sharpen the curvelet fusion result. The values of the numerical evaluation metrics for this experiment are tabulated in Tables 9.7 and 9.8. We notice from these metrics that the curvelet fusion achieves the maximum similarity values with high enough PSNR values.

The effect of noise on the fusion results has also been studied. Figure 9.55 shows the variation of the PSNR of the fusion results with the SNR in the images with and without denoising for case 1. Figure 9.56 shows the variation of the similarity of the fusion results with the SNR in the images with and without denoising for the same case. From these figures, we notice that the denoising process is required for wavelet fusion at low SNR cases, while it produces small effect for curvelet fusion.

Another experiment was performed and the results are shown in Figures 9.57 through 9.62. The values of the numerical evaluation metrics for this experiment are also tabulated in Tables 9.7 and 9.8. From these results, we notice that the curvelet fusion approach achieves the maximum similarity values in both cases with high PSNR results. We notice also, as in the previous experiment, that denoising is required for wavelet fusion at low SNR cases and is not required for curvelet fusion.

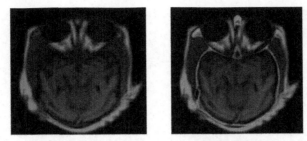

(a) PCA fusion   (b) DWT fusion

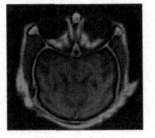   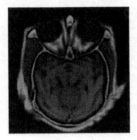

(c) Curvelet fusion   (d) Curvelet fusion with
                           post processing

Figure 9.54 Fusion results for case 1.

Table 9.7 PSNR Results for Fusion of MR and CT Images

Images	PSNR	PCA	Wavelet	Curvelet
Case 1	$PSNR_C$	11.1593	11.1120	11.3848
	$PSNR_M$	34.1244	19.7349	19.9887
Case 2	$PSNR_C$	19.0940	14.5962	15.3955
	$PSNR_M$	13.9911	12.0869	12.8597

Table 9.8 Similarity Results for Fusion of MR and CT Images

Images	Similarity	PCA	Wavelet	Curvelet
Case 1	S_C	0.0056	0.0423	0.0828
	S_M	0.1276	0.1162	0.1445
Case 2	S_C	0.0815	0.0656	0.1437
	S_M	0.0775	0.1093	0.2212

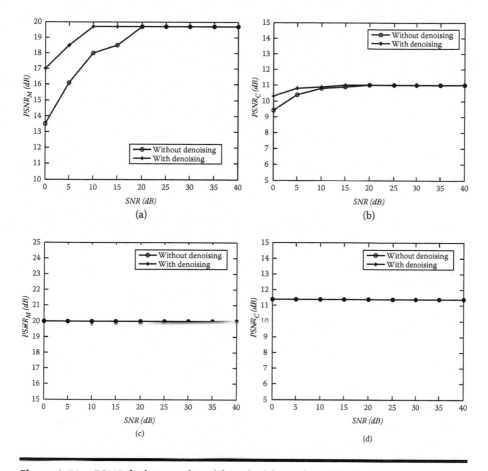

Figure 9.55 PSNR fusion results with and without denoising for case 1. (a) PSNR$_M$ versus SNR for wavelet fusion. (b) PSNR$_C$ versus SNR for wavelet fusion. (c) PSNR$_M$ versus SNR for curvelet fusion. (d) PSNR$_C$ versus SNR for curvelet fusion. (d) Post processing of curvelet fusion result.

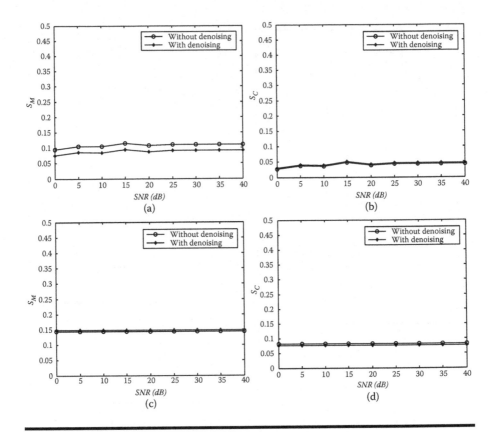

Figure 9.56 Similarity fusion results with and without denoising for case 1. (a) S_M versus SNR for wavelet fusion. (b) S_C versus SNR for wavelet fusion. (c) S_M versus SNR for curvelet fusion. (d) S_C versus SNR for curvelet fusion.

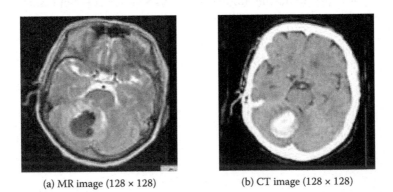

(a) MR image (128 × 128)   (b) CT image (128 × 128)

Figure 9.57 Original images for case 2.

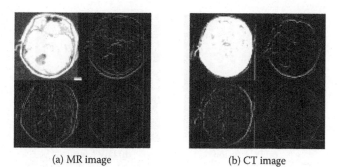

(a) MR image   (b) CT image

Figure 9.58   DWT for images of case 2.

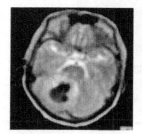 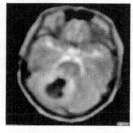 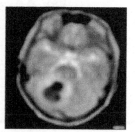

(a) P_1 of MR image   (b) P_2 of MR image   (c) P_3 of MR image

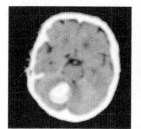 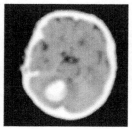 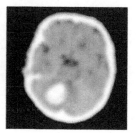

(d) P_1 of CT image   (e) P_2 of CT image   (f) P_3 of CT image

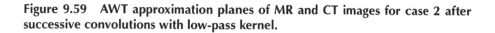

Figure 9.59   AWT approximation planes of MR and CT images for case 2 after successive convolutions with low-pass kernel.

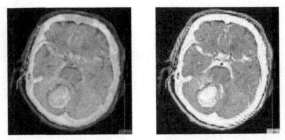

(a) PCA fusion (b) DWT fusion

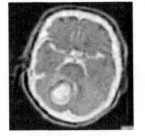

(c) Curvelet fusion (d) Post processing of the curvelet fusion result

Figure 9.60 Fusion results for case 2.

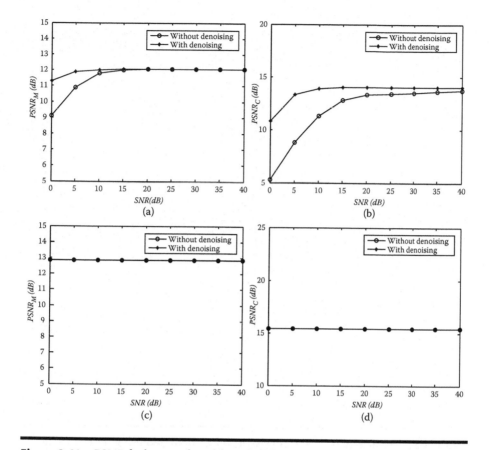

Figure 9.61   PSNR fusion results with and without denoising for case 2. (a) PSNR$_M$ versus SNR for wavelet fusion. (b) PSNR$_C$ versus SNR for wavelet fusion. (c) PSNR$_M$ versus SNR for curvelet fusion. (d) PSNR$_C$ versus SNR for curvelet fusion.

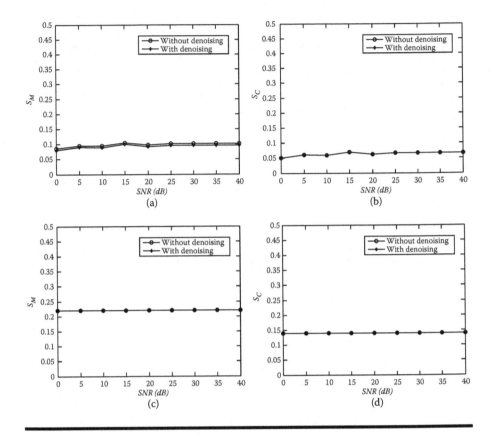

Figure 9.62 **Similarity fusion results with and without denoising for case 2. (a) S_M versus SNR for wavelet fusion. (b) S_C versus SNR for wavelet fusion. (c) S_M versus SNR for curvelet fusion. (d) S_C versus SNR curvelet fusion.**

Chapter 10

Super-Resolution with a *Priori* Information

10.1 Introduction

In the previous chapters, we have dealt with the problem of image interpolation. As defined in Chapter 1, interpolation is intended to estimate a high-resolution (HR) image based on a degraded low-resolution (LR) image. In real life, there are applications in which multiple degraded LR images are available and a single HR image is required. As mentioned previously, this problem is known as super-resolution image reconstruction and it is treated in this chapter. Concentration in the chapter is on super-resolution reconstruction with *a priori* information, where the degradations in the available LR images are assumed to be known or can be estimated through the super-resolution reconstruction process.

The super-resolution reconstruction process is an ill-posed inverse problem involving matrices of very large dimensions. This problem has been treated in the literature [112–123]. The first treatment of the super-resolution reconstruction problem was an iterative frequency domain method since the Fourier transform has superior properties for translational shifts between observations [112–114].

The maximum *a posteriori* (MAP) estimation algorithm has been implemented in the field of image super-resolution [115–117]. Other theoretical approaches have been also presented [114,118]. The special nature of this problem forces most image super-resolution reconstruction algorithms to have iterative natures. These algorithms aim to reduce the computational cost of the matrix inversion processes involved in the solution by using successive approximation methods for the estimation of the HR image.

267

Most of the suggested solutions to this problem are based on the regularization theory [112–123]. The iterative implementation of regularization theory in image super-resolution has been the most popular procedure to solve the problem. Although these algorithms avoid matrix inversion, they are still time consuming and cannot be implemented beyond a certain limit of dimensionality [121,123].

In this chapter, a general framework for image super-resolution based on wavelet fusion is presented [124–127] through discussion of three non-iterative algorithms for image super-resolution: (1) linear minimum mean square error (LMMSE) super-resolution, (2) maximum entropy super-resolution, and (3) regularized super-resolution using wavelet fusion. These algorithms are based on breaking the super-resolution reconstruction problem into four consecutive steps to work on large-dimension images. These steps are (1) registration, (2) multi-channel restoration, (3) wavelet fusion and denoising, and (4) image interpolation. Figure (10.1) illustrates these steps.

The difference between these algorithms lies in the multi-channel restoration step and the interpolation step. The wavelet image fusion process is used in these algorithms as a tool to integrate the information obtained from all the outputs of the multi-channel restoration step into a single image. For the case of low signal-to-noise ratio (SNR) observations, wavelet denoising is applied after the fusion step. The obtained image is then interpolated to give an HR image.

10.2 Multiple Observation LR Degradation Model

In super-resolution image reconstruction algorithms, several degraded LR observations are used to estimate a single HR image. The mathematical model that relates the available LR observations to the required HR image is given by [112-116]:

$$\mathbf{g}_k = \mathbf{D}_k \mathbf{B}_k \mathbf{M}_k \mathbf{f_h} + \mathbf{v}_k \quad 1 \le k \le P \tag{10.1}$$

where P is the number of available observations, $\mathbf{g}_k$ is an $(M^2 \times 1)$ vector representing the kth $(M \times M)$ LR image in lexicographic order, and $\mathbf{f_h}$ is an $(N^2 \times 1)$ vector representing an $(N \times N)$ HR image in lexicographic order. $\mathbf{M}_k$ is the $(N^2 \times N^2)$ registration shift matrix and $\mathbf{B}_k$ is the blur matrix of size $(N^2 \times N^2)$. $\mathbf{D}_k$ is the $(M^2 \times N^2)$ uniform down-sampling matrix and $\mathbf{v}_k$ is the $M^2 \times 1$ noise vector. Equation (10.1) can be rewritten in the following form:

$$\begin{bmatrix} \mathbf{g}_1 \\ \vdots \\ \mathbf{g}_p \end{bmatrix} = \begin{bmatrix} \mathbf{D}_1\mathbf{B}_1\mathbf{M}_1 \\ \vdots \\ \mathbf{D}_p\mathbf{B}_p\mathbf{M}_p \end{bmatrix} \mathbf{f_h} + \begin{bmatrix} \mathbf{v}_1 \\ \vdots \\ \mathbf{v}_p \end{bmatrix} \tag{10.2}$$

Simplifying Equation (10.2) leads to:

$$\mathbf{g} = \mathbf{L}\mathbf{f_h} + \mathbf{v} \tag{10.3}$$

where

$$\mathbf{g} = \begin{bmatrix} \mathbf{g}_1 \\ \vdots \\ \mathbf{g}_p \end{bmatrix}, \quad \mathbf{L} = \begin{bmatrix} \mathbf{D}_1\mathbf{B}_1\mathbf{M}_1 \\ \vdots \\ \mathbf{D}_p\mathbf{B}_p\mathbf{M}_p \end{bmatrix}, \quad \mathbf{n} = \begin{bmatrix} \mathbf{v}_1 \\ \vdots \\ \mathbf{v}_p \end{bmatrix} \tag{10.4}$$

Using the regularization theory to solve Equation (10.3), we get [116,118,123]:

$$\hat{\mathbf{f}}_h = \left[\mathbf{L}^t\mathbf{L} + \tau\mathbf{C}^t\mathbf{C} \right]^{-1} \mathbf{L}^t\mathbf{g} \tag{10.5}$$

where $\mathbf{C}$ is the regularization operator, which is preferred to be a 3-D Laplacian operator to capture the between-channel information in the reconstruction process. The τ is a global regularization parameter.

There is a problem in the implementation of Equation (10.5). The matrix inversion process cannot be performed directly and no approximations such as the Toeplitz-to-circulant approximation can be used to diagonalize the matrices involved in the inversion process because L is not a square matrix. As a result, there is a need for a different implementation of the regularized image super-resolution reconstruction algorithm.

Most previously published work in this field deals with this problem in an iterative manner to avoid the matrix inversion. A large number of iterations is required to get an estimate of the required HR image using the algorithms [121,123]. Another limitation of the iterative solutions is the inability to use LR images of large dimensions beyond 64×64 pixels [121,123]. For example, if three LR images of dimensions 64×64 are used to obtain an HR image of dimensions 128×128, the matrices $\mathbf{L}$ and $\mathbf{C}$ in the iterative algorithms will be of dimensions 12228×49152 and 49152×49152, respectively.

This example shows that the computational cost required to carry out the iterations with matrices of such dimensions is very large. This is clear [123] where the number of iterations required to obtain an HR image from multiple LR images ranges from 13 to 22 iterations, with each iteration requiring 15.5 sec using a Pentium IV 1700 processor. Thus, it is clear that this computation time is very large in iterative algorithms. Working with LR images beyond the limit of 64×64 pixels will require the storage of matrices $\mathbf{L}$ and $\mathbf{C}$ of larger dimensions. In this case, the super-resolution reconstruction algorithm will require sophisticated memory capabilities to store the matrices and carry out the iterations.

A direct solution to this problem could be the segmentation of the LR images into small segments and carrying out the iterations on each segment separately. This solution would produce edge effects. If overlapped segments are used instead to avoid edge effects, the number of segments will increase, requiring much more computational time. In the following section, we present a different treatment to this ill-posed problem that removes these computational limitations.

10.3 Wavelet-Based Image Super-Resolution

Due to the above mentioned limitations, the regularized solution cannot be directly implemented for image super-resolution. In this section, image super-resolution based on wavelet fusion is presented [124–127]. The implementation of this approach is composed of four consecutive steps as mentioned in Section 10.1. In the general solution of the super-resolution reconstruction problem, we deal with three degradation phenomena: general geometric registration warp, blurring, and additive noise. Based on these phenomena, we can break the solution into the following consecutive steps as shown in Figure 10.1:

Step 1 — Image alignment, which means estimating the geometrical registration warp between different images.

Step 2 — Multi-channel image restoration of the registered degraded observations.

Step 3 — Wavelet-based fusion of the multiple images obtained from step 2 to form a single image. Wavelet denoising is performed after the fusion in the case of low SNRs.

Step 4 — Image interpolation of the image resulting from step 3 to obtain an HR image.

We covered the topics of image registration and wavelet fusion and denoising in Chapters 8 and 9, respectively. We also covered image interpolation in Chapters 2 through 7. In the wavelet-based super-resolution reconstruction approach, we use the algorithms discussed in Chapter 7 for interpolation as an inverse problem. Thus, we will devote the rest of this chapter to the multi-channel image restoration step and implementing it using different algorithms with the necessary assumptions. This step requires a simplified image degradation model that is given in the following section.

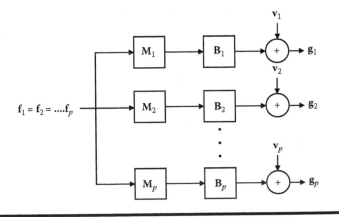

Figure 10.1 Multi-channel image degradation model.

10.4 Simplified Multi-Channel Degradation Model

The multi-channel image restoration step aims at obtaining multiple undegraded images of the same dimensions as those of the available LR images. These obtained images are then used in the next step of image fusion. This step requires a simplified degradation model. This model does not consider the operator **D** of filtering and down-sampling. For a multi-channel imaging system with P channels each of size $M \times M$, the simplified degradation model becomes [124–127]:

$$\mathbf{g}_k = \mathbf{B}_k \mathbf{M}_k \mathbf{f}_k + \mathbf{v}_k \qquad \text{for } k = 1, 2, ...P \tag{10.6}$$

where $\mathbf{g}_k$, $\mathbf{f}_k$ and $\mathbf{v}_k$ are the observed image, the ideal image, and the noise of the kth channel, respectively. $\mathbf{B}_k$ is the degradation matrix of the kth channel. $\mathbf{M}_k$ is the relative registration shift operator of the kth channel. In this chapter, we restrict our work to global translational shifts, which is the case of consideration in most image super-resolution reconstruction algorithms. Other types of motion such as zooming or panning are beyond the scope of this chapter and cannot be modeled using Equation (10.6). Translational motion can be approximated by circular motion except near-edge pixels [124–127]. Using this approximation, the matrices $\mathbf{M}_k$ can be approximated by circulant matrices for all values of k. Equation (10.6) can be written in the following form:

$$\mathbf{g} = \mathbf{BMf} + \mathbf{v} \tag{10.7}$$

where

$$\mathbf{g} = \begin{bmatrix} \mathbf{g}_1 \\ \mathbf{g}_2 \\ \vdots \\ \mathbf{g}_P \end{bmatrix}; \quad \mathbf{f} = \begin{bmatrix} \mathbf{f}_1 \\ \mathbf{f}_2 \\ \vdots \\ \mathbf{f}_P \end{bmatrix}; \quad \mathbf{v} = \begin{bmatrix} \mathbf{v}_1 \\ \mathbf{v}_2 \\ \vdots \\ \mathbf{v}_P \end{bmatrix}; \quad \mathbf{B} = \begin{bmatrix} \mathbf{B}_1 & 0 & \cdots & 0 \\ 0 & \mathbf{B}_2 & \cdots & 0 \\ \vdots & \vdots & \cdots & \vdots \\ 0 & 0 & \cdots & \mathbf{B}_P \end{bmatrix}$$

$$\text{and} \quad \mathbf{M} = \begin{bmatrix} \mathbf{M}_1 & 0 & \cdots & 0 \\ 0 & \mathbf{M}_2 & \cdots & 0 \\ \vdots & \vdots & \cdots & \vdots \\ 0 & 0 & \cdots & \mathbf{M}_P \end{bmatrix} \tag{10.8}$$

If we define:

$$\mathbf{H} = \mathbf{BM} \tag{10.9}$$

the multi-channel image degradation model can be written in the form [124–127]:

$$\mathbf{g} = \mathbf{Hf} + \mathbf{v} \tag{10.10}$$

where $\mathbf{g}$, $\mathbf{f}$, and v are $P \times M^2$ in length. The degradation operator $\mathbf{H}$ of the multi-channel imaging model is of dimensions $(PM^2) \times (PM^2)$. The image degradation model is illustrated in Figure 10.1 for $\mathbf{f}_1 = \mathbf{f}_2 = \ldots = \mathbf{f}_P$. According to the model in Equation (10.10), we can carry the multi-channel restoration process to find the vector $\mathbf{f}$.

10.5 Multi-Channel Image Restoration

Image restoration on a multi-channel basis aims at incorporating all the information existing in all the channels into the restoration process instead of restoring each channel separately. Multi-channel image restoration has been previously studied in the literature for images degraded by both blur and noise only without registration shift [128–136]. Our objective is to carry out multi-channel image restoration in the presence of registration shifts. Multi-channel image restoration can be implemented using several algorithms such as the LMMSE algorithm, the maximum entropy algorithm, and the regularization algorithm. We will discuss the implementation of these algorithms in the presence of registration shifts.

10.5.1 Multi-Channel LMMSE Restoration

The LMMSE algorithm used in Chapter 6 for image interpolation can be used to solve the multi-channel image restoration problem, but with replacing the operator $\mathbf{D}$ by the operator $\mathbf{H}$. The LMMSE solution to Equation (10.10) is thus given by [125,126]:

$$\hat{\mathbf{f}} = \mathbf{R}_\mathbf{f} \mathbf{H}^t [\mathbf{H} \mathbf{R}_\mathbf{f} \mathbf{H}^t + \mathbf{R}_\mathbf{v}]^{-1} \mathbf{g} \tag{10.11}$$

where $\mathbf{R}_\mathbf{f}$ and $\mathbf{R}_\mathbf{n}$ are the multi-channel image and noise correlation matrices. They are of dimensions $(PM^2) \times (PM^2)$, and are given by:

$$\mathbf{R}_\mathbf{f} = \begin{bmatrix} \mathbf{R}_\mathbf{f}^{11} & \mathbf{R}_\mathbf{f}^{12} & \cdots & \mathbf{R}_\mathbf{f}^{1P} \\ \mathbf{R}_\mathbf{f}^{21} & \mathbf{R}_\mathbf{f}^{22} & \cdots & \mathbf{R}_\mathbf{f}^{2P} \\ \vdots & \vdots & \cdots & \vdots \\ \mathbf{R}_\mathbf{f}^{P1} & \mathbf{R}_\mathbf{f}^{P2} & \cdots & \mathbf{R}_\mathbf{f}^{PP} \end{bmatrix}, \quad \mathbf{R}_\mathbf{v} = \begin{bmatrix} \mathbf{R}_\mathbf{v}^{11} & 0 & \cdots & 0 \\ 0 & \mathbf{R}_\mathbf{v}^{22} & \cdots & 0 \\ \vdots & \vdots & \cdots & \vdots \\ 0 & 0 & \cdots & \mathbf{R}_\mathbf{v}^{PP} \end{bmatrix} \tag{10.12}$$

The auto-correlation matrix of the noise $\mathbf{R}_\mathbf{v}$ is defined by assuming that the between-channel noise is uncorrelated. If white noise is assumed for $\mathbf{v}$, then $\mathbf{R}_\mathbf{v}^{kk} = \sigma_{kk}^2 \mathbf{I}$, where σ_{kk}^2 is the variance of $\mathbf{v}_k$. Equation (10.11) can be written in the form [125,126]:

$$\hat{\mathbf{f}} = \mathbf{R}_\mathbf{f} \mathbf{H}^t \overline{\mathbf{R}}^{-1} \mathbf{g} \tag{10.13}$$

where

$$\overline{\mathbf{R}} = \begin{bmatrix} \mathbf{H}_1\mathbf{R}_f^{11}\mathbf{H}_1^t + \mathbf{R}_n^{11} & \mathbf{H}_1\mathbf{R}_f^{12}\mathbf{H}_2^t & \cdots & \mathbf{H}_1\mathbf{R}_f^{1P}\mathbf{H}_P^t \\ \mathbf{H}_2\mathbf{R}_f^{21}\mathbf{H}_1^t & \mathbf{H}_2\mathbf{R}_f^{22}\mathbf{H}_2^t + \mathbf{R}_n^{22} & \cdots & \mathbf{H}_2\mathbf{R}_f^{2P}\mathbf{H}_P^t \\ \vdots & \vdots & \cdots & \vdots \\ \mathbf{H}_P\mathbf{R}_f^{P1}\mathbf{H}_1^t & \mathbf{H}_P\mathbf{R}_f^{P2}\mathbf{H}_2^t & \cdots & \mathbf{H}_P\mathbf{R}_f^{PP}\mathbf{H}_P^t + \mathbf{R}_n^{PP} \end{bmatrix} \tag{10.14}$$

The solution of Equation (10.13) requires the inversion of $\overline{\mathbf{R}}$ which is of dimensions $(PM^2) \times (PM^2)$. This inversion process is impractical. Thus, a simplification is needed for this inversion process. This simplification can be performed using a block Toeplitz-to-block circulant approximation of matrices and the diagonalization property via the 2-D Fourier transform [125,126]. To obtain a solution for the multi-channel LMMSE model, we define the operator [125,126; Appendix B]:

$$\overline{\varphi}^{-1} = \begin{bmatrix} \varphi^{-1} & 0 & \cdots & 0 \\ 0 & \varphi^{-1} & \cdots & 0 \\ \vdots & \vdots & \vdots & \vdots \\ 0 & 0 & \cdots & \varphi^{-1} \end{bmatrix} \tag{10.15}$$

Applying $\boldsymbol{\varphi}^{-1}$ to both sides of Equation (10.11) yields [125,126]:

$$\overline{\varphi}^{-1}\hat{\mathbf{f}} = [\overline{\varphi}^{-1}\mathbf{R}_f\overline{\varphi}][\overline{\varphi}^{-1}\mathbf{H}^t\overline{\varphi}][\overline{\varphi}^{-1}[\mathbf{H}\mathbf{R}_f\mathbf{H}^t + \mathbf{R}_n]\overline{\varphi}]^{-1}\overline{\varphi}^{-1}\mathbf{g} \tag{10.16}$$

where $\overline{\varphi}^{-1}\mathbf{g}$ is the discrete Fourier transform (DFT) of the observed P-channel image and $\overline{\varphi}^{-1}\hat{\mathbf{f}}$ is the DFT of the restored multi-channel image vector. Equation (10.16) can be written in the following form:

$$\hat{\mathbf{F}} = [\overline{\varphi}^{-1}\mathbf{R}_f\overline{\varphi}][\overline{\varphi}^{-1}\mathbf{H}^t\overline{\varphi}][\overline{\varphi}^{-1}[\mathbf{H}\mathbf{R}_f\mathbf{H}^t + \mathbf{R}_n]\overline{\varphi}]^{-1}\mathbf{G} \tag{10.17}$$

where $\mathbf{G}$ and $\hat{\mathbf{F}}$ are the DFTs of $\mathbf{g}$ and $\hat{\mathbf{f}}$, respectively.

For a multi-channel model having $P = 3$ channels, the elements of Equation (10.17) can be written in the following form [125,126]:

$$\overline{\varphi}^{-1}\mathbf{R}_f\overline{\varphi} = \begin{bmatrix} \overline{\varphi}^{-1}\mathbf{R}_f^{11}\overline{\varphi} & \overline{\varphi}^{-1}\mathbf{R}_f^{12}\overline{\varphi} & \overline{\varphi}^{-1}\mathbf{R}_f^{13}\overline{\varphi} \\ \overline{\varphi}^{-1}\mathbf{R}_f^{21}\overline{\varphi} & \overline{\varphi}^{-1}\mathbf{R}_f^{22}\overline{\varphi} & \overline{\varphi}^{-1}\mathbf{R}_f^{23}\overline{\varphi} \\ \overline{\varphi}^{-1}\mathbf{R}_f^{31}\overline{\varphi} & \overline{\varphi}^{-1}\mathbf{R}_f^{32}\overline{\varphi} & \overline{\varphi}^{-1}\mathbf{R}_f^{33}\overline{\varphi} \end{bmatrix} = \begin{bmatrix} \Lambda_f^{11} & \Lambda_f^{12} & \Lambda_f^{13} \\ \Lambda_f^{21} & \Lambda_f^{22} & \Lambda_f^{23} \\ \Lambda_f^{31} & \Lambda_f^{32} & \Lambda_f^{33} \end{bmatrix} \tag{10.18}$$

and

$$\overline{\varphi}^{-1}\mathbf{H}^t\overline{\varphi} = \begin{bmatrix} \overline{\varphi}^{-1}\mathbf{H}_1^t\varphi & \mathbf{0} & \mathbf{0} \\ \mathbf{0} & \overline{\varphi}^{-1}\mathbf{H}_2^t\varphi & \mathbf{0} \\ \mathbf{0} & \mathbf{0} & \overline{\varphi}^{-1}\mathbf{H}_3^t\varphi \end{bmatrix} = \begin{bmatrix} \Lambda_h^{1*} & \mathbf{0} & \mathbf{0} \\ \mathbf{0} & \Lambda_h^{1*} & \mathbf{0} \\ \mathbf{0} & \mathbf{0} & \Lambda_h^{1*} \end{bmatrix} \quad (10.19)$$

where

$$\varphi^{-1}\mathbf{H}_k\varphi = \varphi^{-1}\mathbf{B}_k\varphi\varphi^{-1}\mathbf{M}_k\varphi = \Lambda_b^k\Lambda_m^k = \Lambda_h^k \quad (10.20)$$

and

$$\varphi^{-1}\mathbf{H}_k^t\varphi = \varphi^{-1}\mathbf{M}_k^t\varphi\varphi^{-1}\mathbf{B}_k^t\varphi = \Lambda_m^{k*}\Lambda_b^{k*} = \Lambda_h^{k*} \quad (10.21)$$

Thus:

$$\overline{\varphi}^{-1}[\mathbf{H}\mathbf{R}_f\mathbf{H}^t + \mathbf{R}_v]\overline{\varphi}$$

$$= \begin{bmatrix} \overline{\varphi}^{-1}[\mathbf{H}_1\mathbf{R}_f^{11}\mathbf{H}_1^t + \mathbf{R}_v^{11}]\varphi & \overline{\varphi}^{-1}[\mathbf{H}_1\mathbf{R}_f^{12}\mathbf{H}_2^t]\varphi & \overline{\varphi}^{-1}[\mathbf{H}_1\mathbf{R}_f^{13}\mathbf{H}_3^t]\varphi \\ \overline{\varphi}^{-1}[\mathbf{H}_2\mathbf{R}_f^{21}\mathbf{H}_1^t]\varphi & \overline{\varphi}^{-1}[\mathbf{H}_2\mathbf{R}_f^{22}\mathbf{H}_2^t + \mathbf{R}_v^{22}]\varphi & \overline{\varphi}^{-1}[\mathbf{H}_2\mathbf{R}_f^{23}\mathbf{H}_3^t]\varphi \\ \overline{\varphi}^{-1}[\mathbf{H}_3\mathbf{R}_f^{31}\mathbf{H}_1^t]\varphi & \overline{\varphi}^{-1}[\mathbf{H}_3\mathbf{R}_f^{32}\mathbf{H}_2^t]\varphi & \overline{\varphi}^{-1}[\mathbf{H}_3\mathbf{R}_f^{33}\mathbf{H}_3^t + \mathbf{R}_v^{33}]\varphi \end{bmatrix}$$

$$= \begin{bmatrix} \Lambda_h^1\Lambda_f^{11}\Lambda_h^{1*} + \Lambda_v^{11} & \Lambda_h^1\Lambda_f^{12}\Lambda_h^{2*} & \Lambda_h^1\Lambda_f^{13}\Lambda_h^{3*} \\ \Lambda_h^2\Lambda_f^{21}\Lambda_h^{1*} & \Lambda_h^2\Lambda_f^{22}\Lambda_h^{2*} + \Lambda_v^{22} & \Lambda_h^2\Lambda_f^{23}\Lambda_h^{3*} \\ \Lambda_h^3\Lambda_f^{31}\Lambda_h^{1*} & \Lambda_h^3\Lambda_f^{32}\Lambda_h^{2*} & \Lambda_h^3\Lambda_f^{33}\Lambda_h^{3*} + \Lambda_v^{33} \end{bmatrix} \quad (10.22)$$

Since operations on diagonal matrices yield diagonal matrices, Equation (10.22) can be expressed as follows [125,126]:

$$\overline{\varphi}^{-1}[\mathbf{H}\mathbf{R}_f\mathbf{H}^t + \mathbf{R}_n]\overline{\varphi} = \begin{bmatrix} \Lambda_{11} & \Lambda_{12} & \Lambda_{13} \\ \Lambda_{21} & \Lambda_{22} & \Lambda_{23} \\ \Lambda_{31} & \Lambda_{32} & \Lambda_{33} \end{bmatrix} \quad (10.23)$$

where $\Lambda_{11}, \Lambda_{12}, \Lambda_{13}, \Lambda_{21}, \Lambda_{22}, \Lambda_{23}, \Lambda_{31}, \Lambda_{32}$, and Λ_{33} are all diagonal matrices. The inversion in Equation (10.17) is then possible, since it is a sparse block-diagonal matrix. In general, operations on sparse matrices require small numbers of mathematical operations. The correlation sequences between different channels can be approximated from the undegraded images using the following relation [125,126]:

$$R_f^{ij}(n_1,n_2) \cong \frac{1}{w^2} \sum_{k=1}^{w} \sum_{l=1}^{w} g_i(k,l) g_j(n_1+k,n_2+l) \tag{10.24}$$

where $R_f^{ij}(n_1,n_2)$ is the correlation at position (n_1, n_2) between channels i and j. The signals g_i and g_j are the observed images in channels i and j, respectively.

10.5.2 Multi-Channel Maximum Entropy Restoration

Using the concept of entropy maximization, the solution of the simplified multi-channel image degradation model is represented as follows [126,127]:

$$\hat{\mathbf{f}} \cong \left(\mathbf{H}^t\mathbf{H} + \eta\mathbf{I}\right)^{-1}\mathbf{H}^t\mathbf{g} \tag{10.25}$$

Equation (10.25) can be written in the form [126,127]:

$$\hat{\mathbf{f}} = \bar{\mathbf{R}}^{-1}\mathbf{H}^t\mathbf{g} \tag{10.26}$$

where

$$\bar{\mathbf{R}} = \begin{bmatrix} \mathbf{H}_1^t\mathbf{H}_1 + \eta\mathbf{I}_0 & 0 & \cdots & 0 \\ 0 & \mathbf{H}_2^t\mathbf{H}_2 + \eta\mathbf{I}_0 & \cdots & 0 \\ \vdots & \vdots & \cdots & \vdots \\ 0 & 0 & \cdots & \mathbf{H}_P^t\mathbf{H}_P + \eta\mathbf{I}_0 \end{bmatrix} \tag{10.27}$$

where $\mathbf{I}_0$ is an $(M^2 \times M^2)$ identity matrix.

The solution of Equation (10.26) requires the inversion of $\bar{\mathbf{R}}$, which is of dimensions $(PM^2) \times (PM^2)$. This inversion process is impractical. Thus, a simplification is needed. This simplification can be performed using a block Toeplitz-to-block circulant approximation of matrices and the diagonalization property via the two-dimensional (2-D) Fourier transform as used in the previous section [126,127]. Applying $\bar{\varphi}^{-1}$ to both sides of Equation (10.25) yields :

$$\bar{\varphi}^{-1}\hat{\mathbf{f}} = [\bar{\varphi}^{-1}[\mathbf{H}^t\mathbf{H} + \eta\mathbf{I}]\bar{\varphi}]^{-1}\bar{\varphi}^{-1}\mathbf{H}^t\varphi\bar{\varphi}^{-1}\mathbf{g} \tag{10.28}$$

where $\bar{\varphi}^{-1}\mathbf{g}$ is the DFT of the observed P-channel image vector and $\bar{\varphi}^{-1}\hat{\mathbf{f}}$ is the DFT of the restored multi-channel image vector. Equation (10.28) can be written in the following form:

$$\hat{\mathbf{F}} = [\bar{\varphi}^{-1}\mathbf{H}^t\varphi\bar{\varphi}^{-1}\mathbf{H}\varphi + \eta\mathbf{I}]^{-1}\bar{\varphi}^{-1}\mathbf{H}^t\varphi\mathbf{G} \tag{10.29}$$

where $\mathbf{G}$ and $\hat{\mathbf{F}}$ are the DFTs of $\mathbf{g}$ and $\hat{\mathbf{f}}$, respectively.

For a multi-channel model having $P = 3$ channels, the elements of Equation (10.29) can be written in the following form:

$$
\hat{\mathbf{F}} = \begin{bmatrix} \Lambda_{\mathbf{h}}^{1*}\Lambda_{\mathbf{h}}^{1} + \eta \mathbf{I}_0 & 0 & 0 \\ 0 & \Lambda_{\mathbf{h}}^{2*}\Lambda_{\mathbf{h}}^{2} + \eta \mathbf{I}_0 & 0 \\ 0 & 0 & \Lambda_{\mathbf{h}}^{3*}\Lambda_{\mathbf{h}}^{3} + \eta \mathbf{I}_0 \end{bmatrix}^{-1} \begin{bmatrix} \Lambda_{\mathbf{h}}^{1*} & 0 & 0 \\ 0 & \Lambda_{\mathbf{h}}^{2*} & 0 \\ 0 & 0 & \Lambda_{\mathbf{h}}^{3*} \end{bmatrix} \mathbf{G}
$$

(10.30)

Thus:

$$
\hat{\mathbf{F}} = \begin{bmatrix} \left(\Lambda_{\mathbf{h}}^{1*}\Lambda_{\mathbf{h}}^{1} + \eta \mathbf{I}_0\right)^{-1}\Lambda_{\mathbf{h}}^{1*} & 0 & 0 \\ 0 & \left(\Lambda_{\mathbf{h}}^{2*}\Lambda_{\mathbf{h}}^{2} + \eta \mathbf{I}_0\right)^{-1}\Lambda_{\mathbf{h}}^{2*} & 0 \\ 0 & 0 & \left(\Lambda_{\mathbf{h}}^{3*}\Lambda_{\mathbf{h}}^{3} + \eta \mathbf{I}_0\right)^{-1}\Lambda_{\mathbf{h}}^{3*} \end{bmatrix} \mathbf{G}
$$

(10.31)

This leads to [126,127]:

$$
\hat{\mathbf{F}} = \begin{bmatrix} \Lambda_{11} & 0 & 0 \\ 0 & \Lambda_{22} & 0 \\ 0 & 0 & \Lambda_{33} \end{bmatrix} \mathbf{G}
$$

(10.32)

where

$$
\Lambda_{kk} = \left(\Lambda_{\mathbf{h}}^{k*}\Lambda_{\mathbf{h}}^{k} + \eta \mathbf{I}_0\right)^{-1}\Lambda_{\mathbf{h}}^{k*}
$$

(10.33)

Λ_{11}, Λ_{22}, and Λ_{33} are diagonal matrices. The inversion in Equation (10.29) is then possible since it is a sparse diagonal matrix inversion process.

10.5.3 Multi-Channel Regularized Restoration

Multi-channel regularized restoration has been previously studied in relation to blurred images. Our objective in this section is to extend it to the case of multiple images degraded by both blurring and registration shift in the presence of noise. Let us denote the degradation function associated with the kth channel by the $M^2 \times (PM^2)$ matrix [124,126]:

$$
\bar{\mathbf{H}}_k = [0, \mathbf{H}_k, \ldots\ldots, 0]
$$

(10.34)

Thus, the degradation model defined by Equation (10.10) can be written as [124,126]:

$$\mathbf{g}_k = \bar{\mathbf{H}}_k \mathbf{f} + \mathbf{v}_k \quad k = 1,2,\dots P \tag{10.35}$$

A set theoretic approach can be used to solve Equation (10.35). In this approach, *a priori* knowledge about **f** is assumed, which restricts the solution to lie in a set, that is [124,126]:

$$\mathbf{f} \in \mathbf{S_f} \tag{10.36}$$

where $\mathbf{S_f}$ is a PM^2 dimensional space. The noise is also assumed to belong to a set $\mathbf{S}_{\mathbf{v}_k}$. Since $\mathbf{v}_k$ must lie in a set, it follows that a given observation $\mathbf{g}_k$ combines with the set to define a new set that must contain **f**. Thus, the observation $\mathbf{g}_k$ specifies a set $\mathbf{S}_{\mathbf{f}/\mathbf{g}_k}$ that must contain **f**. This can be expressed as follows [124,126]:

$$\mathbf{f} \in \mathbf{S}_{\mathbf{f}/\mathbf{g}_k} = \left[\mathbf{f} : \left(\bar{\mathbf{H}}\mathbf{f} - \mathbf{g}_k \right) \in \mathbf{S}_{\mathbf{v}_k} \right] \tag{10.37}$$

Now, all the sets $S_f, \mathbf{S}_{\mathbf{f}/\mathbf{g}_1}, \dots\dots\mathbf{S}_{\mathbf{f}/\mathbf{g}_N}$ contain **f**. Therefore **f** must lie in the intersection of these sets as follows:

$$\mathbf{S}_{\hat{\mathbf{f}}} = \mathbf{S_f} \cap \mathbf{S}_{\mathbf{f}/\mathbf{g}_1} \cap \dots\dots \cap \mathbf{S}_{\mathbf{f}/\mathbf{g}_N} \tag{10.38}$$

The $\mathbf{S}_{\hat{\mathbf{f}}}$ is the smallest set that must contain **f**. All the sets can be represented by ellipsoids. The ellipsoids that represent $\mathbf{S_f}$ and $\mathbf{S}_{\mathbf{v}_k}$ are given by [124,126]:

$$\left\| \mathbf{C}\hat{\mathbf{f}} \right\|^2 \leq E^2 \tag{10.39}$$

and

$$\left\| \mathbf{v}_k \right\|^2 \leq e_k^2, \quad k = 1,2,\dots,P. \tag{10.40}$$

where E^2 denotes the maximum amount of high-frequency energy in the required HR image. **C** is the regularization operator, which is a 3-D Laplacian operator given by [124,126]:

$$\mathbf{C} = \begin{bmatrix} \mathbf{C}_{11} & \mathbf{C}_{12} & \cdots & \mathbf{C}_{1P} \\ \mathbf{C}_{21} & \mathbf{C}_{22} & \cdots & \mathbf{C}_{2P} \\ \vdots & \vdots & \cdots & \vdots \\ \mathbf{C}_{P1} & \mathbf{C}_{P2} & \cdots & \mathbf{C}_{PP} \end{bmatrix} \tag{10.41}$$

According to the regularization theory, an $\hat{\mathbf{f}}$ that achieves the following is sought [124,126]:

$$Minimizes \quad \left\| \mathbf{C}\hat{\mathbf{f}} \right\|^2 \tag{10.42}$$

$$subject \quad \left\| \bar{\mathbf{H}}_k \hat{\mathbf{f}} - \mathbf{g}_k \right\|^2 = \left\| \mathbf{n}_k \right\|^2 = \varepsilon_k^2 \quad for \quad k = 1, 2, \ldots\ldots, P \tag{10.43}$$

Using the method of Lagrange multipliers, the solution $\hat{\mathbf{f}}$ is obtained by minimizing the following cost function [124,126]:

$$\Psi(\hat{\mathbf{f}}, \lambda_1, \lambda_2, \ldots\ldots, \lambda_P) = \sum_{k=1}^{P} \lambda_k \left[\left\| \bar{\mathbf{H}}_k \hat{\mathbf{f}} - \mathbf{g}_k \right\|^2 - \left\| \mathbf{n}_k \right\|^2 \right] + \left\| \mathbf{C}\hat{\mathbf{f}} \right\|^2 \tag{10.44}$$

The Lagrangian method is used for solving a system of equations that constitute necessary conditions for optimality. For the problem described by Equation (10.44), these conditions are given by [124,126]:

$$\nabla_{\hat{\mathbf{f}}} \Psi(\hat{\mathbf{f}}, \lambda) = 0 \tag{10.45}$$

$$\nabla_{\lambda} \Psi(\hat{\mathbf{f}}, \lambda) = 0 \tag{10.46}$$

where the vector $\lambda = (\lambda_1, \lambda_2, \ldots\ldots, \lambda_p)$ and ∇_z denote the gradient of a function with respect to vector z. Solving Equations (10.45) and (10.46) leads to [124,126]:

$$\left[\mathbf{H}^t \mathbf{H} + \mathbf{C}^t \mathbf{C} \right] \hat{\mathbf{f}}(\lambda_1, \lambda_2, \ldots\ldots, \lambda_P) = \Lambda \mathbf{H}^t \mathbf{g} \tag{10.47}$$

where Λ is defined as [124,126]:

$$\Lambda = \begin{bmatrix} \lambda_1[\mathbf{I}] & 0 & \cdots & 0 \\ 0 & \lambda_2[\mathbf{I}] & \cdots & 0 \\ \vdots & \vdots & \cdots & \vdots \\ 0 & 0 & \cdots & \lambda_P[\mathbf{I}] \end{bmatrix} \tag{10.48}$$

where the identity matrix $\mathbf{I}$ is of size $M^2 \times M^2$. Finally, solving for $\hat{\mathbf{f}}$ in Equation (10.47) yields [124,126]:

$$\hat{\mathbf{f}} = \left[\mathbf{H}^t \mathbf{H} + \Lambda^{-1} \mathbf{C}^t \mathbf{C} \right]^{-1} \mathbf{H}^t \mathbf{g} \tag{10.49}$$

The evaluation of the optimum vector $\boldsymbol{\lambda} = (\lambda_1, \lambda_2, \ldots, \lambda_p)$ required to carry out Equation (10.48) can be performed using Newton's method [124,126; Appendix C]. This method is iterative and cannot be implemented for the large dimension matrices in the equation. Thus, there is a need to diagonalize the matrices prior to the application of this method. All the matrices $\mathbf{B}_k$, $\mathbf{M}_k$, and $\mathbf{C}_{ij}$ are block Toeplitz matrices that can be approximated by block circulant matrices [124,126]. The circulant matrix can be diagonalized via a 2-D Fourier transform [124,126].

To obtain a solution for the multi-channel regularized restoration model, we utilize the 2-D Fourier transform operator as in the previous sections. Applying $\overline{\varphi}^{-1}$ to both sides of Equation (10.49) yields:

$$\overline{\varphi}^{-1}\hat{\mathbf{f}} = [\overline{\varphi}^{-1}[\mathbf{H}^t\mathbf{H} + \boldsymbol{\Lambda}^{-1}\mathbf{C}^t\mathbf{C}]\overline{\varphi}]^{-1}\overline{\varphi}^{-1}\mathbf{H}^t\overline{\varphi}\overline{\varphi}^{-1}\mathbf{g} \tag{10.50}$$

where $\overline{\varphi}^{-1}\mathbf{g}$ is the DFT of the observed P-channel image vector and $\overline{\varphi}^{-1}\hat{\mathbf{f}}$ is the DFT of the restored multi-channel image vector. Equation (10.50) can be written in the following form:

$$\hat{\mathbf{F}} = [\overline{\varphi}^{-1}\mathbf{H}^t\overline{\varphi}\overline{\varphi}^{-1}\mathbf{H}\overline{\varphi} + \overline{\varphi}^{-1}\boldsymbol{\Lambda}^{-1}\overline{\varphi}\overline{\varphi}^{-1}\mathbf{C}^t\overline{\varphi}\overline{\varphi}^{-1}\mathbf{C}\overline{\varphi}]^{-1}\overline{\varphi}^{-1}\mathbf{H}^t\overline{\varphi}\mathbf{G} \tag{10.51}$$

where $\mathbf{G}$ and $\hat{\mathbf{F}}$ are the DFTs of $\mathbf{g}$ and $\hat{\mathbf{f}}$, respectively. For a multi-channel model having $P = 3$ channels, Equation (10.51) can be written as follows [124,126]:

$$\hat{\mathbf{F}} = \begin{bmatrix} \boldsymbol{\Lambda}_1 & \mathbf{0} & \mathbf{0} \\ \mathbf{0} & \boldsymbol{\Lambda}_2 & \mathbf{0} \\ \mathbf{0} & \mathbf{0} & \boldsymbol{\Lambda}_3 \end{bmatrix}^{-1} \begin{bmatrix} \boldsymbol{\Lambda}_h^{1*} & \mathbf{0} & \mathbf{0} \\ \mathbf{0} & \boldsymbol{\Lambda}_h^{2*} & \mathbf{0} \\ \mathbf{0} & \mathbf{0} & \boldsymbol{\Lambda}_h^{3*} \end{bmatrix} \mathbf{G} \tag{10.52}$$

where

$$\boldsymbol{\Lambda}_1 = \boldsymbol{\Lambda}_h^{1*}\boldsymbol{\Lambda}_h^1 + \lambda_1^{-1}\left(\boldsymbol{\Lambda}_c^{11*}\boldsymbol{\Lambda}_c^{11} + \boldsymbol{\Lambda}_c^{21*}\boldsymbol{\Lambda}_c^{21} + \boldsymbol{\Lambda}_c^{31*}\boldsymbol{\Lambda}_c^{31}\right) \tag{10.53}$$

$$\boldsymbol{\Lambda}_2 = \boldsymbol{\Lambda}_h^{2*}\boldsymbol{\Lambda}_h^2 + \lambda_2^{-1}\left(\boldsymbol{\Lambda}_c^{12*}\boldsymbol{\Lambda}_c^{12} + \boldsymbol{\Lambda}_c^{22*}\boldsymbol{\Lambda}_c^{22} + \boldsymbol{\Lambda}_c^{32*}\boldsymbol{\Lambda}_c^{32}\right) \tag{10.54}$$

$$\boldsymbol{\Lambda}_3 = \boldsymbol{\Lambda}_h^{3*}\boldsymbol{\Lambda}_h^3 + \lambda_1^{-1}\left(\boldsymbol{\Lambda}_c^{13*}\boldsymbol{\Lambda}_c^{13} + \boldsymbol{\Lambda}_c^{23*}\boldsymbol{\Lambda}_c^{23} + \boldsymbol{\Lambda}_c^{33*}\boldsymbol{\Lambda}_c^{33}\right) \tag{10.55}$$

$$\varphi^{-1}\mathbf{C}_{ij}^t\varphi = \boldsymbol{\Lambda}_c^{ij} \tag{10.56}$$

$$\varphi^{-1}\mathbf{C}_{ij}^t\varphi = \boldsymbol{\Lambda}_c^{ij*} \tag{10.57}$$

The matrices $\Lambda_h^k, \Lambda_h^{k^*}, \Lambda_b^k, \Lambda_b^{k^*}, \Lambda_m^k, \Lambda_m^{k^*}, \Lambda_c^{ij}$ and $\Lambda_c^{ij^*}$ are all diagonal matrices. The inversion in Equation (10.52) is then possible since it is a sparse matrix composed of block diagonal sparse matrices.

10.6 Simulation Examples

The super-resolution image reconstruction algorithms were tested on different noisy degraded LR observations with different SNRs. Several experiments were conducted. In the first experiment, three LR degraded observations of Lenna images of size 128×128 were used to obtain a single HR image of size 256×256. The general degradation model of Equation (10.1) was used to generate the degraded observations. The degradation in each observation comprises a relative translational shift with the reference observation, out-of-focus blurring, and additive noise with

Figure 10.2   Original Lenna image.

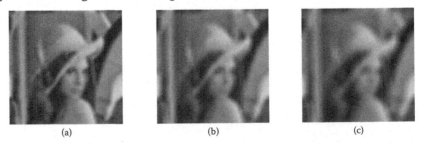

(a)   (b)   (c)

Figure 10.3   Degraded Lenna image. Available observations with SNR = 40 dB. (a) Observation 1 with 5 × 5 blur operator. (b) Observation 2 with 7 × 7 blur operator. (c) Observation 3 with 9 × 9 blur operator.

SNR = 40 dB. The original LR image is shown in Figure 10.2. The degraded observations are given in Figure 10.3.

The LMMSE interpolated version of the original LR image is given in Figure 10.4. The image obtained from the fusion of the multi-channel LMMSE restoration step is given in Figure 10.5a. The HR image obtained using the LMMSE super-resolution algorithm is given in Figure 10.5b. It is clear that the computational time is moderate, but the visual quality and the peak signal-to-noise ratio (PSNR) value obtained are not satisfactory.

Figure 10.4   LMMSE interpolation version of original LR image.

(a)

(b)

Figure 10.5   Results of LMMSE super-resolution algorithm, CPU = 55 sec. (a) Fused image, PSNR = 24 dB. (b) LMSSE interpolation of fused image, PSNR = 24 dB.

Figure 10.6   Original LR image.

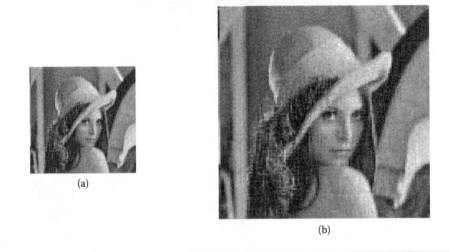

(a)

(b)

Figure 10.7   Results after multi-channel restoration. (a) Fused image. (b) HR image using maximum entropy super-resolution algorithm.

The maximum entropy interpolated version of the original LR image is given in Figure 10.6. The image obtained from the fusion of the multi-channel maximum entropy restoration step is given in Figure 10.7a. The HR image obtained using the maximum entropy super-resolution algorithm is given in Figure 10.7b. The parameter η used in the multi-channel restoration step and in the interpolation step is 0.001. It is clear from the obtained results that the computational time is reduced significantly and the visual quality and the PSNR value get better.

The regularized interpolated version of the original LR image is given in Figure 10.8. The image obtained from the fusion of the multi-channel regularized

Figure 10.8   Original LR image.

Figure 10.9   Results of regularized super-resolution reconstruction algorithm. (a) Fused image, PSNR = 27.7 dB. (b) Obtained image, PSNR = 27.7 dB, CPU = 105 sec on 1 GHz processor.

restoration step is given in Figure 10.9a. The HR image obtained using the regularized super-resolution algorithm is given in Figure 10.9b. The optimum values of the Lagrangian multipliers used in the multi-channel regularized restoration step are estimated using Newton's method. The parameter λ used in the interpolation step is 0.001. Based on the obtained results, the PSNR value obtained using this algorithm is the highest obtained value but at the cost of much more computational time.

The rule used in image fusion in all the algorithms is the maximum frequency rule. The wavelet fusion process was performed in one decomposition level using the Haar basis function. The PSNR values given in parts *a* of Figures 10.5, 10.7, and 10.9 were calculated using the mean square error (MSE) between the fused image and the original LR image given in Figure 10.2. On the other hand, the PSNR values given in parts *b* of the same figures were calculated using the MSE between the obtained HR image and the interpolated version of the original LR image using the same algorithm that produced the HR image.

Another experiment was carried out on an magnetic resonance (MR) image. The results are shown in Figures 10.10 through 10.17. The PSNR values were estimated as in the first experiment. In the multi-channel restoration steps in both experiments, the matrices $\mathbf{B}$, $\mathbf{M}$, $\mathbf{R}_f$, and $\mathbf{C}$ used in the different algorithms were approximated by circulant matrices that were diagonalized via a 2-D Fourier transform.

The Newton method required to obtain the values of λ_1, λ_2, and λ_3 in the regularized multi-channel restoration step is fast converging and requires two iterations only. This method is used after the diagonalization of matrices to reduce the computational cost and ensure fast convergence. The convergence study of the Newton method in both experiments is illustrated in Figure 10.18.

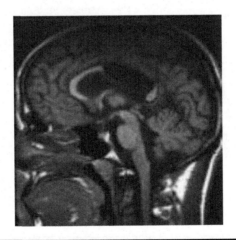

Figure 10.10   Original MR image.

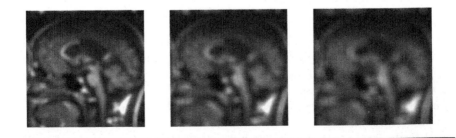

Figure 10.11   Available observations at SNR = 40 dB.

The two experiments carried out in this section are very difficult to implement using the traditional iterative solutions [121,123] due to the large-dimension matrices that cannot be saved using traditional principal components (PCs) as mentioned previously.

It is clear that the super-resolution reconstruction algorithms succeeded in obtaining HR images with good visual quality as compared to the available observations and high PSNR values. The computational cost is acceptable when the quality of the HR image obtained is the most important factor. Regularized super-resolution is the best algorithm from a visual quality perspective, but it has the

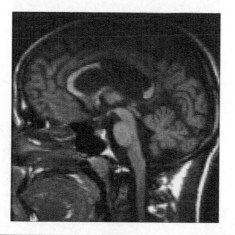

Figure 10.12   Original MR image.

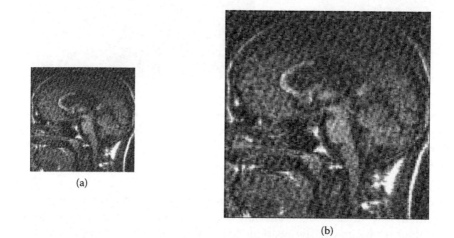

Figure 10.13   Results of suggested LMMSE super-resolution reconstruction algorithm, CPU = 55 sec. (a) Fused image, PSNR = 18.55 dB. (b) LMMSE interpolation of fused image, PSNR = 18.66 dB.

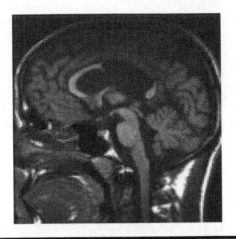

Figure 10.14 Original MR image.

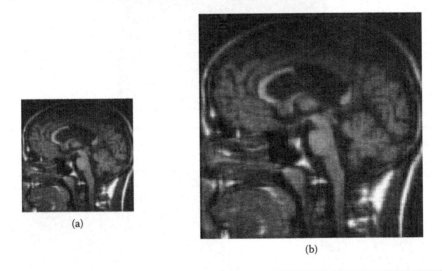

(a)

(b)

Figure 10.15   Results of maximum entropy super-resolution reconstruction algorithm. (a) Fused image, PSNR = 28.71 dB. (b) Obtained HR image, PSNR = 28.74 dB, CPU = 105 sec on 1 GHz processor.

maximum computational time. The success of the super-resolution reconstruction algorithms with three observations only is an indication of the superiority of these algorithms to the previously mentioned algorithms that require large numbers of frames to obtain a single HR image.

The sensitivity of the regularized super-resolution reconstruction algorithm to noise was tested for both Lenna and MR experiments. This algorithm was used on degraded LR images with different SNRs and the results are shown in Figure 10.19.

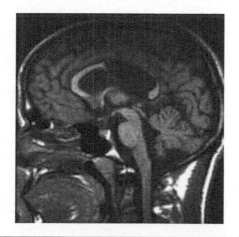

Figure 10.16   Regularized interpolation of original image.

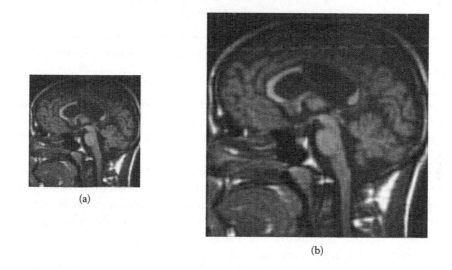

(a)

(b)

Figure 10.17   Results of regularized super-resolution reconstruction algorithm. (a) Fused image, PSNR = 32.44 dB. (b) Obtained HR image, PSNR = 31.49 dB, CPU = 105 sec on 1 GHz processor.

The MSE values in Figures 10.18 and 10.19 were calculated between the obtained HR image and the regularized interpolated image. It is clear from the obtained results that the regularized super-resolution reconstruction algorithm without wavelet denoising works properly for high SNRs. For low SNRs, wavelet denoising is required to reduce the noise effect in the fused image prior to the regularized interpolation step.

An important issue is to decide whether to use denoising. From the experiments on the Lenna and MR images, we have found that the threshold SNR values

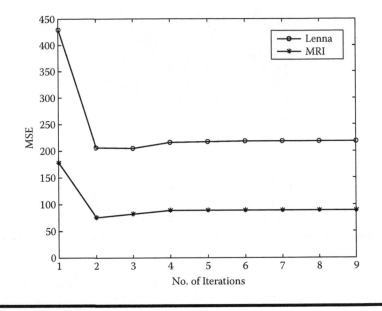

Figure 10.18 **Convergence study of Newton's method used in estimating values of** λ_1, λ_2, **and** λ_3.

below which we can use denoising are 24 and 21 dB, respectively. It is clear that the threshold SNR value is variable from image to image, but we can make a rough estimate of this threshold that can be suitable to most applications. This estimate is 22.5 dB, midway between 21 and 24 dB.

The evaluation of the SNR in the available degraded images is another important issue that is closely related to the noise variance estimation from the degraded image. For the case of independent additive noise, we can approximate the total degraded image variance as the original image variance plus the noise variance. If the noise variance is estimated, the image variance can also be estimated, and thus the SNR value can be approximated.

There are several methods for noise variance estimation. The classical approach to this problem is to use a smooth region of the image and estimate the noise variance as the variance of this region. Other approaches to solve this problem are the maximum likelihood approach and the expectation maximization approach [90]. In the simulation experiments, the Haar basis function has been used for both wavelet fusion and denoising.

The fusion of MR and computed tomography (CT) images was studied in Chapter 9. To obtain HR fusion results, inverse interpolation techniques were used for this purpose. Unfortunately, in image interpolation, there is no reference image to measure the PSNR and the similarity of the obtained fusion results with it. To solve this problem and compare the different interpolation algorithms with numerical metrics, we can decimate the original MR and CT images prior to fusion, and

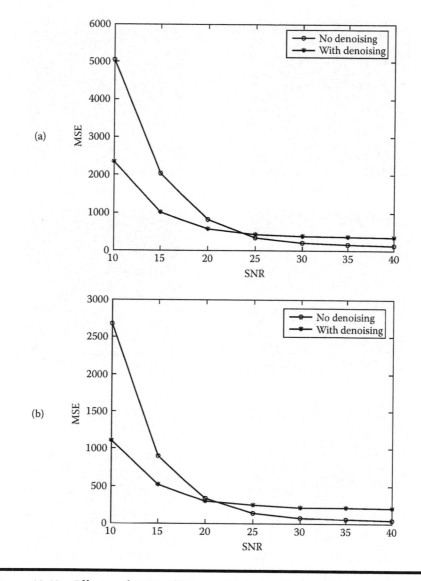

Figure 10.19 Effects of SNR (dB) on LR observations of regularized super-resolution reconstruction algorithm. (a) Lenna image. (b) MR image.

then interpolate the fused images, and calculate the numerical metrics between the interpolated fused images and the original MR and CT images.

For case 1, we performed a decimation of the MR and CT images, and then performed wavelet and curvelet fusions on the decimated images. The fused images were interpolated using LMMSE, maximum entropy, and regularized interpolation methods, and the results are given in Figure 10.20. The values of the numerical

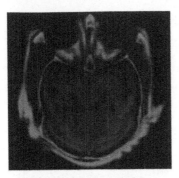

(a) LMMSE interpolation

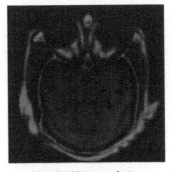

(b) LMMSE interpolation

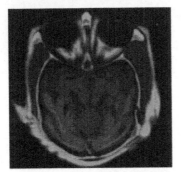

(c) Maximum entropy interpolation

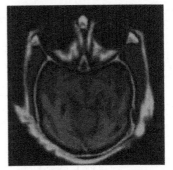

(d) Maximum entropy interpolation

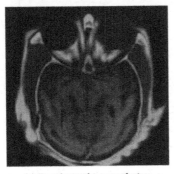

(e) Regularized interpolation

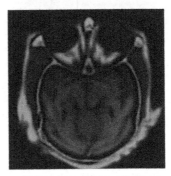

(f) Regularized interpolation

Figure 10.20 **Interpolation of fused images of case 1 using inverse techniques. Left column: Interpolation of wavelet results. Right column: Interpolation of curvelet results.**

Table 10.1 PSNR Results for Interpolation of Fused Images

Images		PSNR	Cubic Spline	Entropy	LMMSE	Regularized
Case (1)	Wavelet	$PSNR_C$	11.1131	11.1840	14.9100	11.1447
		$PSNR_M$	19.7378	19.5364	17.3211	19.4927
	Curvelet	$PSNR_C$	11.2459	11.3939	14.8021	11.3437
		$PSNR_M$	18.4124	19.3386	16.1004	19.3637
Case (2)	Wavelet	$PSNR_C$	14.4399	14.7183	15.9280	14.6995
		$PSNR_M$	11.9996	12.1531	13.3155	12.0685
	Curvelet	$PSNR_C$	14.3624	15.0243	14.6420	15.0474
		$PSNR_M$	13.0389	13.1860	13.6369	13.1018

Table 10.2 Similarity Results for Interpolation of Fused Images

Images		Similarity	Cubic Spline	Entropy	LMMSE	Regularized
Case (1)	Wavelet	S_C	0.0206	0.0197	0.0167	0.0216
		S_M	0.0277	0.0448	0.0300	0.0448
	Curvelet	S_C	0.0530	0.0484	0.0569	0.0504
		S_M	0.0653	0.0829	0.1192	0.0825
Case (2)	Wavelet	S_C	0.0307	0.0331	0.0256	0.0303
		S_M	0.0415	0.0517	0.0476	0.0515
	Curvelet	S_C	0.1112	0.1009	0.1066	0.0973
		S_M	0.1444	0.1764	0.2079	0.1693

evaluation metrics for these results are tabulated in Tables 10.1 and 10.2. It is clear from these images and metrics that the curvelet fusion achieves the maximum similarity values with high enough PSNR values.

Another experiment (case 2) was performed on two MR and CT scans of the human brain and its results are tabulated in Tables 10.1 and 10.2. The numerical results of this experiment coincide with those obtained from the first experiment.

Chapter 11

Blind Super-Resolution Reconstruction of Images

11.1 Introduction

In the previous chapter, we considered the problem of super-resolution reconstruction of images with *a priori* information. In the general solution of the super-reolution reconstruction problem, we deal with three degradation phenomena: general geometric registration warp, blurring, and additive noise. We assumed that these degradation phenomena were known *a priori*. There are applications in which the degradation phenomena are unknown and difficult to estimate within the super-resolution reconstruction process. Obtaining a high-resolution (HR) image from degraded low-resolution (LR) images with unknown degradation phenomena is known as blind super-resolution reconstruction of images. Few methods deal only with blind super-resolution reconstruction [114–123].

In dealing with the problem of blind image super-resolution, the first step is to register the images—estimating the geometrical registration warps between different images. We assume that the registration process has already been done and focus on the blind super-resolution of several blurred noisy images degraded with different unknown blurring operators [132].

When multiple distorted versions of the same scene are available, it is possible to restore the original image from these distorted versions without prior knowledge of the distortion functions [134–136]. The problem of reconstructing an image from two distorted observations in a high signal-to-noise ratio (SNR) environment was investigated using a two-dimensional (2-D) greatest common divisor (GCD) algorithm that is very sensitive to noise [137].

In the z-domain, the original image can be regarded as the GCD among the distorted observations if noise is neglected and the distortion filters are assumed as finite impulse responses (FIRs) and are relatively co-prime [137]. Because small variations in the point spread function (PSF) of the imaging system used in capturing the observations lead to blurring operators of different parameters that are thus co-prime in the z-domain, the co-prime assumption of blurring operators is realistic.

Our interest in this chapter is to benefit from multiple observations of this 2-D GCD algorithm by making use of all observations rather than only two observations to obtain a single undegraded image that is then interpolated under certain constraints [138–140].

11.2 Problem Formulation

The super-resolution reconstruction problem to be solved is the estimation of an HR image from several images degraded by both blur and noise. An image degraded by both blur and noise can be modeled by the following equation [138–140]:

$$y(m,n) = x(m,n)*b(m,n) + v(m,n) \tag{11.1}$$

where $x(m,n)$ is the original image, $b(m,n)$ is the degradation PSF, and $v(m,n)$ is an additive zero mean white Gaussian noise. We then assume that we have two degraded observations of the same scene given by the following equation:

$$y_k(m,n) = x(m,n)*b_k(m,n) + v_k(m,n), \qquad k = 1,2 \tag{11.2}$$

where $b_1(m,n)$ and $b_2(m,n)$ are co-prime filters. The co-primeness assumption is justified since any slight variation of the blurring operators affecting the same observation leads to the co-primeness of these operators. In the z-domain, this equation translates to [138–140]:

$$Y_k(z_1,z_2) = X(z_1,z_2)B_k(z_1,z_2) + V_k(z_1,z_2), \qquad k = 1,2 \tag{11.3}$$

If the noise $v_k(m,n)$ is neglected, the $V(z_1,z_2)$ term vanishes and hence the above equation becomes:

$$Y_k(z_1,z_2) = X(z_1,z_2)B_k(z_1,z_2) \qquad k = 1,2 \tag{11.4}$$

From the above equation, if the two distortion transfer functions $B_1(z_1,z_2)$ and $B_2(z_1,z_2)$ are co-prime, then $X_1(z_1,z_2)$ is the GCD of $Y_1(z_1,z_2)$ and $Y_2(z_1,z_2)$. This can be written mathematically as [138–140]:

$$if \tag{11.5}$$

$$GCD \ \{B_1(z_1,z_2), \ B_2(z_1,z_2)\} = 1$$

then

(11.6)

$$GCD\ \{Y_1(z_1,z_2),\ Y_1(z_1,z_2)\} = X(z_1,z_2)$$

The GCD between two 2-D functions using the 2-D GCD algorithm is discussed in the next section [137–140].

11.3 Two-Dimensional GCD Algorithm

In the GCD algorithm [96], it is assumed that the two blurred images $y_1(m,n)$ and $y_2(m,n)$ of the original image $x(m,n)$ are both $P \times P$ matrices. Substituting $z_1 = e^{-j2\pi m/P}$, $m = 0,1,\dots P-1$ into both $Y_1(z_1,z_2)$ and $Y_2(z_1,z_2)$ for each m yields two 1-D polynomials :

$$Y_k(e^{-j(2\pi m/P)},z_2) = X(e^{-j(2\pi m/P)},z_2).B_k(e^{-j(2\pi m/P)},z_2) \qquad k = 1,2 \qquad (11.7)$$

Thus, the one-dimensional (1-D) GCD between these two polynomials yields the scaled quantity $c_0(e^{-j(2\pi m/P)})X(e^{-j(2\pi m/P)},z_2)$. For each value of m, we further substitute $z_2 = e^{-j2\pi n/P}$, $n = 0,1,\dots P-1$ in this GCD to form a matrix of discrete Fourier transform elements [137–140]:

$$A(m,n)a(m) = X(e^{-j(2\pi m/P)}, e^{-j(2\pi n/P)}) \qquad (11.8)$$

Carrying out similar operations on columns, we get [137–140]:

$$L(m,n)l(m) = X(e^{-j(2\pi m/P)}, e^{-j(2\pi n/P)}) \qquad (11.9)$$

where $a(m)$ and $l(n)$ are scalar quantities that must be determined. From Equations (11.8) and (11.9), we have [137–140]:

$$A(m,n)a(m) - L(m,n)l(n) = 0 \qquad (11.10)$$

The evaluation of $a(m)$ and $l(n)$ can be made on a least-squares basis. Thus, the estimated Fourier transform of the original image is then calculated as [137–140]:

$$X(e^{-j(2\pi m/P)},e^{-j(2\pi n/P)}) = \frac{1}{2}[A(m,n)a(m)+L(m,n)l(n)] \qquad (11.11)$$

The inverse Fourier transform is then used to obtain an estimate of the original image.

11.4 Blind Super-Resolution Reconstruction Approach

Assume that we have K degraded observations of the same scene given by the following equation [137–140]:

$$y_k(m,n) = x(m,n)*b_k(m,n) + v_k(m,n), \qquad k = 1,2,...,K \qquad (11.12)$$

The application of the 2-D GCD algorithm described in the previous section involving only two observations at a time is not an efficient method of restoration because it does not incorporate the information from all observations into the restoration process. To incorporate the information from each observation into the restoration, we suggest generating a new observation image y_{K+1} given by the following equation [137–140]:

$$y_{K+1}(m,n) = \sum_{k=1}^{K} w_k y_k(m,n) \qquad (11.13)$$

where w_k values are scalars chosen according to an estimation of the SNR in each image based on noise variance estimation. Another restriction on the values of w_k is the normalization condition as follows [137–140]:

$$\sum_{k=1}^{K} w_k = 1 \qquad (11.14)$$

Substituting from Equation (11.12) into Equation (11.13), we get:

$$y_{K+1}(m,n) = \sum_{k=1}^{K} w_k \left[x(m,n) * b_k(m,n) + v_k(m,n) \right] \qquad (11.15)$$

Thus:

$$y_{K+1}(m,n) = x(m,n) * \left[\sum_{k=1}^{K} w_k b_k(m,n) \right] + \sum_{k=1}^{K} w_k v_k(m,n) \qquad (11.16)$$

This equation can be written in the form [137–140]:

$$y_{K+1}(m,n) = x(m,n)*b_{K+1}(m,n) + v_{K+1}(m,n) \qquad (11.17)$$

where

$$b_{K+1}(m,n) = \left[\sum_{k=1}^{K} w_k b_k(m,n) \right] \qquad (11.18)$$

and

$$v_{K+1}(m, n) = \sum_{k=1}^{K} w_k v_k(m, n). \tag{11.19}$$

In z-domain, this leads to:

$$B_{K+1}(z_1, z_2) = \sum_{k-1}^{K} w_k B_k(z_1, z_2) \tag{11.20}$$

It can be proved that $B_{K+1}(z_1, z_2)$ is co-prime with all $B_K(z_1, z_2)$ for $k \leq K$. Dividing $B_{K+1}(z_1, z_2)$ by $B_K(z_1, z_2)$, where $k \leq K$ gives:

$$\frac{B_{K+1}(z_1, z_2)}{B_k(z_1, z_2)} = \frac{w_1 B_1(z_1, z_2)}{B_k(z_1, z_2)} + \frac{w_2 B_2(z_1, z_2)}{B_k(z_1, z_2)} + \ldots \ldots$$

$$+ w_k + \ldots \ldots + \frac{w_K B_K(z_1, z_2)}{B_k(z_1, z_2)} \tag{11.21}$$

Since $B_i(z_1, z_2)$ and $B_k(z_1, z_2)$ are co-prime functions for $i \neq k$ and $i, k \leq K$, we have

$$\frac{B_i(z_1, z_2)}{B_k(z_1, z_2)} = Q_i(z_1, z_2) + \frac{R_i(z_1, z_2)}{B_k(z_1, z_2)} \tag{11.22}$$

where $Q_i(z_1, z_2)$ is the quotient and $R_i(z_1, z_2)$ is the remainder and $R_i(z_1, z_2) \neq 0$. Substituting from Equation (11.22) into Equation (11.21):

$$\frac{B_{K+1}(z_1, z_2)}{B_k(z_1, z_2)} = w_k + \sum_{\substack{i=1, \\ i \neq k}}^{K} w_i \left(Q_i(z_1, z_2) + \frac{R_i(z_1, z_2)}{B_k(z_1, z_2)} \right) \tag{11.23}$$

Thus

$$\frac{B_{K+1}(z_1, z_2)}{B_k(z_1, z_2)} = Q_r(z_1, z_2) + \frac{R_r(z_1, z_2)}{B_k(z_1, z_2)} \tag{11.24}$$

where

$$Q_r(z_1, z_2) = w_k + \sum_{\substack{i=1, \\ i \neq k}}^{K} w_i \left(Q_i(z_1, z_2) \right) \tag{11.25}$$

and

$$R_t(z_1, z_2) = \sum_{\substack{i=1, \\ i \neq k}}^{K} w_i R_i(z_1, z_2) \neq 0 \tag{11.26}$$

This leads to the conclusion that $B_{K+1}(z_1, z_2)$ is co-prime with all $B_K(z_1, z_2)$ for $k \leq K$ [138–140]. Thus, the 2-D GCD algorithm can be carried out between $y_{K+1}(m,n)$ and any of $y_k(m,n)$, where $k \leq K$ to give good estimates of $X(e^{-j(2\pi m/P)}, e^{-j(2\pi n/P)})$. The relation $v_{K+1}(m,n) = \sum_{k=1}^{K} w_k v_k(m,n)$ leads to an image with noise variance σ_{K+1}^2 given by:

$$\sigma_{K+1}^2 = \sum_{k=1}^{K} w_k^2 \sigma_k^2. \tag{11.27}$$

For equal weight averaging we have $w_1 = w_2 = ... = w_K = 1/K$

$$\sigma_{K+1}^2 = \sum_{k=1}^{K} \frac{\sigma_k^2}{K^2} \tag{11.28}$$

Assuming all observations are taken in the same noisy environment leads to [138–140]:

$$\sigma_{K+1}^2 = \frac{\sigma_k^2}{K} \tag{11.29}$$

The above equation leads to an improvement in the SNR of the image $y_{K+1}(m,n)$ by a factor of K. This increase in SNR enables a robust application of the 2-D GCD algorithm between $y_{K+1}(m,n)$ and any other observation $y_k(m,n)$, since the 2-D GCD algorithm is very sensitive to the presence of noise.

We summarize the steps of the blind super-resolution reconstruction algorithm in the following steps [138–140]:

1. Begin with multiple observations blurred by relatively co-prime blurring operators in the presence of noise.
2. Generate a new image by a weighted averaging process of the available observations.
3. Carry out the 2-D GCD algorithm between the generated image and each observation.
4. Carry out an image fusion process on the results obtained in step 3.
5. Carry out a regularized image interpolation process on the obtained result of step 4.

11.5 Simulation Examples

The GCD blind super-resolution image reconstruction algorithm has been tested. Three degraded observations blurred with 5 × 5 relatively co-prime operators were used to test this algorithm. The SNR for each observation was 60 dB. The original image and the degraded observations are given in Figure 11.1. The co-prime blurring operators used are as follows:

$$b_1 = \begin{bmatrix} 0.03 & 0.035 & 0.04 & 0.045 & 0.05 \\ 0.05 & 0.025 & 0.055 & 0.03 & 0.04 \\ 0.033 & 0.043 & 0.053 & 0.025 & 0.055 \\ 0.04 & 0.04 & 0.038 & 0.042 & 0.04 \\ 0.045 & 0.05 & 0.065 & 0.025 & 0.015 \end{bmatrix},$$

$$b_2 = \begin{bmatrix} 0.03 & 0.05 & 0.033 & 0.04 & 0.045 \\ 0.035 & 0.025 & 0.043 & 0.04 & 0.05 \\ 0.04 & 0.055 & 0.053 & 0.038 & 0.065 \\ 0.045 & 0.03 & 0.025 & 0.042 & 0.025 \\ 0.05 & 0.04 & 0.055 & 0.04 & 0.015 \end{bmatrix}$$

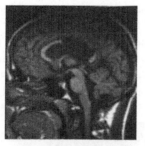

(a) Original image

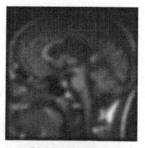

(b) Observation (1)

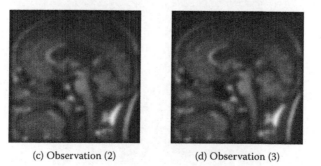

(c) Observation (2)   (d) Observation (3)

Figure 11.1   Original image and available observations.

and

$$
b_3 = \begin{bmatrix}
0.05 & 0.045 & 0.04 & 0.035 & 0.03 \\
0.04 & 0.03 & 0.055 & 0.025 & 0.05 \\
0.055 & 0.025 & 0.053 & 0.043 & 0.033 \\
0.04 & 0.042 & 0.038 & 0.04 & 0.04 \\
0.015 & 0.025 & 0.065 & 0.05 & 0.045
\end{bmatrix}
$$

In the first experiment, the original image was interpolated using a regularized interpolation scheme with a regularization parameter of 0.001. The result is given in Figure 11.2. In the second experiment, the 2-D GCD was estimated between observations 1 and 2 and was interpolated using the regularized interpolation approach. Figure 11.3 shows the result. A similar experiment was performed using observation 1 with observation 3 and observation 2 with observation 3. The results are shown in Figures 11.4 and 11.5, respectively. The fused image of the results in Figures 11.3 through 11.5 appears in Figure 11.6. Based on the obtained results, the experiments failed to produce an HR image.

Another experiment was conducted to test the GCD blind super-resolution reconstruction algorithm. A combinational image was generated from the available observations by an equal-weight averaging process since the available observations had the same SNR. The 2-D GCD was estimated between the combinational image and each observation. The results of each 2-D GCD operation were fused together using a simple averaging process. The combinational image and the resultant image are given in Figure 11.7. The resultant image from the fusion process was interpolated using the same regularized interpolation scheme used to obtain the image in Figure 11.2 and the obtained HR image appears in Figure 11.8.

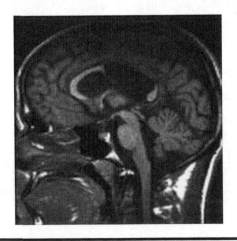

Figure 11.2   Interpolation of original image.

Figure 11.3   Interpolation of two-dimensional GCD between observations 1 and 2, PSNR = 14.6 dB.

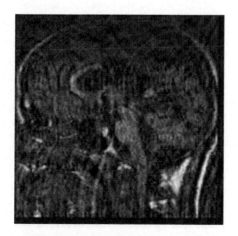

Figure 11.4   Interpolation of two-dimensional GCD between observations 1 and 3, PSNR = 20.9 dB.

Figure 11.5 Interpolation of two-dimensional GCD between observations 2 and 3, PSNR = 15.8 dB.

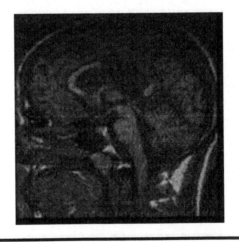

Figure 11.6 Fusion of results from Figures 11.3 to 11.5, PSNR = 20 dB.

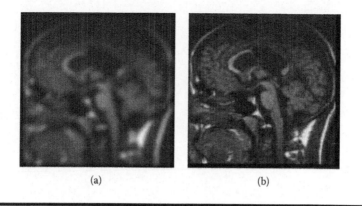

(a)   (b)

Figure 11.7 **(a) Combinational image. (b) Image obtained after fusion process.**

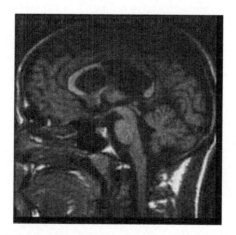

Figure 11.8 **HR image obtained using blind GCD super-resolution reconstruction algorithm, PSNR = 30 dB.**

It is clear that the blind super-resolution reconstruction algorithm succeeded in generating an HR image with a good visual quality and a high peak signal-to-noise ratio (PSNR). The robustness of the algorithm is guaranteed since the co-primeness of the blurring operator in the generated combinational image with that of each observation is guaranteed. The sensitivity of the 2-D GCD to noise is reduced by the averaging process used to generate the combinational image.

Appendix A: Discrete B-Splines

We define [15]:

$$b_m^n(x_k) = \beta^n(x/m)\Big|_{x=x_k} \qquad (A.1)$$

with the following starting condition:

$$b_m^0(x_k) = \begin{cases} 1 & \text{for } -m/2 \le k \le m/2 \\ 0 & \text{otherwise} \end{cases} \qquad (A.2)$$

For a discrete B-spline with up-sampling integer m greater than 1, a convolution property can be established as follows [15]:

(a) *m* odd:

$$b_m^n(x_k) = \frac{1}{m^n} \underbrace{(b_m^0 * b_m^0 * \ldots\ldots * b_m^0)}_{(n+1) \text{ times}} b_1^n(x_k) \qquad (A.3)$$

(b) *n* odd and *m* even:

$$b_m^n(x_k) = \frac{1}{m^n} \delta_{(n+1)/2} \underbrace{(b_m^0 * b_m^0 * \ldots\ldots * b_m^0)}_{(n+1) \text{ times}} b_1^n(x_k) \qquad (A.4)$$

where $\delta_i(x_k)$ is the shift operator (e.g., $\delta_i * a(k) = a(k-i)$).

(c) *n* even and *m* even:

$$b_m^n(x_k) = \frac{1}{m^n} \delta_{(n+2)/2} \underbrace{(b_m^0 * b_m^0 * \ldots\ldots * b_m^0)}_{(n+1) \text{ times}} b_1^n(x_k + 1/2) \qquad (A.5)$$

The z-transform representations of the above equations can be easily obtained.

The up-sampling B-spline of order 0 is a rectangular window of length m given by [15]:

$$B_m^0(z) = \sum_{k=-[m/2]}^{[(m-1)/2]} z^{[m/2]}\left(\frac{1-z^{-m}}{1-z^{-1}}\right) \tag{A.6}$$

where [x] denotes the truncation of the variable x to the smaller integer. The basic (m = 1) sampled discrete B-spline is a symmetric function characterized by [15]:

$$B_1^n(z) = \sum_{k=-[n/2]}^{[n/2]} b_1^n(x_k)z^{-k} \tag{A.7}$$

Approximations for $B_1^n(z)$ for different values of n are given in Table A.1. Using the convolution property, the z-transform of $b_m^n(x_k)$, when n and m are not both even, is given by [15]:

$$B_m^n(z) = \frac{z^{\alpha}}{m^n}\left(\frac{1-z^{-m}}{1-z^{-1}}\right)^{n+1}B_1^n(z) \tag{A.8}$$

where $\alpha = (m-1)(n+1)/2$.

Table A.1 z-Transform of Basic Symmetric B-Splines of n = 0 to 5

n	$B_1^n(z)$
0	1
1	1
2	$\dfrac{z+6+z^{-1}}{8}$
3	$\dfrac{z+4+z^{-1}}{6}$
4	$\dfrac{z^2+76z+230+76z^{-1}+z^{-2}}{384}$
5	$\dfrac{z^2+26z+66+26z^{-1}+z^{-2}}{120}$

Appendix B: Toeplitz-to-Circulant Approximations

B.1 Toeplitz-to-Circulant Approximation

Let **H** be an $S \times S$ Toeplitz matrix of the following form:

$$\mathbf{H} = \begin{bmatrix} h(0) & \cdots & h(-l) & & \mathbf{0} \\ \vdots & \ddots & & \ddots & \\ h(k) & & \ddots & & h(-l) \\ & \ddots & & \ddots & \vdots \\ \mathbf{0} & & h(k) & \cdots & h(0) \end{bmatrix} \tag{B.1}$$

which is approximated by an $S \times S$ circulant matrix $\mathbf{H^c}$ defined as [77]:

$$\mathbf{H^c} = \begin{bmatrix} h(0) & \cdots & \cdots & h(-l) & 0 & \cdots & h(k) & \cdots & h(1) \\ \vdots & \ddots & & & \ddots & \ddots & \cdots & \ddots \\ \vdots & & \ddots & & & \ddots & \ddots & \cdots & h(k) \\ h(k) & & & \ddots & & & \ddots & \ddots & \cdots \\ 0 & \ddots & & & \ddots & & & \ddots & 0 \\ \vdots & \ddots & \ddots & & & \ddots & & & h(-l) \\ h(-l) & \vdots & \ddots & \ddots & & & \ddots & & \vdots \\ \vdots & \ddots & \vdots & \ddots & \ddots & & & \ddots & \vdots \\ h(-1) & \cdots & h(-l) & \vdots & 0 & h(k) & \cdots & \cdots & h(0) \end{bmatrix}$$

$$\tag{B.2}$$

where each row is a circular shift of the row above and the first row is a circular shift of the last row. The primary difference between the matrices $\mathbf{H}$ and $\mathbf{H}^c$ is in the elements added in the upper right and lower left corners to produce the cyclic structure in the rows. If S is large and the number of non-zero elements on the main diagonals compared to the number of zero elements is small (i.e., the matrix is sparse), the elements added to the upper right and lower left corners do not affect the matrix because the number of these elements is small compared to the number of the main diagonal elements. It can be shown from the Eigenvalue distribution of both matrices that $\mathbf{H}$ and $\mathbf{H}^c$ are asymptotically equivalent.

It is known that an $S \times S$ circulant matrix $\mathbf{H}^c$ is diagonalized by [77]:

$$\Lambda = \boldsymbol{\varphi}^{-1}\, \mathbf{H}^c\, \boldsymbol{\varphi} \tag{B.3}$$

where Λ is an $S \times S$ diagonal matrix whose elements $\lambda(s,s)$ are the Eigenvalues of $\mathbf{H}^c$, and $\boldsymbol{\varphi}$ is an $S \times S$ unitary matrix of the Eigenvectors of $\mathbf{H}^c$. Thus, we have:

$$\boldsymbol{\varphi}\boldsymbol{\varphi}^{*t} = \boldsymbol{\varphi}^{*t} = \mathbf{I} \tag{B.4}$$

The elements $\varphi(s_1, s_2)$ of $\boldsymbol{\varphi}$ are given by [77]:

$$\varphi(s_1, s_2) = \exp\left[\frac{j2\pi s_1 s_2}{S}\right] \tag{B.5}$$

for $s_1, s_2 = 0, 1, \ldots, S-1$ and $j^2 = -1$.

The Eigenvalues $\lambda(s,s)$ can be named $\lambda(s)$. For these Eigenvalues, the following relation holds [77]:

$$\lambda(s) = h(0) + \sum_{m=1}^{k} h(m)\exp\left[\frac{-j2\pi ms}{S}\right] + \sum_{m=-l}^{-1} h(m)\exp\left[\frac{-j2\pi ms}{S}\right] \tag{B.6}$$

$s = 0, 1, \ldots, S-1$. Because of the cyclic nature of $\mathbf{H}^c$, we define:

$$h(S-m) = h(-m) \tag{B.7}$$

and thus Equation (B.6) can be written in the form [77]:

$$\lambda(s) = \sum_{m=0}^{S-1} h(m)\exp\left[\frac{-j2\pi ms}{S}\right] \tag{B.8}$$

for $s = 0, 1, \ldots, S-1$. Thus, the circulant matrix can be simply diagonalized by computing the discrete Fourier transform (DFT) of the cyclic sequence $h(0), h(1), \ldots, h(S-1)$.

B.2 Block Toeplitz-to-Block Circulant Approximation

Block Toeplitz matrices can also be approximated by block circulant matrices and diagonalized via the two-dimensional (2-D) Fourier transform. Let $[\mathbf{H}]$ be a block Toeplitz matrix of size $S^2 \times S^2$ of the form [77]:

$$[\mathbf{H}] = \begin{bmatrix} [\mathbf{h}_0] & \cdots & [\mathbf{h}_{-1}] & & [\mathbf{0}] \\ \vdots & \ddots & & \ddots & \\ [\mathbf{h}_k] & & \ddots & & [\mathbf{h}_{-1}] \\ & \ddots & & \ddots & \vdots \\ [\mathbf{0}] & & [\mathbf{h}_k] & \cdots & [\mathbf{h}_0] \end{bmatrix} \qquad (B.9)$$

where the matrix $\mathbf{h}_s$ is a Toeplitz matrix of size $S \times S$ given by [77]:

$$[\mathbf{h}_s] = \begin{bmatrix} h(0) & \cdots & h(-l) & & \mathbf{0} \\ \vdots & \ddots & & \ddots & \\ h(k) & & \ddots & & h(-l) \\ & \ddots & & \ddots & \vdots \\ \mathbf{0} & & h(k) & \cdots & h(0) \end{bmatrix} \qquad (B.10)$$

$s = 0,1,\ldots\ldots,S-1$. The matrix $[\mathbf{H}]$ can be approximated by a matrix $[\mathbf{H}^c]$ of dimensions $S^2 \times S^2$ in the form:

$$[\mathbf{H}^c] = \begin{bmatrix} [\mathbf{h}_0^c] & \cdots & \cdots & [\mathbf{h}_{-1}^c] & [\mathbf{0}] & \cdots & [\mathbf{h}_k^c] & \cdots & [\mathbf{h}_1^c] \\ \vdots & \ddots & & & \ddots & \ddots & \cdots & \ddots \\ \vdots & & \ddots & & & \ddots & \ddots & \cdots & [\mathbf{h}_k^c] \\ [\mathbf{h}_k^c] & & & \ddots & & & \ddots & \ddots & \cdots \\ [\mathbf{0}] & \ddots & & & \ddots & & & \ddots & [\mathbf{0}] \\ \vdots & \ddots & \ddots & & & \ddots & & & [\mathbf{h}_{-1}^c] \\ [\mathbf{h}_{-1}^c] & \vdots & \ddots & \ddots & & & \ddots & & \vdots \\ \vdots & \ddots & \vdots & \ddots & \ddots & & & \ddots & \vdots \\ [\mathbf{h}_{-1}^c] & \cdots & [\mathbf{h}_{-1}^c] & \vdots & [\mathbf{0}] & [\mathbf{h}_k^c] & \cdots & \cdots & [\mathbf{h}_0^c] \end{bmatrix}$$

$$(B.11)$$

where $[\mathbf{h}_s^c]$ is the circulant approximation of $[\mathbf{h}_s]$, and it is of dimensions $S \times S$.

In this case, we can also define $[\mathbf{h}^c_{S-m}]=[\mathbf{h}^c_{-m}]$ due to the cyclic structure of $[\mathbf{H}^c]$.

The process of digitalization of a block circulant matrix $[\mathbf{H}^c]$ is analogous to that of the circulant matrix, but with the use of the 2-D Fourier transform instead of the one-dimensional (1-D) Fourier transform. This can be given in the following equation [77]:

$$\Lambda = \boldsymbol{\varphi}^{-1}[\mathbf{H}^c]\boldsymbol{\varphi} \qquad (B.12)$$

where Λ is an $S^2 \times S^2$ diagonal matrix, whose main diagonal is formed of the elements $\lambda(s,s)$ taken from the lexicographically ordered version of the 2-D Fourier-transformed circular sequence.

Appendix C: Newton's Method

According to the regularized solution:

$$\text{Minimizes} \quad \left\| \mathbf{C}\hat{\mathbf{f}} \right\| \tag{C.1}$$

$$\text{subject} \quad \left\| \overline{\mathbf{H}}_i\hat{\mathbf{f}} - \mathbf{g}_i \right\|^2 = \left\| \mathbf{n}_i \right\|^2 = \varepsilon_i^2 \quad \text{for} \quad i = 1, 2, \ldots\ldots\ldots, N \tag{C.2}$$

Using the method of Lagrange multipliers, the solution $\hat{\mathbf{f}}$ is obtained by minimizing the following cost function [130]:

$$\Psi(\hat{\mathbf{f}}, \lambda_1, \lambda_2, \ldots\ldots, \lambda_N) = \sum_{i=1}^{N} \lambda_i \left[\left\| \overline{\mathbf{H}}_i\hat{\mathbf{f}} - \mathbf{g}_i \right\|^2 - \left\| \mathbf{n}_i \right\|^2 \right] + \left\| \mathbf{C}\hat{\mathbf{f}} \right\|^2 \tag{C.3}$$

The Lagrangian method is based on the possibility of solving a system of equations that constitute necessary conditions for optimality. For the problem described by (C.1) and (C.2), these conditions are given by [130]:

$$\nabla_{\hat{\mathbf{f}}} \Psi(\hat{\mathbf{f}}, \lambda) = 0 \tag{C.4}$$

$$\nabla_{\lambda} \Psi(\hat{\mathbf{f}}, \lambda) = 0 \tag{C.5}$$

where the vector $\lambda = (\lambda_1, \lambda_2, \ldots\ldots, \lambda_N)$ and $\nabla_{\mathbf{z}}$ denotes the gradient of a function with respect to vector $\mathbf{z}$. The Lagrange multiplier method consists of a sequence of unconstrained problems of Equations (C.4) and (C.5). The solution $\hat{\mathbf{f}}$ satisfying Equation (C.3) is a function of $(\lambda_1, \lambda_2, \ldots\ldots, \lambda_N)$ and is denoted by $\hat{\mathbf{f}}(\lambda_1, \lambda_2, \ldots\ldots, \lambda_N)$. Thus, we have:

$$\left[\mathbf{\Lambda}\mathbf{H}^t\mathbf{H} + \mathbf{C}^t\mathbf{C} \right]\hat{\mathbf{f}}(\lambda_1, \lambda_2, \ldots\ldots, \lambda_N) = \mathbf{\Lambda}\mathbf{H}^t\mathbf{g} \tag{C.6}$$

while $\mathbf{H}$ is the multi-channel degradation and $\boldsymbol{\Lambda}$ is defined as [130]:

$$\boldsymbol{\Lambda} = \begin{bmatrix} \lambda_1[\mathbf{I}] & \mathbf{0} & \cdots & \mathbf{0} \\ \mathbf{0} & \lambda_2[\mathbf{I}] & \cdots & \mathbf{0} \\ \vdots & \vdots & \cdots & \vdots \\ \mathbf{0} & \mathbf{0} & \cdots & \lambda_N[\mathbf{I}] \end{bmatrix} \tag{C.7}$$

where the identity matrix $\mathbf{I}$ is of size $M^2 \times M^2$. Finally, solving for $\hat{\mathbf{f}}$ in Equation (C.6) yields [130]:

$$\hat{\mathbf{f}}(\lambda_1, \lambda_2, \ldots \ldots, \lambda_N) = \left[\mathbf{H}^{\mathrm{t}}\mathbf{H} + \boldsymbol{\Lambda}^{-1}\mathbf{C}^{\mathrm{t}}\mathbf{C}\right]^{-1}\mathbf{H}^{\mathrm{t}}\mathbf{g} \tag{C.8}$$

The solution $\hat{\mathbf{f}}(\lambda_1, \lambda_2, \ldots \ldots, \lambda_N)$ must satisfy the constraint in Equation (C.2), which can be written as:

$$Z_i(\lambda_1, \lambda_2, \ldots \ldots, \lambda_N) = \left[\left\|\bar{\mathbf{H}}_i\mathbf{f} - \mathbf{g}_i\right\|^2 - \left\|\mathbf{n}_i\right\|^2\right] \quad \textit{for} \quad i = 1, 2, \ldots, N \tag{C.9}$$

Finding the roots of the functions $Z_i(\lambda_1, \lambda_2, \ldots \ldots, \lambda_N)$ simultaneously yields the desired λ_i. The functions $Z_i(\lambda_1, \lambda_2, \ldots \ldots, \lambda_N)$ are nonlinear; therefore Newton's method is used to find λ_i numerically. The (i,j) element of the Jacobean of the function $Z_i(\lambda_1, \lambda_2, \ldots \ldots, \lambda_N)$ is given by [130]:

$$\frac{\partial Z_i(\lambda_1, \lambda_2, \ldots \ldots, \lambda_N)}{\partial(\lambda_j)^{-1}} = J_{ij} = \frac{\partial\left\|\bar{\mathbf{H}}_i\hat{\mathbf{f}} - \mathbf{g}_i\right\|^2}{\partial(\lambda_j)^{-1}} \tag{C.10}$$

From Equation (C.8), we can write [130]:

$$\hat{\mathbf{f}} = \mathbf{A}^{-1}\mathbf{H}^{\mathrm{t}}\mathbf{g} \tag{C.11}$$

where

$$\mathbf{A} = \left[\mathbf{H}^{\mathrm{t}}\mathbf{H} + \boldsymbol{\Lambda}^{-1}\mathbf{C}^{\mathrm{t}}\mathbf{C}\right] \tag{C.12}$$

Using Equation (C.11) and the property [130]:

$$\frac{\partial \mathbf{A}^{-1}}{\partial x} = -\mathbf{A}^{-1}\frac{\partial \mathbf{A}}{\partial x}\mathbf{A}^{-1} \tag{C.13}$$

The derivative of Equation (C.10) can be written as:

$$\frac{\partial \left\| \bar{\mathbf{H}}_i \hat{\mathbf{f}} - \mathbf{g}_i \right\|^2}{\partial (\lambda_j)^{-1}} = -\mathbf{g}^t \mathbf{H} \mathbf{A}^{-1} \frac{\partial \mathbf{A}^t}{\partial (\lambda_j)^{-1}} \mathbf{A}^{-1} \bar{\mathbf{H}}_i^t \bar{\mathbf{H}}_i \mathbf{A}^{-1} \mathbf{H}^t \mathbf{g}$$

$$- \mathbf{g}^t \mathbf{H} \mathbf{A}^{-1} \bar{\mathbf{H}}_i^t \bar{\mathbf{H}}_i \mathbf{A}^{-1} \frac{\partial \mathbf{A}}{\partial (\lambda_j)^{-1}} \mathbf{A}^{-1} \mathbf{H}^t \mathbf{g} \qquad (C.14)$$

$$+ 2 \mathbf{g}_i \bar{\mathbf{H}}_i \mathbf{A}^{-1} \frac{\partial \mathbf{A}}{\partial (\lambda_j)^{-1}} \mathbf{A}^{-1} \mathbf{H}^t \mathbf{g}$$

Equation (C.12) can be equivalently written as:

$$\mathbf{A} = \left[\mathbf{H}^t \mathbf{H} + \sum_{l=1}^{N} (\lambda_l)^{-1} \mathbf{I}_{ll} \mathbf{C}^t \mathbf{C} \right] \qquad (C.15)$$

with

$$\mathbf{I}_{ll} = \begin{bmatrix} [\mathbf{0}] & [\mathbf{0}] & \cdots & [\mathbf{0}] & \cdots & [\mathbf{0}] \\ [\mathbf{0}] & [\mathbf{0}] & \cdots & [\mathbf{0}] & \cdots & [\mathbf{0}] \\ \vdots & \vdots & \cdots & [\mathbf{I}] & \cdots & \vdots \\ [\mathbf{0}] & [\mathbf{0}] & \cdots & [\mathbf{0}] & \cdots & [\mathbf{0}] \end{bmatrix} \qquad (C.16)$$

where $\mathbf{I}_{ll}$ is an $NM^2 \times NM^2$ matrix, $[\mathbf{0}]$ represents an $M^2 \times M^2$ zero matrix and $[\mathbf{I}]$ an $M^2 \times M^2$ identity matrix at location (l, l). Using this notation, we have:

$$\frac{\partial \mathbf{A}}{\partial (\lambda_j)^{-1}} = \mathbf{I}_{jj} \mathbf{C}^t \mathbf{C} \qquad (C.17)$$

Thus, the $(i,j)th$ element of the Jacobean is equal to [130]:

$$\frac{\partial \left\| \bar{\mathbf{H}}_i \hat{\mathbf{f}} - \mathbf{g}_i \right\|^2}{\partial (\lambda_j)^{-1}} = -2 (\bar{\mathbf{H}}_i \hat{\mathbf{f}} - \mathbf{g}_i)^t \bar{\mathbf{H}}_i \mathbf{A}^{-1} \mathbf{I}_{jj} \mathbf{C}^t \mathbf{C} \hat{\mathbf{f}} \qquad (C.18)$$

The block Toeplitz-to-block circulant approximation can be used to carry out the computations in Equation (C.8) in the frequency domain.

Appendix D:
MATLAB® Codes

D.1 Polynomial Interpolation

```
clear all;
f1 = imread('1111.jpg');
f1 = f1(:,:,1);
f1 = double(f1)/255;
[M,N] = size(f1);
h = [.5,.5];
g = filter(h,1,f1');
g = g(1:2:N,:);
g = filter(h,1,g');
g = g(1:2:M,:);
SNR = input('Enter the value of SNR in dB');
gg = im2col(g,[M/2,N/2],'distinct');
n_var = var(gg)/10^(SNR/10)
g = imnoise(g,'gaussian',0,n_var);
key0 = input('Press 1 for Bilinear, 2 for Bicubic and 3 for
Cubic Spline and 4 for cubic o-Moms');
key1 = input('press 1 for warped-distance and 2 for no
warping');
key2 = input('press 1 for adaptive weights and 2 for no
adaptation');
tic
f = g;
if (key0 = =3)|(key0 = =4)
f = spline_coeff(g);
fs = g;
end;
a = -1/2;
[M,N] = size(f);
ff = zeros(M,N);
x = f(:,N-1:N);
x = rot90(x,2);
```

```
y = f(:,1);
f = [y,f,x];
L = 256;
k = 3;
k1 = 1.5;
s = 1/2;
l1 = 1;
l2 = 1;
l3 = 1;
l4 = 1;
for i = 1:M
for j = 2:N+1
A = (abs(f(i,j+1)-f(i,j-1))-abs(f(i,j+2)-f(i,j)))/(L-1);
switch key1
case 1
s = s-k*A*s*(s-1);
case 2
s = 1/2;
end;
switch key2
case 1
l1 = l1-k1*A;
l2 = l2-k1*A;
l3 = l3+k1*A;
l4 = l4+k1*A;
case 2
l1 = 1;
l2 = 1;
l3 = 1;
l4 = 1;
end;
switch key0
case 1
ff(i,j-1) = l2*f(i,j)*(1-s)+l3*f(i,j+1)*s;%bilinear
case 2
ff(i,j-1) = l1*f(i,j-1)*(a*s^3-2*a*s^2+a*s)+l2*f(i,j)*((a+2)
*s^3-(3+a)*s^2+1)+l3*f(i,j+1)*(-(a+2)*s^3+(2*a+3)*s^2-a*s)+l4*
f(i,j+2)*(-a*s^3+a*s^2); %Bicubic
case 3
ff(i,j-1) = l1*f(i,j-1)*((3+s)^3-4*(2+s)^3+6*(1+s)^3-4*s^3)/6+
l2*f(i,j)*((2+s)^3-4*(1+s)^3+6*s^3)/6+l3*f(i,j+1)*((1+s)^3-4*
s^3)/6+l4*f(i,j+2)*s^3/6;% B-Spline
case 4
ff(i,j-1) = l1*f(i,j-1)*((-1/6)*(1+s)^3+(1+s)^2+(-85/42)*(1+s)
+(29/21))+l2*f(i,j)*(0.5*s^3-s^2+(1/14)*s+13/21)+l3*f(i,j+1)*
(0.5*(1-s)^3-(1-s)^2+(1/14)*s+13/21)+l4*f(i,j+2)*((-1/6)*(2-s)
^3+(2-s)^2-(85/42)*(2-s)+29/21);% Cubic o-Moms
end;
```

```
end;
end;
ff = ff(:,1:N);
fff(1:M,1:2:2*N) = f(1:M,2:N+1);
if (key0 = =3)|(key0 = =4)
fff(1:M,1:2:2*N) = fs(1:M,1:N);
end;
fff(1:M,2:2:2*N) = ff(1:M,1:N);
fff = (fff> = 0).*fff;
f = fff';
imshow(f)
figure
if (key0 = =3)|(key0 = =4)
fs = f;
f = spline_coeff(f);
end;
ss = f;
clear ff,fff;
a = -1/2;
[M,N] = size(f)
ff = zeros(M,N);
x = f(:,N-1:N);
x = rot90(x,2);
y = f(:,1);
f = [y,f,x];
L = 256;
k = 3;
k1 = 1.5;
s = 1/2;
l1 = 1;
l2 = 1;
l3 = 1;
l4 = 1;
for i = 1:M
for j = 2:N+1
A = (abs(f(i,j+1)-f(i,j-1))-abs(f(i,j+2)-f(i,j)))/(L-1);
switch key1
case 1
s = s-k*A*s*(s-1);
case 2
s = 1/2;
end;
switch key2
case 1
l1 = l1-k1*A;
l2 = l2-k1*A;
l3 = l3+k1*A;
l4 = l4+k1*A;
```

```
case 2
l1 = 1;
l2 = 1;
l3 = 1;
l4 = 1;
end;
switch key0
case 1
ff(i,j-1) = l2*f(i,j)*(1-s)+l3*f(i,j+1)*s;%bilinear
case 2
ff(i,j-1) = l1*f(i,j-1)*(a*s^3-2*a*s^2+a*s)+l2*f(i,j)*
((a+2)*s^3-(3+a)*s^2+1)+l3*f(i,j+1)*(-(a+2)*s^3+(2*a+3)*s^2
-a*s)+l4*f(i,j+2)*(-a*s^3+a*s^2); %Bicubic
case 3
ff(i,j-1) = l1*f(i,j-1)*((3+s)^3-4*(2+s)^3+6*(1+s)^3-4*s^3)/
6+l2*f(i,j)*((2+s)^3-4*(1+s)^3+6*s^3)/6+l3*f(i,j+1)*((1+s)^3-4
*s^3)/6+l4*f(i,j+2)*s^3/6;% B-Spline
case 4
ff(i,j-1) = l1*f(i,j-1)*((-1/6)*(1+s)^3+(1+s)^2+(-85/42)*(1+s)
+(29/21))+l2*f(i,j)*(0.5*s^3-s^2+(1/14)*s+13/21)+l3*f(i,j+1)
*(0.5*(1-s)^3-(1-s)^2+(1/14)*s+13/21)+l4*f(i,j+2)*((-1/6)*
(2-s)^3+(2-s)^2-(85/42)*(2-s)+29/21);% Cubic o-Moms
end;
end;
end;
ss = ff;
clear fff;
ff = ff(:,1:N);
fff(1:M,1:2:2*N) = f(1:M,2:N+1);
if (key0 = =3)|(key0 = =4)
fff(1:M,1:2:2*N) = fs(1:M,1:N);
end;
fff(1:M,2:2:2*N) = ff(1:M,1:N);
toc
fff = (fff> = 0).*fff;
fff = min(fff,1);
error = fff'-f1;
[M1,M2] = size(f1);
MSE = sum(sum(error(3:M1-2,3:M2-2).^2))*255^2/((M1-4)*(M2-4))
PSNR2 = 10*log(sum(sum(ones(size(error)))))/
sum(sum(error.^2)))/log(10)
imshow(fff');
ct2 = fff';
save datct2 ct2
```

D.2 Adaptive Polynomial Interpolation

```
Clear all;
load woman
f1 = X;
f1 = double(f1(:,:,1))/255;
[M,N] = size(f1);
h = [.5,.5];
g = filter(h,1,f1');
g = g(1:2:N,:);
g = filter(h,1,g');
g = g(1:2:M,:);
SNR = input('Enter the value of SNR in dB');
gg = im2col(g,[M/2,N/2],'distinct');
n_var = var(gg)/10^(SNR/10)
g = imnoise(g,'gaussian',0,n_var);
key0 = input('Press 1 for Bilinear, 2 for Bicubic and 3 for
Cubic Spline and 4 for cubic Moms');
f = g;
if (key0 = =3)|(key0 = =4)
f = spline_coeff(g);
fs = g;
end;
a = -1/2;
[M,N] = size(f);
ff = zeros(M,N);
x = f(:,N-1:N);
x = rot90(x,2);
y = f(:,1);
f = [y,f,x];
L = 256;
k = 1.5;
k1 = 1.5;
s = 1/2;
s = linspace(0,1,11);
l1 = 1;
l2 = 1;
l3 = 1;
l4 = 1;
temp1 = zeros(11,2*N);
h = [.5,.5];
for i = 1:M
for c = 1:11
for j = 2:N+1
A = (abs(f(i,j+1)-f(i,j-1))-abs(f(i,j+2)-f(i,j)))/(L-1);
switch key0
case 1
temp(c,j-1) = l2*f(i,j)*(1-s(c))+l3*f(i,j+1)*s(c);
```

```
case 2
temp(c,j-1) = l1*f(i,j-1)*(a*s(c)^3-2*a*s(c)^2+a*s(c))+l2*f
(i,j)*((a+2)*s(c)^3-(3+a)*s(c)^2+1)+l3*f(i,j+1)*(-(a+2)*s(c)
^3+(2*a+3)*s(c)^2-a*s(c))+l4*f(i,j+2)*(-a*s(c)^3+a*s(c)^2);
%Bicubic
case 3
temp(c,j-1) = l1*f(i,j-1)*((3+s(c))^3-4*(2+s(c))^3+6*(1+s(c))
^3-4*s(c)^3)/6+l2*f(i,j)*((2+s(c))^3-4*(1+s(c))^3+6*s(c)^3)/
6+l3*f(i,j+1)*((1+s(c))^3-4*s(c)^3)/6+l4*f(i,j+2)*s(c)^3/6;
% B-Spline
case 4
temp(c,j-1) = l1*f(i,j-1)*((-1/6)*(1+s(c))^3+(1+s(c))^2+
(-85/42)*(1+s(c))+(29/21))+l2*f(i,j)*(0.5*s(c)^3-s(c)^2+(1/14)
*s(c)+13/21)+l3*f(i,j+1)*(0.5*(1-s(c))^3-(1-s(c))^2+(1/14)*s(c)
+13/21)+l4*f(i,j+2)*((-1/6)*(2-s(c))^3+(2-s(c))^2-(85/42)
*(2-s(c))+29/21);% Cubic o-Moms
end;
end;
end;
for c = 1:11
temp1(c,1:2:2*N) = f(i,2:N+1);
if (key0 = =3)|(key0 = =4)
temp1(c,1:2:2*N) = fs(i,1:N);
end;
end;
temp1(1:11,2:2:2*N) = temp(1:11,1:N);
temp1 = [temp1(:,1),temp1];
temp2 = filter(h,1,temp1');
temp2 = temp2';
temp3 = temp2(:,1:2:2*N);
for c = 1:11
error(c,:) = abs(f(i,2:N+1)-temp3(c,:));
if (key0 = =3)|(key0 = =4)
error(c,:) = abs(fs(i,1:N)-temp3(c,:));
end;
end;
[min_val,min_idx] = min(error);
for j = 1:N
ff(i,j) = temp(min_idx(j),j);
end;
end;
ff = ff(:,1:N);
fff(1:M,1:2:2*N) = f(1:M,2:N+1);
if (key0 = =3)|(key0 = =4)
fff(1:M,1:2:2*N) = fs(1:M,1:N);
end;
fff(1:M,2:2:2*N) = ff(1:M,1:N);
fff = (fff> = 0).*fff;
```

```
f = fff';
save data f;
clear all
load data f;
key0 = input('Press 1 for Bilinear, 2 for Bicubic and 3 for
Cubic Spline and 4 for cubic o-Moms');
if (key0 = =3)|(key0 = =4)
fs = f;
f = spline_coeff(f);
end;
a = -1/2;
[M,N] = size(f);
ff = zeros(M,N);
x = f(:,N-1:N);
x = rot90(x,2);
y = f(:,1);
f = [y,f,x];
L = 256;
k = 1.5;
k1 = 1.5;
s = 1/2;
s = linspace(0,1,11);
%s = ones(1,11)/2;
l1 = 1;
l2 = 1;
l3 = 1;
l4 = 1;
temp1 = zeros(11,2*N);
h = [.5,.5];
for i = 1:M
for c = 1:11
for j = 2:N+1
A = (abs(f(i,j+1)-f(i,j-1))-abs(f(i,j+2)-f(i,j)))/(L-1);
switch key0
case 1
temp(c,j-1) = l2*f(i,j)*(1-s(c))+l3*f(i,j+1)*s(c);
case 2
temp(c,j-1) = l1*f(i,j-1)*(a*s(c)^3-2*a*s(c)^2+a*s(c))+l2*f
(i,j)*((a+2)*s(c)^3-(3+a)*s(c)^2+1)+l3*f(i,j+1)*(-(a+2)*s(c)^3
+(2*a+3)*s(c)^2-a*s(c))+l4*f(i,j+2)*(-a*s(c)^3+a*s(c)^2);
%Bicubic
case 3
temp(c,j-1) = l1*f(i,j-1)*((3+s(c))^3-4*(2+s(c))^3+6*(1+s(c))
^3-4*s(c)^3)/6+l2*f(i,j)*((2+s(c))^3-4*(1+s(c))^3+6*s(c)^3)/
6+l3*f(i,j+1)*((1+s(c))^3-4*s(c)^3)/6+l4*f(i,j+2)*s(c)^3/6;
%B-Spline
case 4
```

```
temp(c,j-1) = l1*f(i,j-1)*((-1/6)*(1+s(c))^3+(1+s(c))^2+
(-85/42)*(1+s(c))+(29/21))+l2*f(i,j)*(0.5*s(c)^3-s(c)^2+(1/14)*
s(c)+13/21)+l3*f(i,j+1)*(0.5*(1-s(c))^3-(1-s(c))^2+(1/14)*s(c)+
13/21)+l4*f(i,j+2)*((-1/6)*(2-s(c))^3+(2-s(c))^2-(85/42)*
(2-s(c))+29/21);% Cubic o-Moms
end;
end;
end;
for c = 1:11
temp1(c,1:2:2*N) = f(i,2:N+1);
if (key0 = =3)|(key0 = =4)
temp1(c,1:2:2*N) = fs(i,1:N);
end;
end;
temp1(1:11,2:2:2*N) = temp(1:11,1:N);
temp1 = [temp1(:,1),temp1];
temp2 = filter(h,1,temp1');
temp2 = temp2';
temp3 = temp2(:,1:2:2*N);
for c = 1:11
error(c,:) = abs(f(i,2:N+1)-temp3(c,:));
if (key0 = =3)|(key0 = =4)
error(c,:) = abs(fs(i,1:N)-temp3(c,:));
end;
end;
[min_val,min_idx] = min(error);
for j = 1:N
ff(i,j) = temp(min_idx(j),j);
end;
end;
ff = ff(:,1:N);
fff(1:M,1:2:2*N) = f(1:M,2:N+1);
if (key0 = 3)|(key0 = =4)
fff(1:M,1:2:2*N) = fs(1:M,1:N);
end;
fff(1:M,2:2:2*N) = ff(1:M,1:N);
f = fff';
load woman
f1 = X;
f1 = double(f1(:,:,1))/255;
fff = (fff> = 0).*fff;
error = fff'-f1;
[M1,M2] = size(f1);
MSE = sum(sum(error(3:M1-2,3:M2-2).^2))*255^2/((M1-4)*
(M2-4))
imshow(fff');
save data f;
```

D.3 Spline Coefficients

```
function f = spline_coeff(s)
[M,N] = size(s);
z1 = -2+sqrt(3);
kk = 0:N-1;
d = z1.^kk;
f = s;
for i = 1:M
g(i,1) = sum(d.*s(i,:));
for j = 2:N
g(i,j) = s(i,j)+z1*g(i,j-1);
end;
end;
for i = 1:M
f(i,N) = abs((z1/(1-z1^2)))*(g(i,N)+z1*g(i,N-1));
for j = N-1:-1:1
f(i,j) = abs(z1*(f(i,j+1)-g(i,j)));
end;
end;
f = f*6;
```

D.4 Neural Modeling of Polynomial Interpolation

```
clear all;
f1 = imread('bldg.jpg');
f1 = f1(:,:,1);
f1 = double(f1)/255;
[M,N] = size(f1);
h = [.5,.5];
g = filter(h,1,f1');
g = g(1:2:N,:);
g = filter(h,1,g');
g = g(1:2:M,:);
SNR = input('Enter the value of SNR in dB');
gg = im2col(g,[M/2,N/2],'distinct');
n_var = var(gg)/10^(SNR/10)
g = imnoise(g,'gaussian',0,n_var);
key0 = input('Press 1 for Bilinear, 2 for Bicubic and 3 for
Cubic Spline and 4 for cubic o-Moms');
key1 = input('press 1 for warped-distance and 2 for no
warping');
key2 = input('press 1 for adaptive weights and 2 for no
adaptation');
tic
if (key0 = =3)
```

```
f = spline_coeff(g);
fs = g;
end;
f = g;
a = -1/2;
[M,N] = size(f);
ff = zeros(M,N);
x = f(:,N-1:N);
x = rot90(x,2);
y = f(:,1);
f = [y,f,x];
L = 256;
l1 = 1;
l2 = 1;
l3 = 1;
l4 = 1;
k = 3;
k1 = 1.5;
s = 1/2;
net_ip = zeros(2,2000);
if (key0 = =2)|(key0 = =3)
net_ip = zeros(4,2000);
end
net_op = zeros(1,2000);
counter = 1;
for i = 1:M
for j = 2:N+1
A = (abs(f(i,j+1)-f(i,j-1))-abs(f(i,j+2)-f(i,j)))/(L-1);
switch key1
case 1
s = s-k*A*s*(s-1);
case 2
s = 1/2;
end;
switch key2
case 1
l1 = l1-k1*A;
l2 = l2-k1*A;
l3 = l3+k1*A;
l4 = l4+k1*A;
case 2
l1 = 1;
l2 = 1;
l3 = 1;
l4 = 1;
end;
switch key0
case 1
```

```
tt = l2*f(i,j)*(1-s)+l3*f(i,j+1)*s;%bilinear
net_ip(:,counter) = [f(i,j);f(i,j+1)];
case 2
tt = l1*f(i,j-1)*(a*s^3-2*a*s^2+a*s)+l2*f(i,j)*((a+2)*s^3
-(3+a)*s^2+1)+l3*f(i,j+1)*(-(a+2)*s^3+(2*a+3)*s^2-a*s)+l4*f
(i,j+2)*(-a*s^3+a*s^2); %Bicubic
net_ip(:,counter) = [f(i,j-1);f(i,j);f(i,j+1);f(i,j+2)];
case 3
tt = l1*f(i,j-1)*((3+s)^3-4*(2+s)^3+6*(1+s)^3-4*s^3)/6+l2*f
(i,j)*((2+s)^3-4*(1+s)^3+6*s^3)/6+l3*f(i,j+1)*((1+s)^3-4*s^3)/
6+l4*f(i,j+2)*s^3/6;% B-Spline
net_ip(:,counter) = [f(i,j-1);f(i,j);f(i,j+1);f(i,j+2)];
end;
net_op(:,counter) = tt;
counter = counter+1;
end;
end;
net_ip = net_ip(:,1:counter-1);
net_op = net_op(:,1:counter-1);
net = newff(minmax(net_ip),[2 1],{'logsig' 'logsig'},
'trainscg');
net = init(net);
net.performFcn = 'mse';
net.trainparam.epochs = 20000;
net.trainparam.show = 10;
net.trainparam.goal = 1e-10;
net.trainparam.min_grad = 1e-7;
net = train(net,net_ip,net_op)
for i = 1:M
for j = 2:N+1
switch key0
case 1
y = [f(i,j);f(i,j+1)];
case 2
y = [f(i,j-1);f(i,j);f(i,j+1);f(i,j+2)];
case 3
y = [f(i,j-1);f(i,j);f(i,j+1);f(i,j+2)];
end
ff(i,j-1) = sim(net, y);
end;
end;
M
N
ff = ff(:,1:N);
fff(1:M,1:2:2*N) = f(1:M,2:N+1);
if (key0 = =3)
fff(1:M,1:2:2*N) = fs(1:M,1:N);
end;
```

```
fff(1:M,2:2:2*N) = ff(1:M,1:N);
fff = (fff> = 0).*fff;
f = fff';
imshow(f)
figure
if (key0 = =3)
f = spline_coeff(g);
fs = g;
end;
ss = f;
clear ff,fff;
a = -1/2;
[M,N] = size(f)
ff = zeros(M,N);
x = f(:,N-1:N);
x = rot90(x,2);
y = f(:,1);
f = [y,f,x];
L = 256;
s = 1/2;
k = 3;
k1 = 1.5;
l1 = 1;
l2 = 1;
l3 = 1;
l4 = 1;
N
net_ip = zeros(2,2000);
if (key0 = =2)|(key0 = =3)
net_ip = zeros(4,2000);
end
net_op = zeros(1,2000);
counter = 1;
for i = 1:M
for j = 2:N+1
A = (abs(f(i,j+1)-f(i,j-1))-abs(f(i,j+2)-f(i,j)))/(L-1);
switch key1
case 1
s = s-k*A*s*(s-1);
case 2
s = 1/2;
end;
switch key2
case 1
l1 = l1-k1*A;
l2 = l2-k1*A;
l3 = l3+k1*A;
l4 = l4+k1*A;
```

```
case 2
l1 = 1;
l2 = 1;
l3 = 1;
l4 = 1;
end;
switch key0
case 1
tt = l2*f(i,j)*(1-s)+l3*f(i,j+1)*s;%bilinear
net_ip(:,counter) = [f(i,j);f(i,j+1)];
case 2
tt = l1*f(i,j-1)*(a*s^3-2*a*s^2+a*s)+l2*f(i,j)*((a+2)*s^3
-(3+a)*s^2+1)+l3*f(i,j+1)*(-(a+2)*s^3+(2*a+3)*s^2-a*s)+l4*
f(i,j+2)*(-a*s^3+a*s^2); %Bicubic
net_ip(:,counter) = [f(i,j-1);f(i,j);f(i,j+1);f(i,j+2)];
case 3
tt = l1*f(i,j-1)*((3+s)^3-4*(2+s)^3+6*(1+s)^3-4*s^3)/6+l2*
f(i,j)*((2+s)^3-4*(1+s)^3+6*s^3)/6+l3*f(i,j+1)*((1+s)^3-4*s^3)/
6+l4*f(i,j+2)*s^3/6;% B-Spline
net_ip(:,counter) = [f(i,j-1);f(i,j);f(i,j+1);f(i,j+2)];
end;
net_op(:,counter) = tt;
counter = counter+1;
end;
end;
net_ip = net_ip(:,1:counter-1);
net_op = net_op(:,1:counter-1);
net = newff(minmax(net_ip),[2 1],{'logsig'
'logsig'},'trainscg');
net = init(net);
net.performFcn = 'mse';
net.trainparam.epochs = 20000;
net.trainparam.show = 10;
net.trainparam.goal = 1e-10;
net.trainparam.min_grad = 1e-7;
net = train(net,net_ip,net_op)
for i = 1:M
for j = 2:N+1
switch key0
case 1
y = [f(i,j);f(i,j+1)];
case 2
y = [f(i,j-1);f(i,j);f(i,j+1);f(i,j+2)];
case 3
y = [f(i,j-1);f(i,j);f(i,j+1);f(i,j+2)];
end;
ff(i,j-1) = sim(net, y);
end;
```

```
end;
ss = ff;
clear fff;
ff = ff(:,1:N);
fff(1:M,1:2:2*N) = f(1:M,2:N+1);
if (key0 = =3)
fff(1:M,1:2:2*N) = fs(1:M,1:N);
end;
fff(1:M,2:2:2*N) = ff(1:M,1:N);
toc
fff = (fff> = 0).*fff;
fff = min(fff,1);
size(fff')
size(f1)
error = fff'-f1;
[M1,M2] = size(f1);
MSE = sum(sum(error(3:M1-2,3:M2-2).^2))*255^2/((M1-4)*(M2-4))
PSNR = 10*log(sum(sum(ones(size(error)))))/sum(sum(error.^2)))/
log(10)
imshow(fff');
figure
imshow(1-error);
```

D.5 Color Image Interpolation

```
im = imread('flowers.bmp');
zzz1 = im(:,:,1);
[x y z] = size(im);
im = double(im);
tic
DH = zeros(x,y,1);
DV = zeros(x,y,1);
for i = 4:2:x-2
for j = 3:2:y-2
D1 = abs(2*im(i,j,3)-im(i,j-2,3)-im(i,j+2,3))+abs(im(i,j-1,2)-
im(i,j+1,2));
D2 = abs(2*im(i,j,3)-im(i-2,j,3)-im(i+2,j,3))+abs(im(i-1,j,2)-
im(i+1,j,2));
if D1<D2
DH(i,j) = 1;
elseif D1 = =D2
DH(i,j) = 0.5;
else
DH(i,j) = 0;
end
if D1>D2
DV(i,j) = 1;
```

```
elseif D1 = =D2
DV(i,j) = 0.5;
else
DV(i,j) = 0;
end
end
end
for i = 3:2:x-2
for j = 4:2:y-2
D1 = abs(2*im(i,j,1)-im(i,j-2,1)-im(i,j+2,1))+abs(im(i,j-1,2)-
im(i,j+1,2));
D2 = abs(2*im(i,j,1)-im(i-2,j,1)-im(i+2,j,1))+abs(im(i-1,j,2)-
im(i+1,j,2));
if D1<D2
DH(i,j) = 1;
elseif D1 = =D2
DH(i,j) = 0.5;
else
DH(i,j) = 0;
end
if D1>D2
DV(i,j) = 1;
elseif D1 = =D2
DV(i,j) = 0.5;
else
DV(i,j) = 0;
end
end
end
netRG_ip = zeros(2,2000);
netRG_op = zeros(1,2000);
counterRG = 1;
netGR_ip = zeros(10,2000);
netGR_op = zeros(1,2000);
counterGR = 1;
netBR_ip = zeros(2,2000);
netBR_op = zeros(1,2000);
counterBR = 1;
netRB_ip = zeros(2,2000);
netRB_op = zeros(1,2000);
counterRB = 1;
netGB_ip = zeros(10,2000);
netGB_op = zeros(1,2000);
counterGB = 1;
netBG_ip = zeros(2,2000);
netBG_op = zeros(1,2000);
counterBG = 1;
%%%%%%%%%%%%%%%%%%%%%%%%%%%%%%%%%%%%%%%%%%%%%%%%%%%
```

```
%Interpolating missing Green values at Blue Pixels
for i = 4:2:x-2
for j = 3:2:y-2
Dx = DH(i,j-2)+DH(i,j)+DH(i,j+2);
Dy = DV(i-2,j)+DV(i,j)+DV(i+2,j+2);
if Dx> 2.5
im(i,j,2) = 0.5*(2*im(i,j,3)-im(i,j-2,3)-im(i,j+2,3))+0.5*
(im(i,j-1,2)+im(i,j+1,2));
elseif Dy> = 2.5
im(i,j,2) = 0.5*(2*im(i,j,3)-im(i-2,j,3)-im(i+2,j,3))+0.5*
(im(i-1,j,2)+im(i+1,j,2));
else
im(i,j,2) = 0.25*(im(i-1,j,2)+im(i+1,j,2)+im(i,j-1,2)+
im(i,j+1,2))+0.125*(4*im(i,j,3)-im(i,j-2,3)-im(i,j+2,3)-
im(i-2,j,3)-im(i+2,j,3));
end
end
end
%Interpolating missing Green values at Red Pixels
for i = 3:2:x-2
for j = 4:2:y-2
Dx = DH(i,j-2)+DH(i,j)+DH(i,j+2);
Dy = DV(i-2,j)+DV(i,j)+DV(i+2,j);
if Dx>2.5
im(i,j,2) = 0.5*(2*im(i,j,1)-im(i,j-2,1)-im(i,j+2,1))+0.5*
(im(i,j-1,2)+im(i,j+1,2));
elseif Dy> = 2.5
im(i,j,2) = 0.5*(2*im(i,j,1)-im(i-2,j,1)-im(i+2,j,1))+0.5*
(im(i-1,j,2)+im(i+1,j,2));
else
im(i,j,2) = 0.25*(im(i-1,j,2)+im(i+1,j,2)+im(i,j-1,2)+
im(i,j+1,2))+0.125*(4*im(i,j,1)-im(i,j-2,1)-im(i,j+2,1)-
im(i-2,j,1)-im(i+2,j,1));
end
end
end
%%%%%%%%%%%%TRAIN%%%%%%%%%%%%%%%%%%%%%%%%%%%%%%%%%
tim = im(:,:,1)+im(:,:,2)+im(:,:,3);
v = [0 -1 0 1 -1 -2 -2 -1 1 2 2 1];
h = [-1 0 1 0 -2 -1 1 2 2 1 -1 -2];
% The gamma is defined here
k = [1 1 1 1 0.6 0.6 0.6 0.6 0.6 0.6 0.6 0.6];
v1 = [-1 -1 1 1];
h1 = [-1 1 1 -1];
%Interpolating missing Blue at Red Pixels
I = [];w = []; Kb = [];
for i = 5:2:x-4
for j = 6:2:y-5
```

```
s = 0;
for n = 1:4
I(n) = 1*(abs(tim(i+v1(n),j+h1(n))-tim(i-v1(n),j-h1(n)))+abs
(tim(i+2*v1(n),j+2*h1(n))-tim(i,j)));
s = s+1/(1+I(n));
end
w = (1./(1+I))./s;
s = 0;
for n = 1:4
Kb(n) = im(i+v1(n),j+h1(n),2)-im(i+v1(n),j+h1(n),3); netBR_
ip(:,counterBR) = [im(i+v1(n),j+h1(n),2);im(i+v1(n),j+h
1(n),3)]; end
s = sum(w.*Kb);
im(i,j,3) = im(i,j,2)-s;
netBR_op(:,counterBR) = im(i,j,3); %NEURALBR OUTPUT
counterBR = counterBR+1;
end
end
%Interpolating missing Red at Blue Pixels
I = [];w = []; Kr = [];
for i = 6:2:x-5
for j = 5:2:y-4
s = 0;
for n = 1:4
I(n) = 1*(abs(tim(i+v1(n),j+h1(n))-tim(i-v1(n),j-h1(n)))+abs
(tim(i+2*v1(n),j+2*h1(n))-tim(i,j)));
s = s+1/(1+I(n));
end
w = (1./(1+I))./s;
s = 0;
for n = 1:4
Kr(n) = im(i+v1(n),j+h1(n),2)-im(i+v1(n),j+h1(n),1);
netRB_ip(:,counterRB) = [im(i+v1(n),j+h1(n),2);im(i+v1(n),j+h
1(n),1)]; end
s = sum(w.*Kr);
im(i,j,1) = im(i,j,2)-s;
netRB_op(:,counterRB) = im(i,j,1); %NEURALRB OUTPUT
counterRB = counterRB+1;
end
end
%Interpolating missing Red at Green Pixels
I = [];w = []; Kr = [];
for i = 5:2:x-4
for j = 5:2:y-4
s = 0;
for n = 1:12
I(n) = k(n)*(abs(tim(i+v(n),j+h(n))-tim(i-v(n),j-h(n)))+abs
(tim(i+2*v(n),j+2*h(n))-tim(i,j)));
```

```
s = s+1/(1+I(n));
end
w = (1./(1+I))./s;
s = 0;
for n = 1:12
Kr(n) = im(i+v(n),j+h(n),2)-im(i+v(n),j+h(n),1);
netRG_ip(:,counterRG) = [im(i+v(n),j+h(n),2);im(i+v(n),
j+h(n),1)];
end
s = sum(w.*Kr);
im(i,j,1) = im(i,j,2)-s;
netRG_op(:,counterRG) = im(i,j,1); %NEURALRG OUTPUT
counterRG = counterRG+1;
end
end
I = [];w = []; Kr = [];
for i = 6:2:x-5
for j = 6:2:y-5
s = 0;
for n = 1:12
I(n) = k(n)*(abs(tim(i+v(n),j+h(n))-tim(i-v(n),j-h(n)))+abs
(tim(i+2*v(n),j+2*h(n))-tim(i,j)));
s = s+1/(1+I(n));
end
w = (1./(1+I))./s;
s = 0;
for n = 1:12
Kr(n) = im(i+v(n),j+h(n),2)-im(i+v(n),j+h(n),1);
netRG_ip(:,counterRG) = [im(i+v(n),j+h(n),2);im(i+v(n),
j+h(n),1)];
end
s = sum(w.*Kr);
im(i,j,1) = im(i,j,2)-s;
netRG_op(:,counterRG) = im(i,j,1); %NEURALRG OUTPUT
counterRG = counterRG+1;
end
end
%Interpolating missing Blue at Green Pixels
I = [];w = []; Kb = [];
for i = 5:2:x-4
for j = 5:2:y-4
s = 0;
for n = 1:12
I(n) = k(n)*(abs(tim(i+v(n),j+h(n))-tim(i-v(n),j-h(n)))+abs
(tim(i+2*v(n),j+2*h(n))-tim(i,j)));
s = s+1/(1+I(n));
end
w = (1./(1+I))./s;
```

```
s = 0;
for n = 1:12
Kb(n) = im(i+v(n),j+h(n),2)-im(i+v(n),j+h(n),3);
netBG_ip(:,counterBG) = [im(i+v(n),j+h(n),2);im(i+v(n),
j+h(n),3)];
end
s = sum(w.*Kb);
im(i,j,3) = im(i,j,2)-s;
netBG_op(:,counterBG) = im(i,j,3); %NEURALBG OUTPUT
counterBG = counterBG+1;
end
end
I = [];w = []; Kb = [];
for i = 6:2:x-5
for j = 6:2:y-5
s = 0;
for n = 1:12
I(n) = k(n)*(abs(tim(i+v(n),j+h(n))-tim(i-v(n),j-h(n)))+abs
(tim(i+2*v(n),j+2*h(n))-tim(i,j)));
s = s+1/(1+I(n));
end
w = (1./(1+I))./s;
s = 0;
for n = 1:12
Kb(n) = im(i+v(n),j+h(n),2)-im(i+v(n),j+h(n),3);
netBG_ip(:,counterBG) = [im(i+v(n),j+h(n),2);im(i+v(n),
j+h(n),3)];
end
s = sum(w.*Kb);
im(i,j,3) = im(i,j,2)-s;
netBG_op(:,counterBG) = im(i,j,3); %NEURALBG OUTPUT
counterBG = counterBG+1;
end
end
%Interpolating again Green at Blue Pixels to reduce aliasing
for i = 6:2:x-5
for j = 5:2:y-4
s = 0;
for n = 1:12
I(n) = k(n)*(abs(tim(i+v(n),j+h(n))-tim(i-v(n),j-h(n)))+abs
(tim(i+2*v(n),j+2*h(n))-tim(i,j)));
s = s+1/(1+I(n));
end
w = (1./(1+I))./s;
s = 0;
for n = 1:12
G = im(i+v(n),j+h(n),2);
B = im(i+v(n),j+h(n),3);
```

```
Kb(n) = G-B;
end
s = sum(w.*Kb);
im(i,j,2) = im(i,j,3)+s;
end
end
%Interpolating again Green at Red Pixels to reduce aliasing
I = [];w = [];Kr = [];
for i = 5:2:x-4
for j = 6:2:y-5
s = 0;
for n = 1:12
I(n) = k(n)*(abs(tim(i+v(n),j+h(n))-tim(i-v(n),j-h(n)))+abs
(tim(i+2*v(n),j+2*h(n))-tim(i,j)));
s = s+1/(1+I(n));
end
w = (1./(1+I))./s;
s = 0;
for n = 1:12
G = im(i+v(n),j+h(n),2);
R = im(i+v(n),j+h(n),1);
Kr(n) = G-R;
end
s = sum(w.*Kr);
im(i,j,2) = im(i,j,1)+s;
end
end
%Interpolating Blue at Red Pixels
I = [];w = []; Kb = [];
for i = 5:2:x-4
for j = 6:2:y-5
s = 0;
for n = 1:4
I(n) = 1*(abs(tim(i+v1(n),j+h1(n))-tim(i-v1(n),j-h1(n)))+abs
(tim(i+2*v1(n),j+2*h1(n))-tim(i,j)));
s = s+1/(1+I(n));
end
w = (1./(1+I))./s;
s = 0;
for n = 1:4
Kb(n) = im(i+v1(n),j+h1(n),2)-im(i+v1(n),j+h1(n),3);
netBR_ip(:,counterBR) = [im(i+v1(n),j+h1(n),2);im(i+v1(n),
j+h1(n),3)]; end
s = sum(w.*Kb);
im(i,j,3) = im(i,j,2)-s;
netBR_op(:,counterBR) = im(i,j,3); %NEURALBR OUTPUT
counterBR = counterBR+1;
end
```

```
end
%Interpolating Red at Blue Pixels
I = [];w = []; Kr = [];
for i = 6:2:x-5
for j = 5:2:y-4
s = 0;
for n = 1:4
I(n) = 1*(abs(tim(i+v1(n),j+h1(n))-tim(i-v1(n),j-h1(n)))+abs
(tim(i+2*v1(n),j+2*h1(n))-tim(i,j)));
s = s+1/(1+I(n));
end
w = (1./(1+I))./s;
s = 0;
for n = 1:4
Kr(n) = im(i+v1(n),j+h1(n),2)-im(i+v1(n),j+h1(n),1);
netRB_ip(:,counterRB) = [im(i+v1(n),j+h1(n),2);im(i+v1(n),
j+h1(n),1)]; end
s = sum(w.*Kr);
im(i,j,1) = im(i,j,2)-s;
netRB_op(:,counterRB) = im(i,j,1); %NEURALRB OUTPUT
counterRB = counterRB+1;
end
end
%Interpolating Red at Green Pixels
I = [];w = []; Kr = [];
for i = 5:2:x-4
for j = 5:2:y-4
s = 0;
for n = 1:12
I(n) = k(n)*(abs(tim(i+v(n),j+h(n))-tim(i-v(n),j-h(n)))+abs
(tim(i+2*v(n),j+2*h(n))-tim(i,j)));
s = s+1/(1+I(n));
end
w = (1./(1+I))./s;
s = 0;
for n = 1:12
Kr(n) = im(i+v(n),j+h(n),2)-im(i+v(n),j+h(n),1);
netRG_ip(:,counterRG) = [im(i+v(n),j+h(n),2);im(i+v(n),
j+h(n),1)]; end
s = sum(w.*Kr);
im(i,j,1) = im(i,j,2)-s;
netRG_op(:,counterRG) = im(i,j,1); %NEURALRG OUTPUT
counterRG = counterRG+1;
end
end
I = [];w = []; Kr = [];
for i = 6:2:x-5
for j = 6:2:y-5
```

```
s = 0;
for n = 1:12
I(n) = k(n)*(abs(tim(i+v(n),j+h(n))-tim(i-v(n),j-h(n)))+abs
(tim(i+2*v(n),j+2*h(n))-tim(i,j)));
s = s+1/(1+I(n));
end
w = (1./(1+I))./s;
s = 0;
for n = 1:12
Kr(n) = im(i+v(n),j+h(n),2)-im(i+v(n),j+h(n),1);
netRG_ip(:,counterRG) = [im(i+v(n),j+h(n),2);im(i+v(n),
j+h(n),1)];
end
s = sum(w.*Kr);
im(i,j,1) = im(i,j,2)-s;
netRG_op(:,counterRG) = im(i,j,1); %NEURALRG OUTPUT
counterRG = counterRG+1;
end
end
%Interpolating Blue at Green Pixels
I = [];w = []; Kb = [];
for i = 5:2:x-4
for j = 5:2:y-4
s = 0;
for n = 1:12
I(n) = k(n)*(abs(tim(i+v(n),j+h(n))-tim(i-v(n),j-h(n)))+abs
(tim(i+2*v(n),j+2*h(n))-tim(i,j)));
s = s+1/(1+I(n));
end
w = (1./(1+I))./s;
s = 0;
for n = 1:12
Kb(n) = im(i+v(n),j+h(n),2)-im(i+v(n),j+h(n),3);
netBG_ip(:,counterBG) = [im(i+v(n),j+h(n),2);im(i+v(n),
j+h(n),3)];
end
s = sum(w.*Kb);
im(i,j,3) = im(i,j,2)-s;
netBG_op(:,counterBG) = im(i,j,3); %NEURALBG OUTPUT
counterBG = counterBG+1;
end
end
I = [];w = []; Kb = [];
for i = 6:2:x-5
for j = 6:2:y-5
s = 0;
for n = 1:12
```

```
I(n) = k(n)*(abs(tim(i+v(n),j+h(n))-tim(i-v(n),j-h(n)))+abs(ti
m(i+2*v(n),j+2*h(n))-tim(i,j)));
s = s+1/(1+I(n));
end
w = (1./(1+I))./s;
s = 0;
for n = 1:12
Kb(n) = im(i+v(n),j+h(n),2)-im(i+v(n),j+h(n),3);          netBG_ip
(:,counterBG) = [im(i+v(n),j+h(n),2);im(i+v(n),j+h(n),3)];
end
s = sum(w.*Kb);
im(i,j,3) = im(i,j,2)-s;
netBG_op(:,counterBG) = im(i,j,3); %NEURALBG OUTPUT
counterBG = counterBG+1;
end
end
%************************TRAIN********************************
***
netRG_ip = netRG_ip(:,1:counterRG-1);
netRG_op = netRG_op(:,1:counterRG-1);
netRB_ip = netRB_ip(:,1:counterRB-1);
netRB_op = netRB_op(:,1:counterRB-1);
netBG_ip = netBG_ip(:,1:counterBG-1);
netBG_op = netBG_op(:,1:counterBG-1);
netBR_ip = netBR_ip(:,1:counterBR-1);
netBR_op = netBR_op(:,1:counterBR-1);
netRG = newff(minmax(netRG_ip),[2 1],
{'logsig' 'logsig'},'trainscg');
netRB = newff(minmax(netRB_ip),[2 1],
{'logsig' 'logsig'},'trainscg');
netBG = newff(minmax(netBG_ip),[2 1],
{'logsig' 'logsig'},'trainscg');
netBR = newff(minmax(netBR_ip),[2 1],
{'logsig' 'logsig'},'trainscg');
netRG = init(netRG);
netRB = init(netRB);
netBG = init(netBG);
netBR = init(netBR);
netRG.performFcn = 'mse';
netRG.trainparam.epochs = 10000;
netRG.trainparam.show = 10;
netRG.trainparam.goal = 1e-10;
netRG.trainparam.min_grad = 1e-7;
netRG = train(netRG,netRG_ip,netRG_op);
netRB.performFcn = 'mse';
netRB.trainparam.epochs = 10000;
netRB.trainparam.show = 10;
netRB.trainparam.goal = 1e-10;
```

```matlab
netRB.trainparam.min_grad = 1e-7;
netRB = train(netRB,netRB_ip,netRB_op);
netBG.performFcn = 'mse';
netBG.trainparam.epochs = 10000;
netBG.trainparam.show = 10;
netBG.trainparam.goal = 1e-10;
netBG.trainparam.min_grad = 1e-7;
netBG = train(netBG,netBG_ip,netBG_op);
netBR.performFcn = 'mse';
netBR.trainparam.epochs = 10000;
netBR.trainparam.show = 10;
netBR.trainparam.goal = 1e-10;
netBR.trainparam.min_grad = 1e-7;
netBR = train(netBR,netBR_ip,netBR_op);
%%%%%%%%%%%%%%%%%%%%%%%%%%%TEST%%%%%%%%%%%%%%%%%%%%%%%%%%%%%
for i = 5:2:x-4
for j = 6:2:y-8%5
for n = 1:4
y = [im(i+v1(n),j+h1(n),2);im(i+v1(n),j+h1(n),3)];
im(i,j,3) = sim(netBR, y);
end;
end;
end;
for i = 6:2:x-5
for j = 5:2:y-4
for n = 1:4
y = [im(i+v1(n),j+h1(n),2);im(i+v1(n),j+h1(n),1)];
im(i,j,1) = sim(netRB, y);
end;
end;
end;
for i = 5:2:x-4
for j = 5:2:y-4
for n = 1:12
y = [im(i+v(n),j+h(n),2);im(i+v(n),j+h(n),1)];
im(i,j,1) = sim(netRG, y);
end;
end;
end
for i = 5:2:x-4
for j = 5:2:y-4
for n = 1:12
y = [im(i+v(n),j+h(n),2);im(i+v(n),j+h(n),3)];
im(i,j,3) = sim(netBG, y);
end;
end;
end
%%%%%%%%%%%%%%%%%%%%%%%%%%%%%%%%%%%%%%%%%%%%%%%%%%%%%%%%%%%%%
```

```
extime = toc;
Img = uint8(im);
imshow(Img);
figure
zz1 = Img(:,:,1);
imshow(zz1-zzz1);
figure
imshow(edge(zz1,'canny'));
figure
imshow(edge(zzz1,'canny'));
figure
imshow(edge(zz1,'canny')-edge(zzz1,'canny'));
MM = edge(zz1,'canny')-edge(zzz1,'canny');
```

D.6 Interpolation for Pattern Recognition

D.6.1 Train Program

```
clear all
disp('Enter s = 0 for features from signal');
disp('Enter s = 1 for features from the wavelet transformed
signal');
disp('Enter s = 2 for features from the signal plus wavelet
transformed signal');
disp('Enter s = 3 for features from DCT of signal');
disp('Enter s = 4 for features from signal plus DCT of
signal');
disp('Enter s = 5 for features from DST of Signal');
disp('Enter s = 6 for features from signal plus DST of
signal');
for s = 0:6
THRESH = 0;
for i = 1:20
if i = =1
filename = 'T1.jpg';
elseif i = =2
filename = 'T2.jpg';
elseif i = =3
filename = 'T3.jpg';
elseif i = =4
filename = 'T4.jpg';
elseif i = =5
filename = 'T5.jpg';
elseif i = =6
filename = 'T6.jpg';
elseif i = =7
filename = 'T7.jpg';
```

```
elseif i = =8
filename = 'T8.jpg';
elseif i = =9
filename = 'T9.jpg';
elseif i = =10
filename = 'T10.jpg';
elseif i = =11
filename = 'P1.jpg';
elseif i = =12
filename = 'P2.jpg';
elseif i = =13
filename = 'P3.jpg';
elseif i = =14
filename = 'P4.jpg';
elseif i = =15
filename = 'P5.jpg';
elseif i = =16
filename = 'P6.jpg';
elseif i = =17
filename = 'P7.jpg';
elseif i = =18
filename = 'P8.jpg';
elseif i = =19
filename = 'P9.jpg';
elseif i = =20
filename = 'P10.jpg';
end;
im = imread(filename);
im = double(im)/255;
im = im(:,:,1);
[d1,d2] = size(im);
vec = im2col(im,[d1 d2],[d1 d2],'distinct');
size(vec)
current_mat(1:length(vec),i) = vec;
plot(vec)
feature_vec = features(current_mat(:,i),s,THRESH);
I(1:length(feature_vec),i) = feature_vec;
end;
size(I)
save current_file current_mat
save Idata_wavelet I
T = eye (20,20);
switch s
case 0
rand0 = rand(size(I));
I = I+rand0;
save rand_data0 rand0
case 1
```

```
rand1 = rand(size(I));
I = I+rand1;
save rand_data1 rand1
case 2
rand2 = rand(size(I));
I = I+rand2;
save rand_data2 rand2
case 3
rand3 = rand(size(I));
I = I+rand3;
save rand_data3 rand3
case 4
rand4 = rand(size(I));
I = I+rand4;
save rand_data4 rand4
case 5
rand5 = rand(size(I));
I = I+rand5;
save rand_data5 rand5
case 6
rand6 = rand(size(I));
I = I+rand6;
save rand_data6 rand6
end;
net_mfcc_cepspoly = newff(minmax(I),[125 20],{'logsig'
'logsig'},'trainscg');
net_mfcc_cepspoly = init(net_mfcc_cepspoly);
net_mfcc_cepspoly.performFcn = 'mse';
net_mfcc_cepspoly.trainparam.epochs = 10000;
net_mfcc_cepspoly.trainparam.show = 10;
net_mfcc_cepspoly.trainparam.goal = 1e-10;
net_mfcc_cepspoly.trainparam.min_grad = 1e-7;
net_mfcc_cepspoly = train(net_mfcc_cepspoly,I,T)
if s = =0
net_mfcc_cepspoly0 = net_mfcc_cepspoly;
save data0 net_mfcc_cepspoly0
elseif s = =1
net_mfcc_cepspoly1 = net_mfcc_cepspoly;
save data1 net_mfcc_cepspoly1
elseif s = =2
net_mfcc_cepspoly2 = net_mfcc_cepspoly;
save data2 net_mfcc_cepspoly2
elseif s = =3
net_mfcc_cepspoly3 = net_mfcc_cepspoly;
save data3 net_mfcc_cepspoly3
elseif s = =4
net_mfcc_cepspoly4 = net_mfcc_cepspoly;
save data4 net_mfcc_cepspoly4
```

```
elseif s = =5
net_mfcc_cepspoly5 = net_mfcc_cepspoly;
save data5 net_mfcc_cepspoly5
elseif s = =6
net_mfcc_cepspoly6 = net_mfcc_cepspoly;
save data6 net_mfcc_cepspoly6
end;
clear all;
end;
```

D.6.2 Test Program

```
clear all
disp('Enter s = 0 for features from signal');
disp('Enter s = 1 for features from the wavelet transformed
signal');
disp('Enter s = 2 for features from the signal plus wavelet
transformed signal');
disp('Enter s = 3 for features from DCT of signal');
disp('Enter s = 4 for features from signal plus DCT of
signal');
disp('Enter s = 5 for features from DST of Signal');
disp('Enter s = 6 for features from signal plus DST of
signal');
de = 1;
s = 0;
for SNRdB = 0:5:50
SNRdB
for i = 1:100
THRESH = 0;
if (i> = 1) & (i<6)
filename = 'T1.jpg';
elseif (i> = 6) & (i<11)
filename = 'T2.jpg';
elseif (i> = 11) & (i<16)
filename = 'T3.jpg';
elseif (i> = 16) & (i<21)
filename = 'T4.jpg';
elseif (i> = 21) & (i<26)
filename = 'T5.jpg';
elseif (i> = 26) & (i<31)
filename = 'T6.jpg';
elseif (i> = 31) & (i<36)
filename = 'T7.jpg';
elseif (i> = 36) & (i<41)
filename = 'T8.jpg';
elseif (i> = 41) & (i<46)
```

```
filename = 'T9.jpg';
elseif (i> = 46) & (i<51)
filename = 'T10.jpg';
elseif (i> = 51) & (i<56)
filename = 'P1.jpg';
elseif (i> = 56) & (i<61)
filename = 'P2.jpg';
elseif (i> = 61) & (i<66)
filename = 'P3.jpg';
elseif (i> = 66) & (i<71)
filename = 'P4.jpg';
elseif (i> = 71) & (i<76)
filename = 'P5.jpg';
elseif (i> = 76) & (i<81)
filename = 'P6.jpg';
elseif (i> = 81) & (i<86)
filename = 'P7.jpg';
elseif (i> = 86) & (i<91)
filename = 'P8.jpg';
elseif (i> = 91) & (i<96)
filename = 'P9.jpg';
elseif (i> = 96) & (i<101)
filename = 'P10.jpg';
end;
im = imread(filename);
im = double(im)/255;
im = im(:,:,1);
if de = =1
[d1,d2] = size(im);
gg = im2col(im,[d1,d2],'distinct');
n_var = var(gg)/10^(SNRdB/10);
im = imnoise(im,'gaussian',0,n_var);
elseif de = =2
[d1,d2] = size(im);
im = imnoise(im,'Salt & Pepper',SNRdB/5000);
elseif de = =3
[d1,d2] = size(im);
gg = im2col(im,[d1,d2],'distinct');
n_var = var(gg)/10^(SNRdB/10);
h = ones(5,5)/25;
im = filter2(h,im);
im = imnoise(im,'gaussian',0,n_var);
elseif de = =4
hh = ones(5,5)/25;
im = filter2(h,im);
im = imnoise(im,'Salt & Pepper',SNRdB/1000);
end;
vec = im2col(im,[d1 d2],[d1 d2],'distinct');
```

```
size(vec)
current_mat(1:length(vec),i) = vec;
plot(vec)
feature_vec = features(current_mat(:,i),s,THRESH);
Is(1:length(feature_vec),i) = feature_vec;
end;
In = Is;
error_count = 0;
if s = =0
load data0 net_mfcc_cepspoly0
net_mfcc_cepspoly = net_mfcc_cepspoly0;
elseif s = =1
load data1 net_mfcc_cepspoly1
net_mfcc_cepspoly = net_mfcc_cepspoly1;
elseif s = =2
load data2 net_mfcc_cepspoly2
net_mfcc_cepspoly = net_mfcc_cepspoly2;
elseif s = =3
load data3 net_mfcc_cepspoly3
net_mfcc_cepspoly = net_mfcc_cepspoly3;
elseif s = =4
load data4 net_mfcc_cepspoly4
net_mfcc_cepspoly = net_mfcc_cepspoly4;
elseif s = =5
load data5 net_mfcc_cepspoly5
net_mfcc_cepspoly = net_mfcc_cepspoly5;
elseif s = =6
load data6 net_mfcc_cepspoly6
net_mfcc_cepspoly = net_mfcc_cepspoly6;
end;
switch s
case 0
load rand_data0 rand0
for count = 1:100
In(:,count) = In(:,count)+rand0(:,ceil(count/5));
end;
case 1
load rand_data1 rand1
for count = 1:100
In(:,count) = In(:,count)+rand1(:,ceil(count/5));
end;
case 2
load rand_data2 rand2
for count = 1:100
In(:,count) = In(:,count)+rand2(:,ceil(count/5));
end;
case 3
load rand_data3 rand3
```

```
for count = 1:100
In(:,count) = In(:,count)+rand3(:,ceil(count/5));
end;
case 4
load rand_data4 rand4
for count = 1:100
In(:,count) = In(:,count)+rand4(:,ceil(count/5));
end;
case 5
load rand_data5 rand5
for count = 1:100
In(:,count) = In(:,count)+rand5(:,ceil(count/5));
end;
case 6
load rand_data6 rand6
for count = 1:100
In(:,count) = In(:,count)+rand6(:,ceil(count/5));
end;
end;
for i = 1:100;
a = sim(net_mfcc_cepspoly,In(:,i));
if max(a)~ = a(ceil(i/5))
error_count = error_count+1;
end;
clear Is;
end
SNRdB
success_rate(SNRdB/5+1) = (1-(error_count/100))*100;
end;
if s = =0
success_rate0 = success_rate;
save datas01 success_rate0
elseif s = =1
success_rate1 = success_rate;
save datas11 success_rate1
elseif s = =2
success_rate2 = success_rate;
save datas21 success_rate2
elseif s = =3
success_rate3 = success_rate;
save datas31 success_rate3
elseif s = =4
success_rate4 = success_rate;
save datas41 success_rate4
elseif s = =5
success_rate5 = success_rate;
save datas51 success_rate5
elseif s = =6
```

```
success_rate6 = success_rate;
save datas61 success_rate6
end;
clear all
disp('Enter s = 0 for features from signal');
disp('Enter s = 1 for features from the wavelet transformed
signal');
disp('Enter s = 2 for features from the signal plus wavelet
transformed signal');
disp('Enter s = 3 for features from DCT of signal');
disp('Enter s = 4 for features from signal plus DCT of
signal');
disp('Enter s = 5 for features from DST of Signal');
disp('Enter s = 6 for features from signal plus DST of
signal');
de = 1;
s = 1;
for SNRdB = 0:5:50
SNRdB
for i = 1:100
THRESH = 0;
if (i> = 1) & (i<6)
filename = 'T1.jpg';
elseif (i> = 6) & (i<11)
filename = 'T2.jpg';
elseif (i> = 11) & (i<16)
filename = 'T3.jpg';
elseif (i> = 16) & (i<21)
filename = 'T4.jpg';
elseif (i> = 21) & (i<26)
filename = 'T5.jpg';
elseif (i> = 26) & (i<31)
filename = 'T6.jpg';
elseif (i> = 31) & (i<36)
filename = 'T7.jpg';
elseif (i> = 36) & (i<41)
filename = 'T8.jpg';
elseif (i> = 41) & (i<46)
filename = 'T9.jpg';
elseif (i> = 46) & (i<51)
filename = 'T10.jpg';
elseif (i> = 51) & (i<56)
filename = 'P1.jpg';
elseif (i> = 56) & (i<61)
filename = 'P2.jpg';
elseif (i> = 61) & (i<66)
filename = 'P3.jpg';
elseif (i> = 66) & (i<71)
```

```
filename = 'P4.jpg';
elseif (i> = 71) & (i<76)
filename = 'P5.jpg';
elseif (i> = 76) & (i<81)
filename = 'P6.jpg';
elseif (i> = 81) & (i<86)
filename = 'P7.jpg';
elseif (i> = 86) & (i<91)
filename = 'P8.jpg';
elseif (i> = 91) & (i<96)
filename = 'P9.jpg';
elseif (i> = 96) & (i<101)
filename = 'P10.jpg';
end;
im = imread(filename);
im = double(im)/255;
if de = =1
[d1,d2] = size(im);
gg = im2col(im,[d1,d2],'distinct');
n_var = var(gg)/10^(SNRdB/10);
im = imnoise(im,'gaussian',0,n_var);
elseif de = =1
[d1,d2] = size(im);
im = imnoise(im,'Salt & Pepper',SNRdB/5000);
elseif de = =3
[d1,d2] = size(im);
gg = im2col(im,[d1,d2],'distinct');
n_var = var(gg)/10^(SNRdB/10);
h = ones(5,5)/25;
im = filter2(h,im);
im = imnoise(im,'gaussian',0,n_var);
elseif de = =4
hh = ones(5,5)/25;
im = filter2(h,im);
im = imnoise(im,'Salt & Pepper',SNRdB/1000);
end;
vec = im2col(im,[d1 d2],[d1 d2],'distinct');
size(vec)
current_mat(1:length(vec),i) = vec;
plot(vec)
feature_vec = features(current_mat(:,i),s,THRESH);
Is(1:length(feature_vec),i) = feature_vec;
end;
In = Is;
error_count = 0;
if s = =0
load data0 net_mfcc_cepspoly0
net_mfcc_cepspoly = net_mfcc_cepspoly0;
```

```
elseif s = =1
load data1 net_mfcc_cepspoly1
net_mfcc_cepspoly = net_mfcc_cepspoly1;
elseif s = =2
load data2 net_mfcc_cepspoly2
net_mfcc_cepspoly = net_mfcc_cepspoly2;
elseif s = =3
load data3 net_mfcc_cepspoly3
net_mfcc_cepspoly = net_mfcc_cepspoly3;
elseif s = =4
load data4 net_mfcc_cepspoly4
net_mfcc_cepspoly = net_mfcc_cepspoly4;
elseif s = =5
load data5 net_mfcc_cepspoly5
net_mfcc_cepspoly = net_mfcc_cepspoly5;
elseif s = =6
load data6 net_mfcc_cepspoly6
net_mfcc_cepspoly = net_mfcc_cepspoly6;
end;
switch s
case 0
load rand_data0 rand0
for count = 1:100
In(:,count) = In(:,count)+rand0(:,ceil(count/5));
end;
case 1
load rand_data1 rand1
for count = 1:100
In(:,count) = In(:,count)+rand1(:,ceil(count/5));
end;
case 2
load rand_data2 rand2
for count = 1:100
In(:,count) = In(:,count)+rand2(:,ceil(count/5));
end;
case 3
load rand_data3 rand3
for count = 1:100
In(:,count) = In(:,count)+rand3(:,ceil(count/5));
end;
case 4
load rand_data4 rand4
for count = 1:100
In(:,count) = In(:,count)+rand4(:,ceil(count/5));
end;
case 5
load rand_data5 rand5
for count = 1:100
```

```
In(:,count) = In(:,count)+rand5(:,ceil(count/5));
end;
case 6
load rand_data6 rand6
for count = 1:100
In(:,count) = In(:,count)+rand6(:,ceil(count/5));
end;
end;
for i = 1:100;
a = sim(net_mfcc_cepspoly,In(:,i));
if max(a)~ = a(ceil(i/5))
error_count = error_count+1;
end;
clear Is;
end
SNRdB
success_rate(SNRdB/5+1) = (1-(error_count/100))*100;
end;
if s = =0
success_rate0 = success_rate;
save datas01 success_rate0
elseif s = =1
success_rate1 = success_rate;
save datas11 success_rate1
elseif s = =2
success_rate2 = success_rate;
save datas21 success_rate2
elseif s = =3
success_rate3 = success_rate;
save datas31 success_rate3
elseif s = =4
success_rate4 = success_rate;
save datas41 success_rate4
elseif s = =5
success_rate5 = success_rate;
save datas51 success_rate5
elseif s = =6
success_rate6 = success_rate;
save datas61 success_rate6
end;
clear all
disp('Enter s = 0 for features from signal');
disp('Enter s = 1 for features from the wavelet transformed
signal');
disp('Enter s = 2 for features from the signal plus wavelet
transformed signal');
disp('Enter s = 3 for features from DCT of signal');
```

```
disp('Enter s = 4 for features from signal plus DCT of
signal');
disp('Enter s = 5 for features from DST of Signal');
disp('Enter s = 6 for features from signal plus DST of
signal');
de = 1;
s = 2;
for SNRdB = 0:5:50
SNRdB
for i = 1:100
THRESH = 0;
if (i> = 1) & (i<6)
filename = 'T1.jpg';
elseif (i> = 6) & (i<11)
filename = 'T2.jpg';
elseif (i> = 11) & (i<16)
filename = 'T3.jpg';
elseif (i> = 16) & (i<21)
filename = 'T4.jpg';
elseif (i> = 21) & (i<26)
filename = 'T5.jpg';
elseif (i> = 26) & (i<31)
filename = 'T6.jpg';
elseif (i> = 31) & (i<36)
filename = 'T7.jpg';
elseif (i> = 36) & (i<41)
filename = 'T8.jpg';
elseif (i> = 41) & (i<46)
filename = 'T9.jpg';
elseif (i> = 46) & (i<51)
filename = 'T10.jpg';
elseif (i> = 51) & (i<56)
filename = 'P1.jpg';
elseif (i> = 56) & (i<61)
filename = 'P2.jpg';
elseif (i> = 61) & (i<66)
filename = 'P3.jpg';
elseif (i> = 66) & (i<71)
filename = 'P4.jpg';
elseif (i> = 71) & (i<76)
filename = 'P5.jpg';
elseif (i> = 76) & (i<81)
filename = 'P6.jpg';
elseif (i> = 81) & (i<86)
filename = 'P7.jpg';
elseif (i> = 86) & (i<91)
filename = 'P8.jpg';
elseif (i> = 91) & (i<96)
```

```
filename = 'P9.jpg';
elseif (i> = 96) & (i<101)
filename = 'P10.jpg';
end;
im = imread(filename);
im = double(im)/255;
im = im(:,:,1);
if de = =1
[d1,d2] = size(im);
gg = im2col(im,[d1,d2],'distinct');
n_var = var(gg)/10^(SNRdB/10);
im = imnoise(im,'gaussian',0,n_var);
elseif de = =2
[d1,d2] = size(im);
im = imnoise(im,'Salt & Pepper',SNRdB/5000);
elseif de = =3
[d1,d2] = size(im);
gg = im2col(im,[d1,d2],'distinct');
n_var = var(gg)/10^(SNRdB/10);
h = ones(5,5)/25;
im = filter2(h,im);
im = imnoise(im,'gaussian',0,n_var);
elseif de = =4
h = ones(5,5)/25;
im = filter2(h,im);
im = imnoise(im,'Salt & Pepper',SNRdB/1000);
end;
vec = im2col(im,[d1 d2],[d1 d2],'distinct');
size(vec)
current_mat(1:length(vec),i) = vec;
plot(vec)
feature_vec = features(current_mat(:,i),s,THRESH);
Is(1:length(feature_vec),i) = feature_vec;
end;
In = Is;
error_count = 0;
if s = =0
load data0 net_mfcc_cepspoly0
net_mfcc_cepspoly = net_mfcc_cepspoly0;
elseif s = =1
load data1 net_mfcc_cepspoly1
net_mfcc_cepspoly = net_mfcc_cepspoly1;
elseif s = =2
load data2 net_mfcc_cepspoly2
net_mfcc_cepspoly = net_mfcc_cepspoly2;
elseif s = =3
load data3 net_mfcc_cepspoly3
net_mfcc_cepspoly = net_mfcc_cepspoly3;
```

```
elseif s = =4
load data4 net_mfcc_cepspoly4
net_mfcc_cepspoly = net_mfcc_cepspoly4;
elseif s = =5
load data5 net_mfcc_cepspoly5
net_mfcc_cepspoly = net_mfcc_cepspoly5;
elseif s = =6
load data6 net_mfcc_cepspoly6
net_mfcc_cepspoly = net_mfcc_cepspoly6;
end;
switch s
case 0
load rand_data0 rand0
for count = 1:100
In(:,count) = In(:,count)+rand0(:,ceil(count/5));
end;
case 1
load rand_data1 rand1
for count = 1:100
In(:,count) = In(:,count)+rand1(:,ceil(count/5));
end;
case 2
load rand_data2 rand2
for count = 1:100
In(:,count) = In(:,count)+rand2(:,ceil(count/5));
end;
case 3
load rand_data3 rand3
for count = 1:100
In(:,count) = In(:,count)+rand3(:,ceil(count/5));
end;
case 4
load rand_data4 rand4
for count = 1:100
In(:,count) = In(:,count)+rand4(:,ceil(count/5));
end;
case 5
load rand_data5 rand5
for count = 1:100
In(:,count) = In(:,count)+rand5(:,ceil(count/5));
end;
case 6
load rand_data6 rand6
for count = 1:100
In(:,count) = In(:,count)+rand6(:,ceil(count/5));
end;
end;
for i = 1:100;
```

```
a = sim(net_mfcc_cepspoly,In(:,i));
if max(a)~ = a(ceil(i/5))
error_count = error_count+1;
end;
clear Is;
end
SNRdB
success_rate(SNRdB/5+1) = (1-(error_count/100))*100;
end;
if s = =0
success_rate0 = success_rate;
save datas01 success_rate0
elseif s = =1
success_rate1 = success_rate;
save datas11 success_rate1
elseif s = =2
success_rate2 = success_rate;
save datas21 success_rate2
elseif s = =3
success_rate3 = success_rate;
save datas31 success_rate3
elseif s = =4
success_rate4 = success_rate;
save datas41 success_rate4
elseif s = =5
success_rate5 = success_rate;
save datas51 success_rate5
elseif s = =6
success_rate6 = success_rate;
save datas61 success_rate6
end;
clear all
disp('Enter s = 0 for features from signal');
disp('Enter s = 1 for features from the wavelet transformed
signal');
disp('Enter s = 2 for features from the signal plus wavelet
transformed signal');
disp('Enter s = 3 for features from DCT of signal');
disp('Enter s = 4 for features from signal plus DCT of
signal');
disp('Enter s = 5 for features from DST of Signal');
disp('Enter s = 6 for features from signal plus DST of
signal');
de = 1;
s = 3;
for SNRdB = 0:5:50
SNRdB
for i = 1:100
```

```
THRESH = 0;
if (i> = 1) & (i<6)
filename = 'T1.jpg';
elseif (i> = 6) & (i<11)
filename = 'T2.jpg';
elseif (i> = 11) & (i<16)
filename = 'T3.jpg';
elseif (i> = 16) & (i<21)
filename = 'T4.jpg';
elseif (i> = 21) & (i<26)
filename = 'T5.jpg';
elseif (i> = 26) & (i<31)
filename = 'T6.jpg';
elseif (i> = 31) & (i<36)
filename = 'T7.jpg';
elseif (i> = 36) & (i<41)
filename = 'T8.jpg';
elseif (i> = 41) & (i<46)
filename = 'T9.jpg';
elseif (i> = 46) & (i<51)
filename = 'T10.jpg';
elseif (i> = 51) & (i<56)
filename = 'P1.jpg';
elseif (i> = 56) & (i<61)
filename = 'P2.jpg';
elseif (i> = 61) & (i<66)
filename = 'P3.jpg';
elseif (i> = 66) & (i<71)
filename = 'P4.jpg';
elseif (i> = 71) & (i<76)
filename = 'P5.jpg';
elseif (i> = 76) & (i<81)
filename = 'P6.jpg';
elseif (i> = 81) & (i<86)
filename = 'P7.jpg';
elseif (i> = 86) & (i<91)
filename = 'P8.jpg';
elseif (i> = 91) & (i<96)
filename = 'P9.jpg';
elseif (i> = 96) & (i<101)
filename = 'P10.jpg';
end;
im = imread(filename);
im = double(im)/255;
im = im(:,:,1);
if de = =1
[d1,d2] = size(im);
gg = im2col(im,[d1,d2],'distinct');
```

```
n_var = var(gg)/10^(SNRdB/10);
im = imnoise(im,'gaussian',0,n_var);
elseif de = =2
[d1,d2] = size(im);
im = imnoise(im,'Salt & Pepper',SNRdB/5000);
elseif de = =3
[d1,d2] = size(im);
gg = im2col(im,[d1,d2],'distinct');
n_var = var(gg)/10^(SNRdB/10);
h = ones(5,5)/25;
im = filter2(h,im);
im = imnoise(im,'gaussian',0,n_var);
elseif de = =4
hh = ones(5,5)/25;
im = filter2(h,im);
im = imnoise(im,'Salt & Pepper',SNRdB/1000);
end;
vec = im2col(im,[d1 d2],[d1 d2],'distinct');
size(vec)
current_mat(1:length(vec),i) = vec;
plot(vec)
feature_vec = features(current_mat(:,i),s,THRESH);
Is(1:length(feature_vec),i) = feature_vec;
end;
In = Is;
error_count = 0;
if s = =0
load data0 net_mfcc_cepspoly0
net_mfcc_cepspoly = net_mfcc_cepspoly0;
elseif s = =1
load data1 net_mfcc_cepspoly1
net_mfcc_cepspoly = net_mfcc_cepspoly1;
elseif s = =2
load data2 net_mfcc_cepspoly2
net_mfcc_cepspoly = net_mfcc_cepspoly2;
elseif s = =3
load data3 net_mfcc_cepspoly3
net_mfcc_cepspoly = net_mfcc_cepspoly3;
elseif s = =4
load data4 net_mfcc_cepspoly4
net_mfcc_cepspoly = net_mfcc_cepspoly4;
elseif s = =5
load data5 net_mfcc_cepspoly5
net_mfcc_cepspoly = net_mfcc_cepspoly5;
elseif s = =6
load data6 net_mfcc_cepspoly6
net_mfcc_cepspoly = net_mfcc_cepspoly6;
end;
```

```
switch s
case 0
load rand_data0 rand0
for count = 1:100
In(:,count) = In(:,count)+rand0(:,ceil(count/5));
end;
case 1
load rand_data1 rand1
for count = 1:100
In(:,count) = In(:,count)+rand1(:,ceil(count/5));
end;
case 2
load rand_data2 rand2
for count = 1:100
In(:,count) = In(:,count)+rand2(:,ceil(count/5));
end;
case 3
load rand_data3 rand3
for count = 1:100
In(:,count) = In(:,count)+rand3(:,ceil(count/5));
end;
case 4
load rand_data4 rand4
for count = 1:100
In(:,count) = In(:,count)+rand4(:,ceil(count/5));
end;
case 5
load rand_data5 rand5
for count = 1:100
In(:,count) = In(:,count)+rand5(:,ceil(count/5));
end;
case 6
load rand_data6 rand6
for count = 1:100
In(:,count) = In(:,count)+rand6(:,ceil(count/5));
end;
end;
for i = 1:100;
a = sim(net_mfcc_cepspoly,In(:,i));
if max(a)~ = a(ceil(i/5))
error_count = error_count+1;
end;
clear Is;
end
SNRdB
success_rate(SNRdB/5+1) = (1-(error_count/100))*100;
end;
if s = =0
```

```
success_rate0 = success_rate;
save datas01 success_rate0
elseif s = =1
success_rate1 = success_rate;
save datas11 success_rate1
elseif s = =2
success_rate2 = success_rate;
save datas21 success_rate2
elseif s = =3
success_rate3 = success_rate;
save datas31 success_rate3
elseif s = =4
success_rate4 = success_rate;
save datas41 success_rate4
elseif s = =5
success_rate5 = success_rate;
save datas51 success_rate5
elseif s = =6
success_rate6 = success_rate;
save datas61 success_rate6
end;
clear all
disp('Enter s = 0 for features from signal');
disp('Enter s = 1 for features from the wavelet transformed
signal');
disp('Enter s = 2 for features from the signal plus wavelet
transformed signal');
disp('Enter s = 3 for features from DCT of signal');
disp('Enter s = 4 for features from signal plus DCT of
signal');
disp('Enter s = 5 for features from DST of Signal');
disp('Enter s = 6 for features from signal plus DST of
signal ');
de = 1;
s = 4;
for SNRdB = 0:5:50
SNRdB
for i = 1:100
THRESH = 0;
if (i> = 1) & (i<6)
filename = 'T1.jpg';
elseif (i> = 6) & (i<11)
filename = 'T2.jpg';
elseif (i> = 11) & (i<16)
filename = 'T3.jpg';
elseif (i> = 16) & (i<21)
filename = 'T4.jpg';
elseif (i> = 21) & (i<26)
```

```
filename = 'T5.jpg';
elseif (i> = 26) & (i<31)
filename = 'T6.jpg';
elseif (i> = 31) & (i<36)
filename = 'T7.jpg';
elseif (i> = 36) & (i<41)
filename = 'T8.jpg';
elseif (i> = 41) & (i<46)
filename = 'T9.jpg';
elseif (i> = 46) & (i<51)
filename = 'T10.jpg';
elseif (i> = 51) & (i<56)
filename = 'P1.jpg';
elseif (i> = 56) & (i<61)
filename = 'P2.jpg';
elseif (i> = 61) & (i<66)
filename = 'P3.jpg';
elseif (i> = 66) & (i<71)
filename = 'P4.jpg';
elseif (i> = 71) & (i<76)
filename = 'P5.jpg';
elseif (i> = 76) & (i<81)
filename = 'P6.jpg';
elseif (i> = 81) & (i<86)
filename = 'P7.jpg';
elseif (i> = 86) & (i<91)
filename = 'P8.jpg';
elseif (i> = 91) & (i<96)
filename = 'P9.jpg';
elseif (i> = 96) & (i<101)
filename = 'P10.jpg';
end;
im = imread(filename);
im = double(im)/255;
im = im(:,:,1);
if de = =1
[d1,d2] = size(im);
gg = im2col(im,[d1,d2],'distinct');
n_var = var(gg)/10^(SNRdB/10);
im = imnoise(im,'gaussian',0,n_var);
elseif de = =2
[d1,d2] = size(im);
im = imnoise(im,'Salt & Pepper',SNRdB/5000);
elseif de = =3
[d1,d2] = size(im);
gg = im2col(im,[d1,d2],'distinct');
n_var = var(gg)/10^(SNRdB/10);
hh = ones(5,5)/25;
```

```
im = filter2(h,im);
im = imnoise(im,'gaussian',0,n_var);
elseif de = =4
h = ones(5,5)/25;
im = filter2(h,im);
im = imnoise(im,'Salt & Pepper',SNRdB/1000);
end;
vec = im2col(im,[d1 d2],[d1 d2],'distinct');
size(vec)
current_mat(1:length(vec),i) = vec;
plot(vec)
feature_vec = features(current_mat(:,i),s,THRESH);
Is(1:length(feature_vec),i) = feature_vec;
end;
In = Is;
error_count = 0;
if s = =0
load data0 net_mfcc_cepspoly0
net_mfcc_cepspoly = net_mfcc_cepspoly0;
elseif s = =1
load data1 net_mfcc_cepspoly1
net_mfcc_cepspoly = net_mfcc_cepspoly1;
elseif s = =2
load data2 net_mfcc_cepspoly2
net_mfcc_cepspoly = net_mfcc_cepspoly2;
elseif s = =3
load data3 net_mfcc_cepspoly3
net_mfcc_cepspoly = net_mfcc_cepspoly3;
elseif s = =4
load data4 net_mfcc_cepspoly4
net_mfcc_cepspoly = net_mfcc_cepspoly4;
elseif s = =5
load data5 net_mfcc_cepspoly5
net_mfcc_cepspoly = net_mfcc_cepspoly5;
elseif s = =6
load data6 net_mfcc_cepspoly6
net_mfcc_cepspoly = net_mfcc_cepspoly6;
end;
switch s
case 0
load rand_data0 rand0
for count = 1:100
In(:,count) = In(:,count)+rand0(:,ceil(count/5));
end;
case 1
load rand_data1 rand1
for count = 1:100
In(:,count) = In(:,count)+rand1(:,ceil(count/5));
```

```
end;
case 2
load rand_data2 rand2
for count = 1:100
In(:,count) = In(:,count)+rand2(:,ceil(count/5));
end;
case 3
load rand_data3 rand3
for count = 1:100
In(:,count) = In(:,count)+rand3(:,ceil(count/5));
end;
case 4
load rand_data4 rand4
for count = 1:100
In(:,count) = In(:,count)+rand4(:,ceil(count/5));
end;
case 5
load rand_data5 rand5
for count = 1:100
In(:,count) = In(:,count)+rand5(:,ceil(count/5));
end;
case 6
load rand_data6 rand6
for count = 1:100
In(:,count) = In(:,count)+rand6(:,ceil(count/5));
end;
end;
for i = 1:100;
a = sim(net_mfcc_cepspoly,In(:,i));
if max(a)~ = a(ceil(i/5))
error_count = error_count+1;
end;
clear Is;
end
SNRdB
success_rate(SNRdB/5+1) = (1-(error_count/100))*100;
end;
if s = =0
success_rate0 = success_rate;
save datas01 success_rate0
elseif s = =1
success_rate1 = success_rate;
save datas11 success_rate1
elseif s = =2
success_rate2 = success_rate;
save datas21 success_rate2
elseif s = =3
success_rate3 = success_rate;
```

```
save datas31 success_rate3
elseif s = =4
success_rate4 = success_rate;
save datas41 success_rate4
elseif s = =5
success_rate5 = success_rate;
save datas51 success_rate5
elseif s = =6
success_rate6 = success_rate;
save datas61 success_rate6
end;
clear all
disp('Enter s = 0 for features from signal');
disp('Enter s = 1 for features from the wavelet transformed
signal');
disp('Enter s = 2 for features from the signal plus wavelet
transformed signal');
disp('Enter s = 3 for features from DCT of signal');
disp('Enter s = 4 for features from signal plus DCT of
signal');
disp('Enter s = 5 for features from DST of Signal');
disp('Enter s = 6 for features from signal plus DST of
signal');
de = 1;
s = 5;
for SNRdB = 0:5:50
SNRdB
for i = 1:100
THRESH = 0;
if (i> = 1) & (i<6)
filename = 'T1.jpg';
elseif (i> = 6) & (i<11)
filename = 'T2.jpg';
elseif (i> = 11) & (i<16)
filename = 'T3.jpg';
elseif (i> = 16) & (i<21)
filename = 'T4.jpg';
elseif (i> = 21) & (i<26)
filename = 'T5.jpg';
elseif (i> = 26) & (i<31)
filename = 'T6.jpg';
elseif (i> = 31) & (i<36)
filename = 'T7.jpg';
elseif (i> = 36) & (i<41)
filename = 'T8.jpg';
elseif (i> = 41) & (i<46)
filename = 'T9.jpg';
elseif (i> = 46) & (i<51)
```

```
filename = 'T10.jpg';
elseif (i> = 51) & (i<56)
filename = 'P1.jpg';
elseif (i> = 56) & (i<61)
filename = 'P2.jpg';
elseif (i> = 61) & (i<66)
filename = 'P3.jpg';
elseif (i> = 66) & (i<71)
filename = 'P4.jpg';
elseif (i> = 71) & (i<76)
filename = 'P5.jpg';
elseif (i> = 76) & (i<81)
filename = 'P6.jpg';
elseif (i> = 81) & (i<86)
filename = 'P7.jpg';
elseif (i> = 86) & (i<91)
filename = 'P8.jpg';
elseif (i> = 91) & (i<96)
filename = 'P9.jpg';
elseif (i> = 96) & (i<101)
filename = 'P10.jpg';
end;
im = imread(filename);
im = double(im)/255;
im = im(:,:,1);
if de = =1
[d1,d2] = size(im);
gg = im2col(im,[d1,d2],'distinct');
n_var = var(gg)/10^(SNRdB/10);
im = imnoise(im,'gaussian',0,n_var);
elseif de = =2
[d1,d2] = size(im);
im = imnoise(im,'Salt & Pepper',SNRdB/5000);
elseif de = =3
[d1,d2] = size(im);
gg = im2col(im,[d1,d2],'distinct');
n_var = var(gg)/10^(SNRdB/10);
hh = ones(5,5)/25;
im = filter2(h,im);
im = imnoise(im,'gaussian',0,n_var);
elseif de = =4
h = ones(5,5)/25;
im = filter2(h,im);
im = imnoise(im,'Salt & Pepper',SNRdB/1000);
end;
vec = im2col(im,[d1 d2],[d1 d2],'distinct');
size(vec)
current_mat(1:length(vec),i) = vec;
```

```
plot(vec)
feature_vec = features(current_mat(:,i),s,THRESH);
Is(1:length(feature_vec),i) = feature_vec;
end;
In = Is;
error_count = 0;
if s = =0
load data0 net_mfcc_cepspoly0
net_mfcc_cepspoly = net_mfcc_cepspoly0;
elseif s = =1
load data1 net_mfcc_cepspoly1
net_mfcc_cepspoly = net_mfcc_cepspoly1;
elseif s = =2
load data2 net_mfcc_cepspoly2
net_mfcc_cepspoly = net_mfcc_cepspoly2;
elseif s = =3
load data3 net_mfcc_cepspoly3
net_mfcc_cepspoly = net_mfcc_cepspoly3;
elseif s = =4
load data4 net_mfcc_cepspoly4
net_mfcc_cepspoly = net_mfcc_cepspoly4;
elseif s = =5
load data5 net_mfcc_cepspoly5
net_mfcc_cepspoly = net_mfcc_cepspoly5;
elseif s = =6
load data6 net_mfcc_cepspoly6
net_mfcc_cepspoly = net_mfcc_cepspoly6;
end;
switch s
case 0
load rand_data0 rand0
for count = 1:100
In(:,count) = In(:,count)+rand0(:,ceil(count/5));
end;
case 1
load rand_data1 rand1
for count = 1:100
In(:,count) = In(:,count)+rand1(:,ceil(count/5));
end;
case 2
load rand_data2 rand2
for count = 1:100
In(:,count) = In(:,count)+rand2(:,ceil(count/5));
end;
case 3
load rand_data3 rand3
for count = 1:100
In(:,count) = In(:,count)+rand3(:,ceil(count/5));
```

```
end;
case 4
load rand_data4 rand4
for count = 1:100
In(:,count) = In(:,count)+rand4(:,ceil(count/5));
end;
case 5
load rand_data5 rand5
for count = 1:100
In(:,count) = In(:,count)+rand5(:,ceil(count/5));
end;
case 6
load rand_data6 rand6
for count = 1:100
In(:,count) = In(:,count)+rand6(:,ceil(count/5));
end;
end;
for i = 1:100;
a = sim(net_mfcc_cepspoly,In(:,i));
if max(a)~ = a(ceil(i/5))
error_count = error_count+1;
end;
clear Is;
end
SNRdB
success_rate(SNRdB/5+1) = (1-(error_count/100))*100;
end;
if s = =0
success_rate0 = success_rate;
save datas01 success_rate0
elseif s = =1
success_rate1 = success_rate;
save datas11 success_rate1
elseif s = =2
success_rate2 = success_rate;
save datas21 success_rate2
elseif s = =3
success_rate3 = success_rate;
save datas31 success_rate3
elseif s = =4
success_rate4 = success_rate;
save datas41 success_rate4
elseif s = =5
success_rate5 = success_rate;
save datas51 success_rate5
elseif s = =6
success_rate6 = success_rate;
save datas61 success_rate6
```

```
end;
clear all
disp('Enter s = 0 for features from signal');
disp('Enter s = 1 for features from the wavelet transformed
signal');
disp('Enter s = 2 for features from the signal plus wavelet
transformed signal');
disp('Enter s = 3 for features from DCT of signal');
disp('Enter s = 4 for features from signal plus DCT of
signal');
disp('Enter s = 5 for features from DST of Signal');
disp('Enter s = 6 for features from signal plus DST of
signal');
de = 1;
s = 6;
for SNRdB = 0:5:50
SNRdB
for i = 1:100
THRESH = 0;
if (i> = 1) & (i<6)
filename = 'T1.jpg';
elseif (i> = 6) & (i<11)
filename = 'T2.jpg';
elseif (i> = 11) & (i<16)
filename = 'T3.jpg';
elseif (i> = 16) & (i<21)
filename = 'T4.jpg';
elseif (i> = 21) & (i<26)
filename = 'T5.jpg';
elseif (i> = 26) & (i<31)
filename = 'T6.jpg';
elseif (i> = 31) & (i<36)
filename = 'T7.jpg';
elseif (i> = 36) & (i<41)
filename = 'T8.jpg';
elseif (i> = 41) & (i<46)
filename = 'T9.jpg';
elseif (i> = 46) & (i<51)
filename = 'T10.jpg';
elseif (i> = 51) & (i<56)
filename = 'P1.jpg';
elseif (i> = 56) & (i<61)
filename = 'P2.jpg';
elseif (i> = 61) & (i<66)
filename = 'P3.jpg';
elseif (i> = 66) & (i<71)
filename = 'P4.jpg';
elseif (i> = 71) & (i<76)
```

```
filename = 'P5.jpg';
elseif (i> = 76) & (i<81)
filename = 'P6.jpg';
elseif (i> = 81) & (i<86)
filename = 'P7.jpg';
elseif (i> = 86) & (i<91)
filename = 'P8.jpg';
elseif (i> = 91) & (i<96)
filename = 'P9.jpg';
elseif (i> = 96) & (i<101)
filename = 'P10.jpg';
end;
im = imread(filename);
im = double(im)/255;
im = im(:,:,1);
if de = =1
[d1,d2] = size(im);
gg = im2col(im,[d1,d2],'distinct');
n_var = var(gg)/10^(SNRdB/10);
im = imnoise(im,'gaussian',0,n_var);
elseif de = =2
[d1,d2] = size(im);
im = imnoise(im,'Salt & Pepper',SNRdB/5000);
elseif de = =3
[d1,d2] = size(im);
gg = im2col(im,[d1,d2],'distinct');
n_var = var(gg)/10^(SNRdB/10);
hh = ones(5,5)/25;
im = filter2(h,im);
im = imnoise(im,'gaussian',0,n_var);
elseif de = =4
h = ones(5,5)/25;
im = filter2(h,im);
im = imnoise(im,'Salt & Pepper',SNRdB/1000);
end;
vec = im2col(im,[d1 d2],[d1 d2],'distinct');
size(vec)
current_mat(1:length(vec),i) = vec;
plot(vec)
feature_vec = features(current_mat(:,i),s,THRESH);
Is(1:length(feature_vec),i) = feature_vec;
end;
In = Is;
error_count = 0;
if s = =0
load data0 net_mfcc_cepspoly0
net_mfcc_cepspoly = net_mfcc_cepspoly0;
elseif s = =1
```

```
load data1 net_mfcc_cepspoly1
net_mfcc_cepspoly = net_mfcc_cepspoly1;
elseif s = =2
load data2 net_mfcc_cepspoly2
net_mfcc_cepspoly = net_mfcc_cepspoly2;
elseif s = =3
load data3 net_mfcc_cepspoly3
net_mfcc_cepspoly = net_mfcc_cepspoly3;
elseif s = =4
load data4 net_mfcc_cepspoly4
net_mfcc_cepspoly = net_mfcc_cepspoly4;
elseif s = =5
load data5 net_mfcc_cepspoly5
net_mfcc_cepspoly = net_mfcc_cepspoly5;
elseif s = =6
load data6 net_mfcc_cepspoly6
net_mfcc_cepspoly = net_mfcc_cepspoly6;
end;
switch s
case 0
load rand_data0 rand0
for count = 1:100
In(:,count) = In(:,count)+rand0(:,ceil(count/5));
end;
case 1
load rand_data1 rand1
for count = 1:100
In(:,count) = In(:,count)+rand1(:,ceil(count/5));
end;
case 2
load rand_data2 rand2
for count = 1:100
In(:,count) = In(:,count)+rand2(:,ceil(count/5));
end;
case 3
load rand_data3 rand3
for count = 1:100
In(:,count) = In(:,count)+rand3(:,ceil(count/5));
end;
case 4
load rand_data4 rand4
for count = 1:100
In(:,count) = In(:,count)+rand4(:,ceil(count/5));
end;
case 5
load rand_data5 rand5
for count = 1:100
In(:,count) = In(:,count)+rand5(:,ceil(count/5));
```

```
end;
case 6
load rand_data6 rand6
for count = 1:100
In(:,count) = In(:,count)+rand6(:,ceil(count/5));
end;
end;
for i = 1:100;
a = sim(net_mfcc_cepspoly,In(:,i));
if max(a)~ = a(ceil(i/5))
error_count = error_count+1;
end;
clear Is;
end
SNRdB
success_rate(SNRdB/5+1) = (1-(error_count/100))*100;
end;
if s = =0
success_rate0 = success_rate;
save datas01 success_rate0
elseif s = =1
success_rate1 = success_rate;
save datas11 success_rate1
elseif s = =2
success_rate2 = success_rate;
save datas21 success_rate2
elseif s = =3
success_rate3 = success_rate;
save datas31 success_rate3
elseif s = =4
success_rate4 = success_rate;
save datas41 success_rate4
elseif s = =5
success_rate5 = success_rate;
save datas51 success_rate5
elseif s = =6
success_rate6 = success_rate;
save datas61 success_rate6
end;
```

D.6.3 Curve Plotter

```
SNRdB = (0:10:100)/10;
load datas02 success_rate0
load datas12 success_rate1
load datas22 success_rate2
load datas32 success_rate3
load datas42 success_rate4
```

```
load datas52 success_rate5
load datas62 success_rate6
figure
plot(SNRdB,success_rate0,'k-o', SNRdB,success_rate1,'b-x',
SNRdB,success_rate2,'r-+', SNRdB,success_
rate3,'g-*',SNRdB,success_rate4,'c-s',SNRdB,success_
rate5,'m-d',SNRdB,success_rate6,'k:v');
xlabel('Error Percentage');
ylabel('Recognition Rate');
axis([0 10 0 100])
legend('Features from signal', 'Features from the DWT of the
signal', 'Features from signal plus DWT of signal','Features
from DCT of signal', 'Features from signal plus DCT of
signal', 'Features from DST of Signal','Features from signal
plus DST of signal');
```

D.6.4 Cepstral Feature Extraction

```
function de = features(x1,s,THRESH)
fs = 10000;
x = x1;
switch s
case 0 % cepstral coefficients from signal Noise free case
de = feature_extraction(x,THRESH);
case 1
[A,B] = dwt(x,'haar');
x = [A;B];
de = feature_extraction(x,THRESH);
case 2
y = x;
[A,B] = dwt(x,'haar');
yy = [A;B];
de1 = feature_extraction(x,THRESH);
x = yy;
de2 = feature_extraction(x,THRESH);
de = [de1;de2];
case 3
de = feature_extraction(dct(x),THRESH);
case 4
de = [feature_extraction(x,THRESH);feature_
extraction(dct(x),THRESH)];
case 5
de = feature_extraction(dst(x),THRESH);
case 6
de = [feature_extraction(x,THRESH);feature_
extraction(dst(x),THRESH)];
end;
function de = feature_extraction(x,THRESH)
```

```
y = x;
len = length(x);
energy = sum(x.*x);
avg_e = energy/len;
segment = y(end-100:end);
abs1 = abs(segment);
max1 = max(abs1);
while(1)
blk = y(1:40);
for o = 1:40
if (blk(o) > = -max1)&&(blk(o)< = max1)
blk(o) = o;
end;
end;
e = sum(blk.*blk)/40;
if(e>THRESH*avg_e)
break;
else
y(1:40) = [];
end;
end;
while (1)
if length(y)> = 40
blk = y(end-39:end);
else
break
end;
for p = 1:40
if (blk(p) > = -max1)&&(blk(p)< = max1)
blk(p) = 0;
end
end
e = sum(blk.*blk)/40;
if(e>THRESH*avg_e)
break;
else
y(end-39:end) = [];
end;
end;
m = length(y);
s = m/16000;
r = 1:1:m;
ri = 1:s:m;
iint = interp1(r,y,ri,'spline');
[ceps,freqresp,fb,fbrecon,freqrecon] = mfcc(iint,10000);
ceps = ceps.';
[M,N] = size(ceps);
sum1 = 0;
```

```
sum2 = 0;
for j = 1:1:9;
p1(j) = j-5;
p2(j) = (j*j)-(10*j)+(55/3);
sum1 = sum1+(p1(j)*p1(j));
sum2 = sum2+(p2(j)*p2(j));
end
[B] = zeros(M,13);
[C] = zeros(M,13);
for i = 1:1:(M-8);
for x = 1:1:13;
sum11 = 0;
sum22 = 0;
for y = 1:1:9;
sum11 = sum11+ceps((i+y-1),x)*p1(y);
sum22 = sum22+ceps((i+y-1),x)*p2(y);
end
B(i,x) = sum11/sum1;
C(i,x) = sum22/sum2;
end
end
for i = (M-7):1:M;
for x = 1:1:13;
sum11 = 0;
sum22 = 0;
for y = 1:1:(M-i-1)
sum11 = sum11+ceps((i+y-1),x)*p1(y);
sum22 = sum22+ceps((i+y-1),x)*p2(y);
end;
B(i,x) = sum11/sum1;
C(i,x) = sum22/sum2;
end
end
[cepsBC] = zeros(M,39);
cepsBC = [ceps,B,C];
[M,N] = size(cepsBC);
de = [cepsBC(:,1);cepsBC(:,2);cepsBC(:,3);cepsBC(:,4);
cepsBC(:,5);cepsBC(:,6);cepsBC(:,7);
cepsBC(:,8);cepsBC(:,9);cepsBC(:,10);cepsBC(:,11);
cepsBC(:,12);cepsBC(:,13);cepsBC(:,14);
cepsBC(:,15);cepsBC(:,16);cepsBC(:,17);cepsBC(:,18);
cepsBC(:,19);cepsBC(:,20);cepsBC(:,21);
cepsBC(:,22);cepsBC(:,23);cepsBC(:,24);cepsBC(:,25);
cepsBC(:,26);cepsBC(:,27);cepsBC(:,28);
cepsBC(:,29);cepsBC(:,30);cepsBC(:,31);cepsBC(:,32);
cepsBC(:,33);cepsBC(:,34);cepsBC(:,35);
cepsBC(:,36);cepsBC(:,37);cepsBC(:,38);cepsBC(:,39)];
```

```
function [ceps,freqresp,fb,fbrecon,freqrecon] = mfcc(input,
samplingRate, frameRate)
global mfccDCTMatrix mfccFilterWeights
[r c] = size(input);
if (r > c)
input = input';
end
lowestFrequency = 0;
linearFilters = 13;
linearSpacing = 66.66666666;
logFilters = 27;
logSpacing = 1.0711703;
fftSize = 512;
cepstralCoefficients = 13;
windowSize = 400;
windowSize = 512; % Standard says 400, but 256 makes more
sense
        % Really should be a function of the sample
        % rate (and the lowestFrequency) and the
        % frame rate.
if (nargin < 2) samplingRate = 16000; end;
if (nargin < 3) frameRate = 100; end;
% Keep this around for later....
totalFilters = linearFilters + logFilters;
% Now figure the band edges. Interesting frequencies are
spaced
% by linearSpacing for a while, then go logarithmic. First
figure
% all the interesting frequencies. Lower, center, and upper
band
% edges are all consequtive interesting frequencies.
freqs = lowestFrequency + (0:linearFilters-1)*linearSpacing;
freqs(linearFilters+1:totalFilters+2) =...
        freqs(linearFilters) * logSpacing.^(1:logFilters+2);
lower = freqs(1:totalFilters);
center = freqs(2:totalFilters+1);
upper = freqs(3:totalFilters+2);
% We now want to combine FFT bins so that each filter has unit
% weight, assuming a triangular weighting function. First
figure
% out the height of the triangle, then we can figure out each
% frequencies contribution
mfccFilterWeights = zeros(totalFilters,fftSize);
triangleHeight = 2./(upper-lower);
fftFreqs = (0:fftSize-1)/fftSize*samplingRate;
for chan = 1:totalFilters
mfccFilterWeights(chan,:) = (fftFreqs > lower(chan) &
fftFreqs < = center(chan)).* triangleHeight(chan).
```

```
*(fftFreqs-lower(chan))/(center(chan)-lower(chan)) + (fftFreqs
> center(chan) & fftFreqs < upper(chan)).*
triangleHeight(chan).*(upper(chan)-fftFreqs)/(upper(chan)-
center(chan));
end
hamWindow = 0.54 - 0.46*cos(2*pi*(0:windowSize-1)/windowSize);
if 0                    % Window it like ComplexSpectrum
windowStep = samplingRate/frameRate;
a =.54;
b = -.46;
wr = sqrt(windowStep/windowSize);
phi = pi/windowSize;
hamWindow = 2*wr/sqrt(4*a*a+2*b*b)*(a +
b*cos(2*pi*(0:windowSize-1)/windowSize + phi));
end
% Figure out Discrete Cosine Transform. We want a matrix
% dct(i,j) which is totalFilters x cepstralCoefficients in
size.
% The i,j component is given by
%                 cos(i * (j+0.5)/totalFilters pi)
% where we have assumed that i and j start at 0.
mfccDCTMatrix = 1/sqrt(totalFilters/2)*cos((0:(cepstralCoeffic
ients-1))' * (2*(0:(totalFilters-1))+1) * pi/2/totalFilters);
mfccDCTMatrix(1,:) = mfccDCTMatrix(1,:) * sqrt(2)/2;
%imagesc(mfccDCTMatrix);
% Filter the input with the preemphasis filter. Also figure how
% many columns of data we will end up with.
if 1
preEmphasized = filter([1 -.97], 1, input);
else
preEmphasized = input;
end
windowStep = samplingRate/frameRate;
cols = fix((length(input)-windowSize)/windowStep)
% Allocate all the space we need for the output arrays.
ceps = zeros(cepstralCoefficients, cols);
if (nargout > 1) freqresp = zeros(fftSize/2, cols); end;
if (nargout > 2) fb = zeros(totalFilters, cols); end;
% Invert the filter bank center frequencies. For each FFT bin
% we want to know the exact position in the filter bank to find
% the original frequency response. The next block of code
finds the
% integer and fractional sampling positions.
if (nargout > 4)
fr = (0:(fftSize/2-1))'/(fftSize/2)*samplingRate/2;
j = 1;
for i = 1:(fftSize/2)
if fr(i) > center(j+1)
```

```
j = j + 1;
end
if j > totalFilters-1
j = totalFilters-1;
end
fr(i) = min(totalFilters-.0001, max(1,j + (fr(i)-center(j))/
(center(j+1)-center(j))));
end
fri = fix(fr);
frac = fr - fri;
freqrecon = zeros(fftSize/2, cols);
end
% For each chunk of data:
% * Window the data with a hamming window,
% * Shift it into FFT order,
% * Find the magnitude of the fft,
% * Convert the fft data into filter bank outputs,
% * Find the log base 10,
% * Find the cosine transform to reduce dimensionality.
for start = 0:cols
first = floor(start*windowStep) + 1;
last = first + windowSize-1;
fftData = zeros(1,fftSize);
fftData(1:windowSize) = preEmphasized(first:last).*hamWindow;
fftMag = abs(fft(fftData));
t = mfccFilterWeights * fftMag';
for i = 1:size(t,1)
for j = 1:size(t,2)
if (t(i,j) = = 0)
t(i,j) = 0.001;
end
end
end
earMag = log10(t);
ceps(:,start+1) = mfccDCTMatrix * earMag;
if (nargout > 1) freqresp(:,start+1) = fftMag(1:fftSize/2)';
end;
if (nargout > 2) fb(:,start+1) = earMag; end
if (nargout > 3)
fbrecon(:,start+1) = mfccDCTMatrix(1:cepstralCoefficients,:)'
* ceps(:,start+1);
end
if (nargout > 4)
f10 = 10.^fbrecon(:,start+1);
freqrecon(:,start+1) = samplingRate/fftSize * (f10(fri).*
(1-frac) + f10(fri+1).*frac);
end
end
```

```
% Let's also reconstruct the original FB
% output. We do this by multiplying the cepstral data by the
transpose
% of the original DCT matrix. This all works because we were
careful to
% scale the DCT matrix so it was orthonormal.
if 1 & (nargout > 3)
fbrecon = mfccDCTMatrix(1:cepstralCoefficients,:)' * ceps;
end;
```

D.7 Least-Squares Interpolation

```
clear all;
load woman
f1 = X;
f1 = f1(:,:,1);
f1 = double(f1)/255;
[M,N] = size(f1);
h = ones(2,2)/4;
g = filter2(h,f1);
g = g(1:2:M,1:2:N);
SNR = input('Enter the value of SNR in dB');
gg = im2col(g,[M/2,N/2],'distinct');
n_var = var(gg)/10^(SNR/10)
g = imnoise(g,'gaussian',0,n_var);
g = [rot90(g(:,1:4),2),g,rot90(g(:,M/2-8:M/2),2)];
g = [rot90(g(1:4,:),2);g;rot90(g(M/2-8:M/2,:),2)];
[L1,L2] = size(f1);
M = 24;
N = 12;
tic
I = speye(M^2);
H1 = sparse(M/2,M);
counter = 1;
for i = 1:M/2
H1(i,counter) = 1;
H1(i,counter+1) = 1;
counter = counter+2;
end;
H1 = H1/2;
H = kron(H1,H1);
HH = H'*H;
beta = 0.125;
landa = 0.001;
Q1 = sparse(M,M);
for i = 1:M
Q1(i,i) = -2;
```

```
end;
for i = 1:M-1
Q1(i,i+1) = 1;
end;
for i = 2:M
Q1(i,i-1) = 1;
end;
Q = kron(Q1,Q1);
QQ = Q'*Q;
ee = zeros(1,20);
iterations = 0;
for ii = 1:L1/8
ii
for jj = 1:L2/8
ii;
f = g(4*ii+1-4:4*(ii+1)+4,4*jj+1-4:4*(jj+1)+4);
z = f;
y = im2col(z,[N N],'distinct');
w = zeros(M^2,N^2);
u = 0.1;
yes = zeros(M^2,1);
ii;
counter = 1;
select = 0;
u = 0.1;
bound = 0.01;
while (norm(y-H*yes)>bound)
ee(counter) = norm(y-H*yes);
if (counter > = 2 & ~select)
if ee(counter)> ee(counter-1)
w = zeros(M^2,N^2);
select = 1;
end;
elseif (counter > = 2 & select)
if ee(counter)> ee(counter-1)
w = zeros(M^2,N^2);
u = 0.01;
select = 0;
end;
end;
w = w+u*H'*(y-H*yes)*y';
yes = w*y;
counter = counter+1;
if counter = = 10
bound = 0.1;
end;
if counter = = 20
w = wp;
```

```
u = 0.1;
end;
end;
wp = w;
counter
x1 = w*y;
x1 = col2im(x1,[M M],[M M],'distinct');
xx1(8*(ii-1)+1:8*(ii),8*(jj-1)+1:8*(jj)) = x1(9:16,9:16);
end;
end;
Avg_iterations = iterations/(L1*L2/64)
toc;
[a,b] = size(xx1)
xx1 = max(xx1,0);
xx1 = min(xx1,1);
error = xx1-f1;
clear sum
MSE = sum(sum(error(3:a-2,3:b-2).^2))*255^2/((a-4)*(b-4))
PSNR = 10*log(sum(sum(ones(size(error))))/sum(sum(error.^2)))/
log(10)
figure
imshow(xx1)
```

D.8 LMMSE Interpolation

```
clear all;
f1 = imread('2222.jpg');
f1 = f1(1:128,1:128,1);
f1 = double(f1)/255;
save data f1
[M,N] = size(f1);
h = ones(2,2)/4;
g = filter2(h,f1);
g = g(1:2:M,1:2:N);
SNR = input('Enter the value of SNR in dB');
selectau = input('Enter 1 for original image in
Auto-correlation 0 otherwise');
if ~selectau
key0 = input('Press 1 for Bilinear, 2 for Bicubic and 3 for
Cubic Spline and 4 for cubic o-Moms');
end;
gg = im2col(g,[M/2,N/2],'distinct');
n_var = var(gg)/10^(SNR/10);
g = imnoise(g,'gaussian',0,n_var);
f = g;
tic
if ~selectau
```

```
tic
a = -1/2;
s = 0.5;
[M,N] = size(f);
ff = zeros(M,N);
x = f(:,N-1:N);
x = rot90(x,2);
y = f(:,1);
f = [y,f,x];
for i = 1:M
for j = 2:N+1
switch key0
case 1
ff(i,j-1) = f(i,j)*(1-s)+f(i,j+1)*s;%bilinear
case 2
ff(i,j-1) = f(i,j-1)*(a*s^3-2*a*s^2+a*s)+f(i,j)*((a+2)*s^3
-(3+a)*s^2+1)+f(i,j+1)*(-(a+2)*s^3+(2*a+3)*s^2-a*s)+f(i,j+2)
*(-a*s^3+a*s^2); % Bicubic
case 3
ff(i,j-1) = f(i,j-1)*((3+s)^3-4*(2+s)^3+6*(1+s)^3-4*s^3)/6+f
(i,j)*((2+s)^3-4*(1+s)^3+6*s^3)/6+f(i,j+1)*((1+s)^3-4*s^3)/6+f
(i,j+2)*s^3/6;% Cubic-Spline
case 4
ff(i,j-1) = f(i,j-1)*((-1/6)*(1+s)^3+(1+s)^2+(-85/42)*(1+s)+
(29/21))+f(i,j)*(0.5*s^3-s^2+(1/14)*s+13/21)+f(i,j+1)*(0.5*
(1-s)^3-(1-s)^2+(1/14)*s+13/21)+f(i,j+2)*((-1/6)*(2-s)^3+
(2-s)^2-(85/42)*(2-s)+29/21);% Cubic o- Moms
end;
end;
end;
ff = ff(:,1:N);
fff(1:M,1:2:2*N) = f(1:M,2:N+1);
fff(1:M,2:2:2*N) = ff(1:M,1:N);
f = fff';
clear ff,fff;
a = -1/2;
s = 0.5;
[M,N] = size(f);
ff = zeros(M,N);
x = f(:,N-1:N);
x = rot90(x,2);
y = f(:,1);
f = [y,f,x];
for i = 1:M
for j = 2:N+1
switch key0
case 1
ff(i,j-1) = f(i,j)*(1-s)+f(i,j+1)*s;%bilinear
```

```
case 2
ff(i,j-1) = f(i,j-1)*(a*s^3-2*a*s^2+a*s)+f(i,j)*((a+2)*s^3-
(3+a)*s^2+1)+f(i,j+1)*(-(a+2)*s^3+(2*a+3)*s^2-a*s)+f(i,j+2)*
(-a*s^3+a*s^2); % Bicubic
case 3
ff(i,j-1) = f(i,j-1)*((3+s)^3-4*(2+s)^3+6*(1+s)^3-4*s^3)/6+f
(i,j)*((2+s)^3-4*(1+s)^3+6*s^3)/6+f(i,j+1)*((1+s)^3-4*s^3)/6+f
(i,j+2)*s^3/6;% Cubic-Spline
case 4
ff(i,j-1) = f(i,j-1)*((-1/6)*(1+s)^3+(1+s)^2+(-85/42)*(1+s)+
(29/21))+f(i,j)*(0.5*s^3-s^2+(1/14)*s+13/21)+f(i,j+1)*(0.5*
(1-s)^3-(1-s)^2+(1/14)*s+13/21)+f(i,j+2)*((-1/6)*(2-s)^3+
(2-s)^2-(85/42)*(2-s)+29/21);% Cubic o- Moms
end;
end;
end;
ff = ff(:,1:N);
fff(1:M,1:2:2*N) = f(1:M,2:N+1);
fff(1:M,2:2:2*N) = ff(1:M,1:N);
fff = fff';
toc
fff = (fff> = 0).*fff;
error = fff-f1;
imshow(fff);
figure
clear sum
[a,b] = size(error);
error = error(3:a-2,3:b-2);
MSE1 = sum(sum(error.^2))*255^2/prod(size(error));
PSNR1 = 10*log(sum(sum(ones(size(error)))))/
sum(sum(error.^2)))/log(10)
else
load data f1
fff = f1;
end;
wlength = 3;
[M1,M1] = size(fff);
fff(M1+wlength,M1+wlength) = 0;
for j = 0:M1-1
for k = 0:M1-1
sum = 0;
for n = 1:wlength
for m = 1:wlength
sum = sum+fff(n,m)*fff(n+j,m+k);
end;
end;
RRR(j+1,k+1) = 1/((wlength)^2)*sum;
end;
```

```
end;
[M M] = size(g);
N = 2*M;
R = 2;
kff = zeros(N,N);
kff = RRR';
kff = im2col(kff,[N N],'distinct');
kff = sparse(1:N^2,1:N^2,kff);
g = g';
g = im2col(g,[M M],'distinct');
I = speye(M^2)/12;
H1 = sparse(M,N);
counter = 1;
for i = 1:M
H1(i,counter) = 1;
H1(i,counter+1) = 1;
counter = counter+2;
end;
H1 = H1/2;
H = kron(H1,H1);
Hz1 = speye(M^2)*n_var;
HH = H*kff*H';
I = speye(size(HH));
Hopt = kff*H'*inv(HH+Hz1);
f = Hopt*g;
f = col2im(f,[N N],[N N],'distinct');
f = (f> = 0).*f;
toc
error = f'-f1;
clear sum
[a,b] = size(error);
error = error(3:a-2,3:b-2);
MSE2 = sum(sum(error.^2))*255^2/prod(size(error));
PSNR2 = 10*log(sum(sum(ones(size(error)))))/
sum(sum(error.^2)))/log(10)
imshow(f');
mr1 = f'
save datamr mr1
```

D.9 Maximum Entropy Interpolation

```
clear all;
f1 = imread('mri.tif');
f1 = f1(1:128,1:128,1);
f1 = double(f1)/255;
f1 = f1(:,:,1);
[M,N] = size(f1);
```

```
[M,N] = size(f1);
h = [.5,.5];
g = filter(h,1,f1');
g = g(1:2:N,:);
g = filter(h,1,g');
g = g(1:2:M,:);
h = ones(2,2)/4;
g = filter2(h,f1);
g = g(1:2:M,1:2:N);
SNR = input('Enter the value of SNR in dB');
gg = im2col(g,[M/2,N/2],'distinct');
n_var = var(gg)/10^(SNR/10);
g = imnoise(g,'gaussian',0,n_var);
imshow(g)
tic
[M M] = size(g);
N = 2*M;
g = g';
g = im2col(g,[M M],'distinct');
I = speye(N^2);
H1 = sparse(M,N);
counter = 1;
for i = 1:M
H1(i,counter) = 1;
H1(i,counter+1) = 1;
counter = counter+2;
end;
H1 = H1/2;
H = kron(H1,H1);
HH = H'*H;
gama = 0.001;
Hopt = inv(HH+gama*I)*H';
f = Hopt*g;
f = col2im(f,[N N],[N N],'distinct');
toc
error = f'-f1;
[a,b] = size(error);
error = error(3:a-2,3:b-2);
MSE2 = sum(sum(error.^2))*255^2/prod(size(error));
PSNR2 = 10*log(sum(sum(ones(size(error)))))/
sum(sum(error.^2)))/log(10)
imshow(f');
```

D.10 Regularized Interpolation (Inverse Technique)

```
clear all;
f1 = imread('1111.jpg');
```

```
f1 = f1(1:128,1:128,1);
f1 = double(f1)/255;
%f1 = double(f1)/255;
[M,N] = size(f1);
h = ones(2,2)/4;
g = filter2(h,f1);
g = g(1:2:M,1:2:N);
SNR = input('Enter the value of SNR in dB');
gg = im2col(g,[M/2,N/2],'distinct');
n_var = var(gg)/10^(SNR/10);
g = imnoise(g,'gaussian',0,n_var);
landa = input('Enter the value of regularization parameter');
g = [rot90(g(:,1:4),2),g,rot90(g(:,M/2-8:M/2),2)];
g = [rot90(g(1:4,:),2);g;rot90(g(M/2-8:M/2,:),2)];
[L1,L2] = size(f1);
M = 24;
N = 12;
tic
I = speye(M^2);
H1 = sparse(M/2,M);
counter = 1;
for i = 1:M/2
H1(i,counter) = 1;
H1(i,counter+1) = 1;
counter = counter+2;
end;
H1 = H1/2;
H = kron(H1,H1);
HH = H'*H;
beta = 0.125;
Q1 = sparse(M,M);
for i = 1:M
Q1(i,i) = -2;
end;
for i = 1:M-1
Q1(i,i+1) = 1;
end;
for i = 2:M
Q1(i,i-1) = 1;
end;
Q = kron(Q1,Q1);
QQ = Q'*Q;
L = inv(HH+landa*QQ);
for ii = 1:L1/8
for jj = 1:L2/8
ii;
f = g(4*ii+1-4:4*(ii+1)+4,4*jj+1-4:4*(jj+1)+4);
z = f;
```

```
y = im2col(z,[N N],'distinct');
x1 = L*H'*y;
x1 = col2im(x1,[M M],[M M],'distinct');
xx1(8*(ii-1)+1:8*(ii),8*(jj-1)+1:8*(jj)) = x1(9:16,9:16);
end;
end;
toc;
[a,b] = size(xx1)
xx1 = max(xx1,0);
xx1 = min(xx1,1);
error = xx1-f1;
clear sum
[a,b] = size(error);
error = error(3:a-2,3:b-2);
MSE2 = sum(sum(error.^2))*255^2/prod(size(error));
PSNR2 = 10*log(sum(sum(ones(size(error)))))/
sum(sum(error.^2)))/log(10)
figure
imshow(xx1)
```

D.11 Regularized Interpolation (Iterative Technique)

```
clear all;
load woman
f1 = X;
f1 = f1(:,:,1);
f1 = double(f1)/255;
[M,N] = size(f1);
h = ones(2,2)/4;
g = filter2(h,f1);
g = g(1:2:M,1:2:N);
SNR = input('Enter the value of SNR in dB');
gg = im2col(g,[M/2,N/2],'distinct');
n_var = var(gg)/10^(SNR/10);
g = imnoise(g,'gaussian',0,n_var);
landa = input('Enter the value of regularization parameter');
beta = input('Enter the value of the convergence parameter');
ni = input('Enter the number of iterations');
f = g;
[N,N] = size(g);
[M,M] = size(f1);
y = im2col(g,[N,N],'distinct');
tic
I = speye(M^2);
H1 = sparse(M/2,M);
counter = 1;
for i = 1:M/2
```

```
H1(i,counter) = 1;
counter = counter+2;
end;
H1 = H1/2;
H = kron(H1,H1);
HH = H'*H;
Q1 = sparse(M,M);
for i = 1:M
Q1(i,i) = -2;
end;
for i = 1:M-1
Q1(i,i+1) = 1;
end;
for i = 2:M
Q1(i,i-1) = 1;
end;
Q = kron(Q1,Q1);
LLL = Q';
QQ = LLL*Q;
x = beta*(H'*y);
th = H'*y;
HHH = sparse(HH+landa*QQ);
for k = 1:ni
k;
x = x+beta*(th-HHH*x);
end;
toc
x = col2im(x,[M M],[M M],'distinct');
x = max(x,0);
x = min(x,1);
error = x-f1;
clear sum
[a,b] = size(error);
error = error(3:a-2,3:b-2);
MSE2 = sum(sum(error.^2))*255^2/prod(size(error));
PSNR2 = 10*log(sum(sum(ones(size(error)))))/
sum(sum(error.^2)))/log(10)
imshow(x)
```

D.12 Wavelet and Curvelet Fusion of MR and CT Images

```
f3 = imread('2222.jpg');
f3 = f3(:,:,1);
f3 = double(f3)/255;
f1 = imread('1111.jpg');
f1 = f1(:,:,1);
f1 = double(f1)/255;
```

```
[M,N] = size(f3);
tic
[a1,hd1,vd1,dd1] = dwt2(f1,'haar');
[a3,hd3,vd3,dd3] = dwt2(f3,'haar');
a = a1.*(abs(a1)> = abs(a3))+a3.*(abs(a1)<abs(a3));
hd = hd1.*(abs(hd1)> = abs(hd3))+hd3.*(abs(hd1)<abs(hd3));
vd = vd1.*(abs(vd1)> = abs(vd3))+vd3.*(abs(vd1)<abs(vd3));
dd = dd1.*(abs(dd1)> = abs(dd3))+dd3.*(abs(dd1)<abs(dd3));
f6 = idwt2(a,hd,vd,dd,'haar');
toc
tic
[x11,x21,x31,w11,w21,w31] = trous_dec(f1);
[x13,x213,x33,w13,w23,w33] = trous_dec(f3);
w1 = curvelet_fusion(w11,w13,8,4);
[M,N] = size(w1);
w2 = curvelet_fusion(w21,w23,8,4);
w3 = curvelet_fusion(w31,w33,8,4);
x3 = curvelet_fusion(x31,x33,8,4);
xx = trous_rec(x11,x21,x3,w1,w2,w3);
toc
PSNR1w = psnr(f1,f6,M,N);
PSNR2w = psnr(f3,f6,M,N);
PSNR1c = psnr(f1,xx,M,N);
PSNR2c = psnr(f3,xx,M,N);
[r_CT_edge_c,r_MR_edge_c, r_CT_edge_w, r_MR_edge_w] =
similarity(f1,f3, f6,xx);
function [x1,x2,x3,w1,w2,w3] = trous_dec(x)
h1 = [1,4,6,4,1;4,16,24,16,4;6,24,36,24,6;4,16,24,16,4;1,4,
     6,4,1];
h = (1/256)*h1;
x1 = filter2(h,x);
x2 = filter2(h,x1);
x3 = filter2(h,x2);
w1 = x-x1;
w2 = x1-x2;
w3 = x2-x3;
function x = trous_rec(x1,x2,x3,w1,w2,w3)
x = w1+w2+w3+x3;
function p = curvelet_fusion(p1,p2,blk_length,redundancy)
[L1,L2] = size(p1);
p = zeros(size(p1));
p1 = [p1(blk_length:-1:1,:);p1;p1(L1:-1:L1-blk_length+1,:)];
p1 = [p1(:,blk_length:-1:1),p1,p1(:,L2:-1:L2-blk_length+1)];
p2 = [p2(blk_length:-1:1,:);p2;p2(L1:-1:L1-blk_length+1,:)];
p2 = [p2(:,blk_length:-1:1),p2,p2(:,L2:-1:L2-blk_length+1)];
[L1,L2] = size(p1);
ii = 2;
jj = 2;
```

```
while ii*blk_length+2 < = L1 & jj*blk_length+2 < = L2
tile1 = p1((ii-1)*blk_length-redundancy+1:ii*blk_
length+redundancy,(jj-1)*blk_length-redundancy+1:jj*blk_
length+redundancy);
[M1,N1] = size(tile1);
tile2 = p2((ii-1)*blk_length-redundancy+1:ii*blk_
length+redundancy,(jj-1)*blk_length-redundancy+1:jj*blk_
length+redundancy);
z = ridgelet_fusion(tile1,tile2);
imshow(z);
p((ii-1)*blk_length+1:ii*blk_length,(jj-1)*blk_
length+1:jj*blk_length) = z(1+redundancy:redundancy+blk_length
,1+redundancy:redundancy+blk_length);
jj = jj+1;
jj1 = jj;
if jj*blk_length+2 > = L2
jj = 2;
ii = ii+1;
end;
end;
p = p(blk_length+1:end,blk_length+1:end);
function [r_CT_edge_c,r_MR_edge_c, r_CT_edge_w, r_MR_edge_w]
= similarity(f1,f3, f6,xx);
Ic = edge(xx,'canny');
Iw = edge(f6,'canny');
I2 = edge(f1,'canny');
I3 = edge(f3,'canny');
[M,N] = size(xx);
counter = 1;
for i = 1:M
for j = 1:N
if Ic(i,j) = =1
z_fused_c(counter) = Ic(i,j);
z_fused_w(counter) = Iw(i,j);
z_CT(counter) = I3(i,j);
z_MR(counter) = I2(i,j);
counter = counter+1;
end;
end;
end;
r_CT_edge_c = sum(and(z_fused_c,z_CT))/sum(z_fused_c)
r_MR_edge_c = sum(and(z_fused_c,z_MR))/sum(z_fused_c)
r_CT_edge_w = sum(and(z_fused_w,z_CT))/sum(z_fused_c)
r_MR_edge_w = sum(and(z_fused_w,z_MR))/sum(z_fused_c)
function I = Ridgelet_fusion(f1,f2);
theta = 0:180;
[R1,xp1] = radon(f1,theta);
[R2,xp2] = radon(f2,theta);
```

```
[m,n] = size(R1);
if mod(n,2) ~ = 0
n = n+1;
theta(n) = 181;
end;
RR = zeros(m,n);
for i = 1:m
r1 = R1(i,:);
[cA1,cD1] = dwt([r1],'haar');
r2 = R2(i,:);
[cA2,cD2] = dwt([r2],'haar');
cA = cA1.*(abs(cA1)> = abs(cA2))+cA2.*(abs(cA1)<abs(cA2));
cD = cD1.*(abs(cD1)> = abs(cD2))+cD2.*(abs(cD1)<abs(cD2));
r = idwt(cA,cD,'haar');
RR(i,:) = r;
end;
I = iradon(RR,theta,'spline','Hann');
function [A] = psnr(image,image_prime,M,N)
image = double(image);
image_prime = double(image_prime);
if ((sum(sum(image-image_prime))) = = 0)
error('Input vectors must not be identical')
else
psnr_num = M*N*max(max(image.^2));          % calculate
                                             numerator
psnr_den = sum(sum((image-image_prime).^2)); % calculate
                                             denominator
A = psnr_num/psnr_den;                       % calculate
                                             PSNR
end
A = 10*log10(A);
return
```

D.13 Interpolation of Fusion Results of MR and CT Images

```
load dataxx xx% Result of curvelet fusion.
load dataf6 f6 % Result of wavelet fusion.
f33 = imread('2222.jpg');
f33 = f33(:,:,1);
f3i = double(f33)/255;
f11 = imread('1111.jpg');
f11 = f11(:,:,1);
f1i = double(f11)/255;
selector = input('Enter the type of interpolation 0 for
spline, 1 for Entropy, 2 for LMMSE, 3 for regularized');
```

```
switch selector
case 0
f6i = cubicspline_interp(f6);
xxi = cubicspline_interp(xx);
case 1
f6i = entropy_interp(f6);
xxi = entropy_interp(xx);
f6i = f6i';
xxi = xxi';
case 2
f6i = LMMSE_interp(f6);
xxi = LMMSE_interp(xx);
case 3
f6i = regularized_interp(f6);
xxi = regularized_interp(xx);
end;
[M,N] = size(f1i);
PSNR1w = psnr(f1i,f6i,M,N)
PSNR2w = psnr(f3i,f6i,M,N)
PSNR1c = psnr(f1i,xxi,M,N)
PSNR2c = psnr(f3i,xxi,M,N)
figure
imshow(f6i)
figure
imshow(xxi)
[r_CT_edge_c,r_MR_edge_c, r_CT_edge_w, r_MR_edge_w] =
similarity(f1i,f3i, f6i,xxi)
function c = cubicspline_interp(xx);
[M,N] = size(xx);
z = 1:2*N;
x = 1:N;
c = zeros(M,2*N);
for i = 1:M
c(i,:) = spline(x,xx(i,:),z/2);
end;
xx = c';
[M,N] = size(xx);
z = 1:2*N;
x = 1:N;
c = zeros(M,2*N);
for i = 1:M
c(i,:) = spline(x,xx(i,:),z/2);
end;
c = c';
function c = LMMSE_interp(xx);
g = xx;
[M,N] = size(g);
n_var = 0.001;
```

```
s = 0.5;
f = g;
[M,N] = size(f);
ff = zeros(M,N);
x = f(:,N-1:N);
x = rot90(x,2);
y = f(:,1);
f = [y,f,x];
for i = 1:M
for j = 2:N+1
ff(i,j-1) = f(i,j)*(1-s)+f(i,j+1)*s;%bilinear
end;
end;
ff = ff(:,1:N);
fff(1:M,1:2:2*N) = f(1:M,2:N+1);
fff(1:M,2:2:2*N) = ff(1:M,1:N);
f = fff';
clear ff,fff;
s = 0.5;
[M,N] = size(f);
ff = zeros(M,N);
x = f(:,N-1:N);
x = rot90(x,2);
y = f(:,1);
f = [y,f,x];
for i = 1:M
for j = 2:N+1
ff(i,j-1) = f(i,j)*(1-s)+f(i,j+1)*s;%bilinear
end;
end;
ff = ff(:,1:N);
fff(1:M,1:2:2*N) = f(1:M,2:N+1);
fff(1:M,2:2:2*N) = ff(1:M,1:N);
fff = fff';
wlength = 3;
[M1,M1] = size(fff);
fff(M1+wlength,M1+wlength) = 0;
for j = 0:M1-1
for k = 0:M1-1
sum = 0;
for n = 1:wlength
for m = 1:wlength
sum = sum+fff(n,m)*fff(n+j,m+k);
end;
end;
RRR(j+1,k+1) = 1/((wlength)^2)*sum;
end;
end;
```

```
[M M] = size(g);
N = 2*M;
R = 2;
kff = zeros(N,N);
kff = RRR';
kff = im2col(kff,[N N],'distinct');
kff = sparse(1:N^2,1:N^2,kff);
g = g';
g = im2col(g,[M M],'distinct');
I = speye(M^2)/12;
H1 = sparse(M,N);
counter = 1;
for i = 1:M
H1(i,counter) = 1;
H1(i,counter+1) = 1;
counter = counter+2;
end;
H1 = H1/2;
H = kron(H1,H1);
Hz1 = speye(M^2)*n_var;
HH = H*kff*H';
I = speye(size(HH));
Hopt = kff*H'*inv(HH+Hz1);
f = Hopt*g;
f = col2im(f,[N N],[N N],'distinct');
f = f';
f = (f> = 0).*f;
imshow(f)
c = f;
function c = entropy_interp(xx);
g = xx;
[M M] = size(g);
N = 2*M;
g = g';
g = im2col(g,[M M],'distinct');
I = speye(N^2);
H1 = sparse(M,N);
counter = 1;
for i = 1:M
H1(i,counter) = 1;
H1(i,counter+1) = 1;
counter = counter+2;
end;
H1 = H1/2;
H = kron(H1,H1);
HH = H'*H;
gama = 0.001;
Hopt = inv(HH+gama*I)*H';
```

```
f = Hopt*g;
f = col2im(f,[N N],[N N],'distinct');
c = f;
function c = regularized_interp(xx);
g = xx;
[M,N] = size(g);
M = 2*M;
N = 2*N;
landa = input('Enter the value of regularization parameter');
g = [rot90(g(:,1:4),2),g,rot90(g(:,M/2-8:M/2),2)];
g = [rot90(g(1:4,:),2);g;rot90(g(M/2-8:M/2,:),2)];
L1 = M;
L2 = N;
M = 24;
N = 12;
I = speye(M^2);
H1 = sparse(M/2,M);
counter = 1;
for i = 1:M/2
H1(i,counter) = 1;
H1(i,counter+1) = 1;
counter = counter+2;
end;
H1 = H1/2;
H = kron(H1,H1);
HH = H'*H;
beta = 0.125;
Q1 = sparse(M,M);
for i = 1:M
Q1(i,i) = -2;
end;
for i = 1:M-1
Q1(i,i+1) = 1;
end;
for i = 2:M
Q1(i,i-1) = 1;
end;
Q = kron(Q1,Q1);
QQ = Q'*Q;
L = inv(HH+landa*QQ);
for ii = 1:L1/8
for jj = 1:L2/8
f = g(4*ii+1-4:4*(ii+1)+4,4*jj+1-4:4*(jj+1)+4);
z = f;
y = im2col(z,[N N],'distinct');
x1 = L*H'*y;
x1 = col2im(x1,[M M],[M M],'distinct');
xx1(8*(ii-1)+1:8*(ii),8*(jj-1)+1:8*(jj)) = x1(9:16,9:16);
```

```
end;
end;
c = xx1;
```

D.14 Fusion of Satellite Images

D.14.1 DWT Fusion

```
t1 = cputime;
F1 = im2double(imread ('ca_pan.bmp','bmp'));
F = im2double(imread ('ca_ms.bmp','bmp'));
%HISTOGRAM MATCHING BETWEEN EVERY BAND IN RGB IMAGE AND THE
PAN IMAGE.
for i = 1:3
F2 = F(:,:,i);
m1 = mean2(F1); m2 = mean2(F2);
s1 = std2(F1); s2 = std2(F2);
g = s2/s1;
of = m2-g*m1;
FH = ((F1)*g+of);
%THE WAVELET TRANSFORM STEP FOR PAN IMAGE and THE MS BANDS
[cA1,cH1,cV1,cD1] = dwt2(FH,'db2');
[cA2,cH2,cV2,cD2] = dwt2(F2,'db2');
%SELECTION OF COEFFICIENT USING Linear combination BETWEEN
both approximations.
A = imlincomb(0.5,cA1,0.5,cA2);
%INVERSE WAVELET TRANSFORM TO GENERATE THE FUSED IMAGE.
Xsyn(:,:,i) = idwt2(A,cH1,cV1,cD1,'db2');
end
figure;imshow(Xsyn)
imwrite(Xsyn,'Xsy_dwt.png');
t2 = cputime;
t = t2-t1
```

D.14.2 DWFT Fusion

```
t1 = cputime;
F1 = im2double(imread ('ca_pan.bmp','bmp'));
F = im2double(imread ('ca_ms.bmp','bmp'));
%HISTOGRAM MATCHING BETWEEN EVERY BAND IN RGB IMAGE AND THE
PAN IMAGE.
for i = 1:3
F2 = F(:,:,i);
m1 = mean2(F1); m2 = mean2(F2);
s1 = std2(F1); s2 = std2(F2);
g = s2/s1;
of = m2-g*m1;
```

```
FH = ((F1)*g+of);
%THE WAVELET FRAME TRANSFORM STEP FOR BOTH IMAGES.
[a1,h1,v1,d1] = swt2(FH,1,'db2');
[a2,h2,v2,d2] = swt2(F2,1,'db2');
%SELECTION OF COEFFICIENT USING Linear combination BETWEEN THE
APPROXIAMTION
A = imlincomb(0.5,a1,0.5,a2);
%INVERSE FRAME WAVELET TRANSFORM TO GENERATE THE FUSED
IMAGE.
Z = iswt2(A,h1,v1,d1,'db2');
Xsyn(:,:,i) = Z;
end
figure;imshow(Xsyn)
imwrite(Xsyn,'Xsy_FWT.png');
t2 = cputime;
t = t2-t1
```

D.14.3 IHS Fusion

```
t1 = cputime;
F1 = im2double(imread ('ca_pan.bmp','bmp'));
F = im2double(imread ('ca_ms.bmp','bmp'));
F2 = rgb2hsv(F);%..........RGB to HSV transform
%Histogram Matching step— — — — — — — — — — — — — — — — —
— — — —
m1 = mean2(F1); m2 = mean2(F2(:,:,3));
s1 = std2(F1); s2 = std2(F2(:,:,3));
g = s2/s1;
of = m2-g*m1;
FH = ((F1)*g+of);
F2(:,:,3) = FH;%...............Intenisty band Replacement
Xsyn = hsv2rgb(F2);%.........HSV to RGB transform
figure;imshow(Xsyn)
imwrite(Xsyn,'Xsy_ihs.png');
t2 = cputime;
t = t2-t1;
```

D.14.4 IHS + DWT Fusion

```
t1 = cputime;
F1 = im2double(imread ('gray1.tif','tiff'));
F = im2double(imread ('rgb.bmp','bmp'));
%IHS and histogram matching
hsv_register = rgb2hsv(F);
F2 = hsv_register(:,:,3);
m1 = mean2(F1); m2 = mean2(F2);
s1 = std2(F1); s2 = std2(F2);
g = s2/s1;
```

```
of = m2-g*m1;
FH = ((F1)*g+of);
% wavelet frame transform for both intensity and pan_matched
image
[a1,h1,v1,d1] = swt2(FH,1,'db2');
[a2,h2,v2,d2] = swt2(F2,1,'db2');
%Linear combination between both of int. & pan
approximation.
L = imlincomb(0.5,a1,0.5,a2);
%inverse wavelet frame transform
Z = iswt2(L,h1,v1,d1,'db2');
%hsv to rgb image conversion
hsv_register(:,:,3) = Z;
Xsyn = hsv2rgb(hsv_register);
figure;imshow(Xsyn)
imwrite(Xsyn,'Xsy_ihsfwt.png');
t2 = cputime;
t = t2-t1
```

D.14.5 IHS + DWFT Fusion

```
t1 = cputime;
F1 = im2double(imread ('gray1.tif','tiff'));
F = im2double(imread ('rgb.bmp','bmp'));
% RGB TO IHS TRANSFORMATION
hsv_register = rgb2hsv(F);
% HISTOGRAM MATCHING BETWEEN THE PAN AND THE INTENISITY IMAGE.
F2 = hsv_register(:,:,3);
m1 = mean2(F1); m2 = mean2(F2);
s1 = std2(F1); s2 = std2(F2);
g = s2/s1;
of = m2-g*m1;
new_pan = ((F1)*g+of);
%WAVELET TRANSFORM FOR BOTH NEW_PAN AND INTENISTY IMAGES
[cA1,cH1,cV1,cD1] = dwt2(new_pan,'db2');
[cA2,cH2,cV2,cD2] = dwt2(hsv_register(:,:,3),'db2');
% COEFFICIENT SELECTION USING Linear combination between
approximations.
A = imlincomb(0.5,cA2,0.5,cA1);
%INVERSE WAVELET TRANSFORM TO GET NEW INTENISTY IMAGE.
new_int = idwt2(A,cH1,cV1,cD1,'db2');
%IHS TO RGB IMAGE BACK.
hsv_register(:,:,3) = new_int;
Xsyn = hsv2rgb(hsv_register);
figure;imshow(Xsyn)
imwrite(Xsyn,'Xsy_ihsdwt.png');
t2 = cputime;
t = t2-t1
```

D.14.6 DWT Fusion with Noise

```
t1 = cputime;
F1 = im2double(imread ('lon_spot_.bmp','bmp'));
F = im2double(imread ('lon_tm_.bmp','bmp'));
counter = 0;
for snr = -20:2:30
n1_var = var(F1(:))/(10^(snr/10));
F1_n = imnoise(F1,'gaussian',0,n1_var);
for i = 1:3
F2 = F(:,:,i);
n2_var = var(F2(:))/(10^(snr/10));
F2_n(:,:,i) = imnoise(F2,'gaussian',0,n2_var);
end
for l = 1:3
%Histogram Matching
F2 = F2_n(:,:,l);
m1 = mean2(F1_n); m2 = mean2(F2);
s1 = std2(F1_n); s2 = std2(F2);
g = s2/s1;
of = m2-g*m1;
FH = ((F1_n)*g+of);
%THE WAVELET TRANSFORM STEP FOR PAN IMAGE and THE MS BANDS
[cA1,cH1,cV1,cD1] = dwt2(FH,'db2');
[cA2,cH2,cV2,cD2] = dwt2(F2,'db2');
%SELECTION OF COEFFICIENT USING Linear combination BETWEEN
the approximations.
A = imlincomb(0.5,cA1,0.5,cA2);
%INVERSE FRAME WAVELET TRANSFORM TO GENERATE THE FUSED IMAGE.
Xsyn(:,:,l) = idwt2(A,cH1,cV1,cD1,'db2');
end
figure;imshow(Xsyn);
imwrite(Xsyn,'Xsy_25_dwt.png')
t2 = cputime;
t = t2-t1
%Discrepancy and spatial calculations
red = Xsyn(:,:,1);green = Xsyn(:,:,2);blue = Xsyn(:,:,3);
counter = counter+1;
%Discrepancy calculation Dr(counter) =
disc(red,F(:,:,1));Dg(counter) = disc(green,F(:,:,2));Db
(counter) = disc(blue,F(:,:,3));
% Spatial correlation using Laplacian filter
high = [-1 -1 -1;-1 8 -1;-1 -1 -1];%HPF FILTER
Fs = imfilter(F1,high,'conv','replicate');%APPLIED FILTER ON
SPOT
Fr = imfilter(red,high,'conv','replicate');%APPLIED FILTER
ON RED
Fg = imfilter(green,high,'conv','replicate');%APPLIED FILTER
ON GREEN
```

```
Fb = imfilter(blue,high,'conv','replicate');%APPLIED FILTER
    ON BLUE
c_red(counter) = corr2(Fs,Fr);c_green(counter) =
corr2(Fs,Fg);c_blue(counter) = corr2(Fs,Fb);
end
```

D.14.7 DWT Fusion with Noise and Denoising

```
t1 = cputime;
F1 = im2double(imread ('lon_spot_.bmp','bmp'));
F = im2double(imread ('lon_tm_.bmp','bmp'));
counter = 0;
for snr = 0:2:30
n1_var = var(F1(:))/(10^(snr/10));
F1_n = imnoise(F1,'gaussian',0,n1_var);
for i = 1:3
F2 = F(:,:,i);
n2_var = var(F2(:))/(10^(snr/10));
F2_n(:,:,i) = imnoise(F2,'gaussian',0,n2_var);
end
for l = 1:3
%Histogram Matching
F2 = F2_n(:,:,l);
m1 = mean2(F1_n); m2 = mean2(F2);
s1 = std2(F1_n); s2 = std2(F2);
g = s2/s1;
of = m2-g*m1;
FH = ((F1_n)*g+of);
%THE WAVELET TRANSFORM STEP FOR PAN IMAGE and THE MS BANDS
[cA1,cH1,cV1,cD1] = dwt2(FH,'db2');
[cA2,cH2,cV2,cD2] = dwt2(F2,'db2');
%SELECTION OF COEFFICIENT USING Linear combination BETWEEN
the approximations.
A = imlincomb(0.5,cA1,0.5,cA2);
%INVERSE FRAME WAVELET TRANSFORM TO GENERATE THE FUSED IMAGE.
Xsyn(:,:,l) = idwt2(A,cH1,cV1,cD1,'db2');
end
%denoising.
for l = 1:3
Xsyn_noise = Xsyn(:,:,l);
[a,h,v,d] = dwt2(Xsyn_noise,'db2');
maxim1 = max(max(abs(h)));
thr1 = 0.8*maxim1;
maxim2 = max(max(abs(v)));
thr2 = 0.8*maxim2;
maxim3 = max(max(abs(d)));
thr3 = 0.8*maxim3;
sorh = 's';
```

```
dwh = wthresh(h,sorh,thr1);
dwv = wthresh(v,sorh,thr2);
dwd = wthresh(d,sorh,thr3);
Xsynn(:,:,l) = idwt2(a,dwh,dwv,dwd,'db2');
end
figure;imshow(Xsynn)
imwrite(Xsynn,'fu_25dB.png');
t2 = cputime;
t = t2-t1
end
```

D.14.8 DWFT Fusion with Noise

```
t1 = cputime
F1 = im2double(imread ('lon_spot_.bmp','bmp'));
F = im2double(imread ('lon_tm_.bmp','bmp'));
counter = 0;
for snr = -20:2:30
n1_var = var(F1(:))/(10^(snr/10));
F1_n = imnoise(F1,'gaussian',0,n1_var);
for i = 1:3
F2 = F(:,:,i);
n2_var = var(F2(:))/(10^(snr/10));
F2_n(:,:,i) = imnoise(F2,'gaussian',0,n2_var);
end
for l = 1:3
%histogram matching
F2 = F2_n(:,:,l);
m1 = mean2(F1_n); m2 = mean2(F2);
s1 = std2(F1_n); s2 = std2(F2);
g = s2/s1;
of = m2-g*m1;
FH = ((F1_n)*g+of);
%THE WAVELET FRAME TRANSFORM STEP FOR BOTH IMAGES.
[a1,h1,v1,d1] = swt2(FH,1,'db2');
[a2,h2,v2,d2] = swt2(F2,1,'db2');
%SELECTION OF COEFFICIENT USING Linear combination BETWEEN
THE APPROXIAMTIONS
A = imlincomb(0.5,a1,0.5,a2);
%INVERSE FRAME WAVELET TRANSFORM TO GENERATE THE FUSED IMAGE.
Z = iswt2(A,h1,v1,d1,'db2');
Xsyn(:,:,l) = Z;
end
figure;imshow(Xsyn);
imwrite(Xsyn,'Xsy_25_fwt.png')
t2 = cputime;
t = t2-t1
%Discrepancy and spatial calculations
```

```
red = Xsyn(:,:,1);green = Xsyn(:,:,2);blue = Xsyn(:,:,3);
counter = counter+1;
%Discrepancy calculation
Dr(counter) = disc(red,F(:,:,1));Dg(counter) = disc(green,
F(:,:,2));Db(counter) = disc(blue,F(:,:,3));
high = [-1 -1 -1;-1 8 -1;-1 -1 -1];%HPF FILTER
Fs = imfilter(F1,high,'conv','replicate');%APPLIED FILTER ON
SPOT
Fr = imfilter(red,high,'conv','replicate');%APPLIED FILTER
ON RED
Fg = imfilter(green,high,'conv','replicate');%APPLIED FILTER
ON GREEN
Fb = imfilter(blue,high,'conv','replicate');%APPLIED FILTER ON
BLUE   c_red(counter) = corr2(Fs,Fr);c_green(counter) =
corr2(Fs,Fg);c_blue(counter) = corr2(Fs,Fb);
end
```

D.14.9 DWFT Fusion with Noise and Denoising

```
t1 = cputime;
F1 = im2double(imread ('lon_spot_.bmp','bmp'));
F = im2double(imread ('lon_tm_.bmp','bmp'));
counter = 0;
snr = 25;
n1_var = var(F1(:))/(10^(snr/10));
F1_n = imnoise(F1,'gaussian',0,n1_var);
for i = 1:3
F2 = F(:,:,i);
n2_var = var(F2(:))/(10^(snr/10));
F2_n(:,:,i) = imnoise(F2,'gaussian',0,n2_var);
end
for l = 1:3
%histogram matching
F2 = F2_n(:,:,l);
m1 = mean2(F1_n); m2 = mean2(F2);
s1 = std2(F1_n); s2 = std2(F2);
g = s2/s1;
of = m2-g*m1;
FH = ((F1_n)*g+of);
%THE WAVELET FRAME TRANSFORM STEP FOR BOTH IMAGES.
[a1,h1,v1,d1] = swt2(FH,1,'db2');
[a2,h2,v2,d2] = swt2(F2,1,'db2');
%SELECTION OF COEFFICIENT USING Linear combination BETWEEN
THE APPROXIAMTIONS
A = imlincomb(0.5,a1,0.5,a2);
%INVERSE FRAME WAVELET TRANSFORM TO GENERATE THE FUSED IMAGE.
Z = iswt2(A,h1,v1,d1,'db2');
Xsyn(:,:,l) = Z;
```

```
end
%denoising.
for l = 1:3
Xsyn_noise = Xsyn(:,:,l);
[a,h,v,d] = swt2(Xsyn_noise,1,'db2');
maxim1 = max(max(abs(h)));
thr1 = 0.8*maxim1;
maxim2 = max(max(abs(v)));
thr2 = 0.8*maxim2;
maxim3 = max(max(abs(d)));
thr3 = 0.8*maxim3;
sorh = 's';
swh = wthresh(h,sorh,thr1);
swv = wthresh(v,sorh,thr2);
swd = wthresh(d,sorh,thr3);
Xsynn(:,:,l) = iswt2(a,swh,swv,swd,'db2');
end
figure;imshow(Xsynn)
imwrite(Xsynn,'C:\MATLAB7\work\paper2_noise_work\london\FWT\
fu_25dB.png');
t2 = cputime;
t = t2-t1
```

D.14.10 IHS Fusion with Noise

```
t1 = cputime;
F1 = im2double(imread ('lon_spot_.bmp','bmp'));
F = im2double(imread ('lon_tm_.bmp','bmp'));
counter = 0;
for snr = -20:2:30
n1_var = var(F1(:))/(10^(snr/10));
F1_n = imnoise(F1,'gaussian',0,n1_var);
for i = 1:3
F2 = F(:,:,i);
n2_var = var(F2(:))/(10^(snr/10));
F2_n(:,:,i) = imnoise(F2,'gaussian',0,n2_var);
end
hsv = rgb2hsv(F2_n);%RGB to HSV
F2 = hsv(:,:,3);
%histogram matching
m1 = mean2(F1_n); m2 = mean2(F2);
s1 = std2(F1_n); s2 = std2(F2);
g = s2/s1;
of = m2-g*m1;
new_pan = ((F1_n)*g+of);
hsv(:,:,3) = new_pan;% replacement
Xsyn = hsv2rgb(hsv);% HSV to RGB
figure;imshow(Xsyn);
```

```
imwrite(Xsyn,'C:\MATLAB7\work\paper2_noise_work\Xsy_25_ihs.
png')
figure;imshow(Xsyn);
t2 = cputime;
t = t2-t1
figure;imshow(F2_n);
figure;imshow(F1_n);
figure;imshow(Xsyn);
%Discrepancy and spatial calculations
red = Xsyn(:,:,1);green = Xsyn(:,:,2);blue = Xsyn(:,:,3);
counter = counter+1;
%Discrepancy calculation
Dr(counter) = disc(red,F(:,:,1));Dg(counter) = disc(green,
F(:,:,2));Db(counter) = disc(blue,F(:,:,3));
% Spatial correlation using Laplacian filter
high = [-1 -1 -1;-1 8 -1;-1 -1 -1];%HPF FILTER
Fs = imfilter(F1,high,'conv','replicate');%APPLIED FILTER ON
SPOT
Fr = imfilter(red,high,'conv','replicate');%APPLIED FILTER
ON RED
Fg = imfilter(green,high,'conv','replicate');%APPLIED FILTER
ON GREEN
Fb = imfilter(blue,high,'conv','replicate');%APPLIED FILTER
ON BLUE
c_red(counter) = corr2(Fs,Fr);c_green(counter) =
corr2(Fs,Fg);c_blue(counter) = corr2(Fs,Fb);
end
```

D.14.11 IHS Fusion with Noise and Denoising

```
t1 = cputime;
F1 = im2double(imread ('lon_spot_.bmp','bmp'));
F = im2double(imread ('lon_tm_.bmp','bmp'));
counter = 0;
for snr = 0:2:30
n1_var = var(F1(:))/(10^(snr/10));
F1_n = imnoise(F1,'gaussian',0,n1_var);
for i = 1:3
F2 = F(:,:,i);
n2_var = var(F2(:))/(10^(snr/10));
F2_n(:,:,i) = imnoise(F2,'gaussian',0,n2_var);
end
hsv = rgb2hsv(F2_n);%RGB to HSV
F2 = hsv(:,:,3);
m1 = mean2(F1_n); m2 = mean2(F2);
s1 = std2(F1_n); s2 = std2(F2);
g = s2/s1;
of = m2-g*m1;
```

```
new_pan = ((F1_n)*g+of);
hsv(:,:,3) = new_pan;% replacement
Xsyn = hsv2rgb(hsv);% HSV to RGB
for l = 1:3
Xsyn_noise = Xsyn(:,:,l);
[a,h,v,d] = dwt2(Xsyn_noise,'db2');
maxim1 = max(max(abs(h)));
thr1 = 0.8*maxim1;
maxim2 = max(max(abs(v)));
thr2 = 0.8*maxim2;
maxim3 = max(max(abs(d)));
thr3 = 0.8*maxim3;
sorh = 's';
dwh = wthresh(h,sorh,thr1);
dwv = wthresh(v,sorh,thr2);
dwd = wthresh(d,sorh,thr3);
Xsynn(:,:,l) = idwt2(a,dwh,dwv,dwd,'db2');
end
figure;imshow(Xsynn)
imwrite(Xsynn,'fu_25dB.png');
t2 = cputime;
t = t2-t1
%Discrepancy and spatial calculations
red_n = Xsyn(:,:,1);green_n = Xsyn(:,:,2);blue_n =
Xsyn(:,:,3);
red_d = Xsynn(:,:,1);green_d = Xsynn(:,:,2);blue_d =
Xsynn(:,:,3);
counter = counter+1;
%Discrepancy calculation
Dr_n(counter) = disc(red_n,F(:,:,1));Dg_n(counter) = disc
(green_n,F(:,:,2));Db_n(counter) = disc(blue_n,F(:,:,3));
Dr_d(counter) = disc(red_d,F(:,:,1));Dg_d(counter) = disc
(green_d,F(:,:,2));Db_d(counter) = disc(blue_d,F(:,:,3));
% Spatial correlation using Laplacian filter
high = [-1 -1 -1;-1 8 -1;-1 -1 -1];%HPF FILTER
Fs = imfilter(F1,high,'conv','replicate');%APPLIED FILTER ON
SPOT
Fr_n = imfilter(red_n,high,'conv','replicate');%APPLIED FILTER
ON RED
Fg_n = imfilter(green_n,high,'conv','replicate');%APPLIED
FILTER ON GREEN
Fb_n = imfilter(blue_n,high,'conv','replicate');%APPLIED
FILTER ON BLUE
Fr_d = imfilter(red_d,high,'conv','replicate');%APPLIED FILTER
ON RED
Fg_d = imfilter(green_d,high,'conv','replicate');%APPLIED
FILTER ON GREEN
```

```
Fb_d = imfilter(blue_d,high,'conv','replicate');%APPLIED FILTER
ON BLUE c_red_n(counter) = corr2(Fs,Fr_n);c_green_n(counter) =
corr2(Fs,Fg_n);c_blue_n(counter) = corr2(Fs,Fb_n);
c_red_d(counter) = corr2(Fs,Fr_d);c_green_d(counter) =
corr2(Fs,Fg_d);c_blue_d(counter) = corr2(Fs,Fb_d);
end
n = 0:2:30;
figure;imshow(Xsyn)
plot(n,Dr_n,'k-.');
hold on
plot(n,Dr_d,'k-*');
xlabel('SNR in (db)');
ylabel('Discrepancy');
title('Discrepancy of the red band ')
legend('Noisy','De-noised','Location','NorthEast')
figure;imshow(Xsyn)
plot(n,Dg_n,'k-.');
hold on
plot(n,Dg_d,'k-*');
xlabel('SNR in (db)');
ylabel('Discrepancy');
title('Discrepancy of the green band ')
legend('Noisy','De-noised','Location','NorthEast')
figure;imshow(Xsyn)
plot(n,Db_n,'k-.');
hold on
plot(n,Db_d,'k-*');
xlabel('SNR in (db)');
ylabel('Discrepancy');
title('Discrepancy of the blue band ')
legend('Noisy','De-noised','Location','NorthEast')
figure; imshow(Xsyn)
plot(n,c_red_n,'k-.');
hold on
plot(n,c_red_d,'k-*');
title('Correlation coefficient for the red band ');
xlabel('SNR in (db)');
ylabel('Correlation Coefficient');
legend('Noisy','De-noised','Location','northwest')
figure;imshow(Xsyn)
plot(n,c_green_n,'k-.');
hold on
plot(n,c_green_d,'k-*');
xlabel('SNR in (db)');
ylabel('Correlation Coefficient');
title('Correlation coefficient for the green band ')
legend('Noisy','De-noised','Location','northwest')
figure;imshow(Xsyn)
```

```
plot(n,c_blue_n,'k-.');
hold on
plot(n,c_blue_d,'k-*');
xlabel('SNR in (db)');
ylabel('Correlation Coefficient');
title('Correlation coefficient for the blue band ')
legend('Noisy','De-noised','Location','northwest')
```

D.14.12 IHS + DWT Fusion with Noise

```
t1 = cputime;
F1 = im2double(imread ('lon_spot_.bmp','bmp'));
F = im2double(imread ('lon_tm_.bmp','bmp'));
counter = 0;
for snr = -20:2:30
n1_var = var(F1(:))/(10^(snr/10));
F1_n = imnoise(F1,'gaussian',0,n1_var);
for i = 1:3
F2 = F(:,:,i);
n2_var = var(F2(:))/(10^(snr/10));
F2_n(:,:,i) = imnoise(F2,'gaussian',0,n2_var);
end
% RGB TO IHS TRANSFORMATION
hsv = rgb2hsv(F2_n);
% HISTOGRAM MATCHING BETWEEN THE PAN AND THE INTENISITY IMAGE.
F2 = hsv(:,:,3);
m1 = mean2(F1_n); m2 = mean2(F2);
s1 = std2(F1_n); s2 = std2(F2);
g = s2/s1;
of = m2-g*m1;
new_pan = ((F1_n)*g+of);
%WAVELET TRANSFORM FOR BOTH NEW_PAN AND INTENISTY IMAGES
[cA1,cH1,cV1,cD1] = dwt2(new_pan,'db2');
[cA2,cH2,cV2,cD2] = dwt2(F2,'db2');
% COEFFICIENT SELECTION
A = imlincomb(0.5,cA2,0.5,cA1);
%-INVERSE WAVELET TRANSFORM TO GET NEW INTENISTY IMAGE.
new_int = idwt2(A,cH1,cV1,cD1,'db2');
% IHS TO RGB IMAGE BACK.
hsv(:,:,3) = new_int;
Xsyn = hsv2rgb(hsv);
figure;imshow(Xsyn);
imwrite(Xsyn,'Xsy_25_ihswt.png')
t2 = cputime;
t = t2-t1
%Discrepancy and spatial calculations
red = Xsyn(:,:,1);green = Xsyn(:,:,2);blue = Xsyn(:,:,3);
counter = counter+1;
```

```
%Discrepancy calculation
Dr(counter) = disc(red,F(:,:,1));Dg(counter) = disc(green,
F(:,:,2));Db(counter) = disc(blue,F(:,:,3));
% Spatial correlation using Laplacian filter
high = [-1 -1 -1;-1 8 -1;-1 -1 -1];%HPF FILTER
Fs = imfilter(F1,high,'conv','replicate');%APPLIED FILTER ON
SPOT
Fr = imfilter(red,high,'conv','replicate');%APPLIED FILTER
ON RED
Fg = imfilter(green,high,'conv','replicate');%APPLIED FILTER
ON GREEN
Fb = imfilter(blue,high,'conv','replicate');%APPLIED FILTER
ON BLUE
c_red(counter) = corr2(Fs,Fr);c_green(counter) =
corr2(Fs,Fg);c_blue(counter) = corr2(Fs,Fb);
end
n = -20:2:30;
figure;imshow(Xsyn)
plot(n,Dr,'k.');
xlabel('SNR in (db)');
ylabel('Discrepancy');
title('Discrepancy of the red band between the fused image and
the orginal MS image ')
legend('IHSWT','Location','NorthEast')
figure;imshow(Xsyn)
plot(n,Dg,'k.');
xlabel('SNR in (db)');
ylabel('Discrepancy');
title('Discrepancy of the green band between the fused image
and the orginal MS image ')
legend('IHSWT','Location','NorthEast')
figure;imshow(Xsyn)
plot(n,Db,'k.');
xlabel('SNR in (db)');
ylabel('Discrepancy');
title('Discrepancy of the blue band between the fused image
and the orginal MS image ')
legend('IHSWT','Location','NorthEast')
figure; imshow(Xsyn)
plot(n,c_red,'k.');
title('Correlation coefficient for the red band between the
fused image and the orginal PAN image ');
xlabel('SNR in (db)');
ylabel('Correlation Coefficient');
legend('IHSWT','Location','southeast')
figure;imshow(Xsyn)
plot(n,c_green,'k.');
xlabel('SNR in (db)');
```

```
ylabel('Correlation Coefficient');
title('Correlation coefficient for the green band between the
fused image and the orginal PAN image ')
legend('IHSWT','Location','southeast')
figure;imshow(Xsyn)
plot(n,c_blue,'k.');
xlabel('SNR in (db)');
ylabel('Correlation Coefficient');
title('Correlation coefficient for the blue band between the
fused image and the orginal PAN image')
legend('IHSWT','Location','southeast')
```

D.14.13 IHS + DWT Fusion with Noise and Denoising

```
t1 = cputime;
F1 = im2double(imread ('lon_spot_.bmp','bmp'));
F = im2double(imread ('lon_tm_.bmp','bmp'));
counter = 0;
for snr = 0:2:30
n1_var = var(F1(:))/(10^(snr/10));
F1_n = imnoise(F1,'gaussian',0,n1_var);
for i = 1:3
F2 = F(:,:,i);
n2_var = var(F2(:))/(10^(snr/10));
F2_n(:,:,i) = imnoise(F2,'gaussian',0,n2_var);
end
% RGB TO IHS TRANSFORMATION
hsv = rgb2hsv(F2_n);
% HISTOGRAM MATCHING BETWEEN THE PAN AND THE INTENISITY
IMAGE.
F2 = hsv(:,:,3);
m1 = mean2(F1_n); m2 = mean2(F2);
s1 = std2(F1_n); s2 = std2(F2);
g = s2/s1;
of = m2-g*m1;
new_pan = ((F1_n)*g+of);
%WAVELET TRANSFORM FOR BOTH NEW_PAN AND INTENISTY IMAGES
[cA1,cH1,cV1,cD1] = dwt2(new_pan,'db2');
[cA2,cH2,cV2,cD2] = dwt2(F2,'db2');
% COEFFICIENT SELECTION
A = imlincomb(0.5,cA2,0.5,cA1);
%INVERSE WAVELET TRANSFORM TO GET NEW INTENISTY IMAGE.
new_int = idwt2(A,cH1,cV1,cD1,'db2');
% IHS TO RGB IMAGE BACK.
hsv(:,:,3) = new_int;
Xsyn = hsv2rgb(hsv);
%denoising.
for l = 1:3
```

```
Xsyn_noise = Xsyn(:,:,1);
[a,h,v,d] = dwt2(Xsyn_noise,'db2');
maxim1 = max(max(abs(h)));
thr1 = 0.8*maxim1;
maxim2 = max(max(abs(v)));
thr2 = 0.8*maxim2;
maxim3 = max(max(abs(d)));
thr3 = 0.8*maxim3;
sorh = 's';
dwh = wthresh(h,sorh,thr1);
dwv = wthresh(v,sorh,thr2);
dwd = wthresh(d,sorh,thr3);
Xsynn(:,:,1) = idwt2(a,dwh,dwv,dwd,'db2');
end
figure;imshow(Xsynn)
imwrite(Xsynn,'fu_25dB.png');
t2 = cputime;
t = t2-t1
```

D.14.14 IHS + DWFT Fusion with Noise

```
t1 = cputime;
F1 = im2double(imread ('lon_spot_.bmp','bmp'));
F = im2double(imread ('lon_tm_.bmp','bmp'));
counter = 0;
for snr = -20:2:30
n1_var = var(F1(:))/(10^(snr/10));
F1_n = imnoise(F1,'gaussian',0,n1_var);
for i = 1:3
F2 = F(:,:,i);
n2_var = var(F2(:))/(10^(snr/10));
F2_n(:,:,i) = imnoise(F2,'gaussian',0,n2_var);
end
%IHS and histogram matching
hsv_register = rgb2hsv(F2_n);
F2 = hsv_register(:,:,3);
m1 = mean2(F1_n); m2 = mean2(F2);
s1 = std2(F1_n); s2 = std2(F2);
g = s2/s1;
of = m2-g*m1;
FH = ((F1_n)*g+of);
% wavelet frame transform for both intensity and pan_matched
image
[a1,h1,v1,d1] = swt2(FH,1,'db2');
[a2,h2,v2,d2] = swt2(F2,1,'db2');
%Coefficient selection between both of int. & pan.
L = imlincomb(0.5,a1,0.5,a2);
%inverse frame wavelet transform
```

```
Z = iswt2(L,h1,v1,d1,'db2');
%hsv to rgb image conversion
hsv_register(:,:,3) = Z;
Xsyn = hsv2rgb(hsv_register);
figure;imshow(Xsyn);
imwrite(Xsyn,'Xsy_25_ihsfw.png')
t2 = cputime;
t = t2-t1;
%Discrepancy and spatial calculations
red = Xsyn(:,:,1);green = Xsyn(:,:,2);blue = Xsyn(:,:,3);
counter = counter+1;
%Discrepancy calculation
Dr(counter) = disc(red,F(:,:,1));Dg(counter) = disc(green,
F(:,:,2));Db(counter) = disc(blue,F(:,:,3));
% Spatial correlation using Laplacian filter
high = [-1 -1 -1;-1 8 -1;-1 -1 -1];%HPF FILTER
Fs = imfilter(F1,high,'conv','replicate');%APPLIED FILTER
ON SPOT
Fr = imfilter(red,high,'conv','replicate');%APPLIED FILTER
ON RED
Fg = imfilter(green,high,'conv','replicate');%APPLIED FILTER
ON GREEN
Fb = imfilter(blue,high,'conv','replicate');%APPLIED FILTER
ON BLUE
c_red(counter) = corr2(Fs,Fr);c_green(counter) =
corr2(Fs,Fg);c_blue(counter) = corr2(Fs,Fb);
end
```

D.14.15 IHS + DWFT Fusion with Noise and Denoising

```
t1 = cputime;
F1 = im2double(imread ('lon_spot_.bmp','bmp'));
F = im2double(imread ('lon_tm_.bmp','bmp'));
counter = 0;
% Input Images with noise for different SNR
for snr = 0:2:30
n1_var = var(F1(:))/(10^(snr/10));
F1_n = imnoise(F1,'gaussian',0,n1_var);
for i = 1:3
F2 = F(:,:,i);
n2_var = var(F2(:))/(10^(snr/10));
F2_n(:,:,i) = imnoise(F2,'gaussian',0,n2_var);
end
%IHS and histogram matching
hsv_register = rgb2hsv(F2_n);
F2 = hsv_register(:,:,3);
m1 = mean2(F1_n); m2 = mean2(F2);
s1 = std2(F1_n); s2 = std2(F2);
```

```
g = s2/s1;
of = m2-g*m1;
FH = ((F1_n)*g+of);
%frame wavelet transform for both intensity and pan_matched
image
[a1,h1,v1,d1] = swt2(FH,1,'db2');
[a2,h2,v2,d2] = swt2(F2,1,'db2');
%Coefficient selection between both of int. & pan.
L = imlincomb(0.5,a1,0.5,a2);
%inverse frame wavelet transform
Z = iswt2(L,h1,v1,d1,'db2');
%hsv to rgb image conversion
hsv_register(:,:,3) = Z;
Xsyn = hsv2rgb(hsv_register);
for l = 1:3
Xsyn_noise = Xsyn(:,:,l);
[a,h,v,d] = swt2(Xsyn_noise,1,'db2');
maxim1 = max(max(abs(h)));
thr1 = 0.8*maxim1;
maxim2 = max(max(abs(v)));
thr2 = 0.8*maxim2;
maxim3 = max(max(abs(d)));
thr3 = 0.8*maxim3;
sorh = 's';
swh = wthresh(h,sorh,thr1);
swv = wthresh(v,sorh,thr2);
swd = wthresh(d,sorh,thr3);
Xsynn(:,:,l) = iswt2(a,swh,swv,swd,'db2');
end
figure;imshow(Xsynn)
imwrite(Xsynn,'fu_25dB.png');
t2 = cputime;
t = t2-t1
```

D.15 Single-Channel LMMSE Restoration

```
f = imread('cameraman.tif');
f = double(f)/255;
l = f(1:30,1:30);
M = 256;
nhood = [7 7];
h = ones(nhood);
h = h/49;
h(M,M) = 0;
H = fft2(h);
F = fft2(f);
```

```
G = F.*H;
g = ifft2(G);
sigma = (std2(1)^2/10^5)^(1/2);
g = imnoise(g,'gaussian',0, sigma^2);
G = fft2(g);
wlength = 12;
g(M+wlength,M+wlength) = 0;
f(M+wlength,M+wlength) = 0;
for j = 0:M-1
for k = 0:M-1
sum = 0;
for n = 1:wlength
for m = 1:wlength
sum = sum+g(n,m)*g(n+j,m+k);
end;
end;
R (j+1,k+1) = 1/((wlength)^2)*sum;
end;
end;
Pf = fft2(R);
I = linspace(1, M^2,M^2);
J = I;
H = H.';
H = im2col(H,[M M],'distinct');
Pf = Pf.';
Pf = im2col(Pf,[M M],'distinct');
G = G.';
G = im2col(G,[M M],'distinct');
Diag1 = sparse(I,J,Pf);
Diag2 = sparse(I,J,H);
Diag3 = conj(Diag2);
Diag4 = sigma^2*speye(M^2);
A = Diag2*Diag1;
A = A*Diag3;
A = A+Diag4;
B = inv(A);
C = Diag1*Diag3;
C = C*B;
FF = C*G;
FF = col2im(FF,[M M],[M M],'distinct');
ff = ifft2(FF);
ff = ff';
ff = real(ff);
imshow(ff)
% End of the single-channel LMMSE restoration.
```

D.16 Multi-Channel LMMSE Restoration

```
f = imread('cameraman.tif');
f = double(f)/255;
l = f(1:30,1:30);
M = 256;
nhood = [5 5];
h1 = ones(nhood);
h1 = h1/25;
h1(M,M) = 0;
H1 = fft2(h1);
F = fft2(f);
G1 = F.*H1;
g1 = ifft2(G1);
sigma1 = (std2(l)^2/10^3)^(1/2);
g1 = imnoise(g1,'gaussian',0, sigma1^2);
nhood = [7 7];
h2 = ones(nhood);
h2 = h2/49;
h2(M,M) = 0;
H2 = fft2(h2);
G2 = F.*H2;
g2 = ifft2(G2);
sigma2 = (std2(l)^2/10^4)^(1/2);
g2 = imnoise(g2,'gaussian',0, sigma2^2);
nhood = [9 9];
h3 = ones(nhood);
h3 = h3/81;
h3(M,M) = 0;
H3 = fft2(h3);
G3 = F.*H3;
g3 = ifft2(G3);
sigma3 = (std2(l)^2/10^5)^(1/2);
g3 = imnoise(g3,'gaussian',0, sigma3^2);
g1 = real(g1);
g2 = real(g2);
g3 = real(g3);
G1 = G1.';
G2 = G2.';
G3 = G3.';
G1 = im2col(G1,[M M],'distinct');
G2 = im2col(G2,[M M],'distinct');
G3 = im2col(G3,[M M],'distinct');
G = [G1;G2;G3];
wlength = 3;
g1(M+wlength,M+wlength) = 0;
g2(M+wlength,M+wlength) = 0;
g3(M+wlength,M+wlength) = 0;
```

```
x = 0+eps*i
Diagonals = repmat(x,M^2,18);
for counter = 1:9
% Begin of formation of correlation matrices.
switch counter          .
case 1
for j = 0:M-1
for k = 0:M-1
sum = 0;
for n = 1:wlength
for m = 1:wlength
sum = sum+g1(n,m)*g1(n+j,m+k);
end;
end;
R (j+1,k+1) = 1/((wlength)^2)*sum;
end;
end;
case 2
for j = 0:M-1
for k = 0:M-1
sum = 0;
for n = 1:wlength
for m = 1:wlength
sum = sum+g2(n,m)*g1(n+j,m+k);
end;
end;
R (j+1,k+1) = 1/((wlength)^2)*sum;
end;
end;
case 3
for j = 0:M-1
for k = 0:M-1
sum = 0;
for n = 1:wlength
for m = 1:wlength
sum = sum+g3(n,m)*g1(n+j,m+k);
end;
end;
R (j+1,k+1) = 1/((wlength)^2)*sum;
end;
end;
case 4
for j = 0:M-1
for k = 0:M-1
sum = 0;
for n = 1:wlength
for m = 1:wlength
sum = sum+g1(n,m)*g2(n+j,m+k);
```

```
end;
end;
R (j+1,k+1) = 1/((wlength)^2)*sum;
end;
end;
case 5
for j = 0:M-1
for k = 0:M-1
sum = 0;
for n = 1:wlength
for m = 1:wlength
sum = sum+g2(n,m)*g2(n+j,m+k);
end;
end;
R (j+1,k+1) = 1/((wlength)^2)*sum;
end;
end;
case 6
for j = 0:M-1
for k = 0:M-1
sum = 0;
for n = 1:wlength
for m = 1:wlength
sum = sum+g3(n,m)*g2(n+j,m+k);
end;
end;
R (j+1,k+1) = 1/((wlength)^2)*sum;
end;
end;
case 7
for j = 0:M-1
for k = 0:M-1
sum = 0;
for n = 1:wlength
for m = 1:wlength
sum = sum+g1(n,m)*g3(n+j,m+k);
end;
end;
R (j+1,k+1) = 1/((wlength)^2)*sum;
end;
end;
case 8
for j = 0:M-1
for k = 0:M-1
sum = 0;
for n = 1:wlength
for m = 1:wlength
sum = sum+g2(n,m)*g3(n+j,m+k);
```

```
end;
end;
R (j+1,k+1) = 1/((wlength)^2)*sum;
end;
end;
case 9
for j = 0:M-1
for k = 0:M-1
sum = 0;
for n = 1:wlength
for m = 1:wlength
sum = sum+g3(n,m)*g3(n+j,m+k);
end;
end;
R (j+1,k+1) = 1/((wlength)^2)*sum;
end;
end;
end;% End of switch.
Pf = fft2(R);
Pf = Pf.';
column = im2col(Pf,[M M],'distinct');
Diagonals(:,counter) = column;
end;
for counter = 1:3
switch counter
case 1
H = H1;
case 2
H = H2;
case 3
H = H3;
end;
H = H.';
column = im2col(H,[M M],'distinct');
Diagonals(:,9+counter) = column;
Diagonals(:,12+counter) = conj(column);
end;
Diagonals(:,16) = sigma1^2;
Diagonals(:,17) = sigma2^2;
Diagonals(:,18) = sigma3^2;
I = linspace(1, M^2,M^2);
J = I;
A1 = sparse(I,J,Diagonals(:,1));
A2 = sparse(I,J+M^2,Diagonals(:,2));
A3 = sparse(I,J+2*M^2,Diagonals(:,3));
A4 = sparse(I+M^2,J,Diagonals(:,4));
A5 = sparse(I+M^2,J+M^2,Diagonals(:,5));
A6 = sparse(I+M^2,J+2*M^2,Diagonals(:,6));
```

```
A7 = sparse(I+2*M^2,J,Diagonals(:,7));
A8 = sparse(I+2*M^2,J+M^2,Diagonals(:,8));
A9 = sparse(I+2*M^2,J+2*M^2,Diagonals(:,9));
A1(3*M^2,3*M^2) = 0;
A2(3*M^2,3*M^2) = 0;
A3(3*M^2,3*M^2) = 0;
A4(3*M^2,3*M^2) = 0;
A5(3*M^2,3*M^2) = 0;
A6(3*M^2,3*M^2) = 0;
A7(3*M^2,3*M^2) = 0;
A8(3*M^2,3*M^2) = 0;
A = A1+A2+A3+A4+A5+A6+A7+A8+A9;
A1 = sparse(I,J,Diagonals(:,13));
A2 = sparse(I+M^2,J+M^2,Diagonals(:,14));
A3 = sparse(I+2*M^2,J+2*M^2,Diagonals(:,15));
A1(3*M^2,3*M^2) = 0;
A2(3*M^2,3*M^2) = 0;
B = A1+A2+A3;
column = Diagonals(:,10).* Diagonals(:,1);
column = column.* Diagonals(:,13);
column = column+ Diagonals(:,16);
Diagonals(:,1) = column;
column = Diagonals(:,10).* Diagonals(:,2);
column = column.* Diagonals(:,14);
Diagonals(:,2) = column;
column = Diagonals(:,10).* Diagonals(:,3);
column = column.* Diagonals(:,15);
Diagonals(:,3) = column;
column = Diagonals(:,11).* Diagonals(:,4);
column = column.* Diagonals(:,13);
Diagonals(:,4) = column;
column = Diagonals(:,11).* Diagonals(:,5);
column = column.* Diagonals(:,14);
column = column+ Diagonals(:,17);
Diagonals(:,5) = column;
column = Diagonals(:,11).* Diagonals(:,6);
column = column.* Diagonals(:,15);
Diagonals(:,6) = column;
column = Diagonals(:,12).* Diagonals(:,7);
column = column.* Diagonals(:,13);
Diagonals(:,7) = column;
column = Diagonals(:,12).* Diagonals(:,8);
column = column.* Diagonals(:,14);
Diagonals(:,8) = column;
column = Diagonals(:,12).* Diagonals(:,9);
column = column.* Diagonals(:,15);
column = column+ Diagonals(:,18);
Diagonals(:,9) = column;
```

```
A1 = sparse(I,J,Diagonals(:,1));
A2 = sparse(I,J+M^2,Diagonals(:,2));
A3 = sparse(I,J+2*M^2,Diagonals(:,3));
A4 = sparse(I+M^2,J,Diagonals(:,4));
A5 = sparse(I+M^2,J+M^2,Diagonals(:,5));
A6 = sparse(I+M^2,J+2*M^2,Diagonals(:,6));
A7 = sparse(I+2*M^2,J,Diagonals(:,7));
A8 = sparse(I+2*M^2,J+M^2,Diagonals(:,8)) ;
A9 = sparse(I+2*M^2,J+2*M^2,Diagonals(:,9));
A1(3*M^2,3*M^2) = 0;
A2(3*M^2,3*M^2) = 0;
A3(3*M^2,3*M^2) = 0;
A4(3*M^2,3*M^2) = 0;
A5(3*M^2,3*M^2) = 0;
A6(3*M^2,3*M^2) = 0;
A7(3*M^2,3*M^2) = 0;
A8(3*M^2,3*M^2) = 0;
C = A1+A2+A3+A4+A5+A6+A7+A8+A9;
C = inv(C);
A = A*B;
C = C*G;
F = A*C;
F1 = F(1:M^2,1);
F2 = F(M^2+1:2*M^2,1);
F3 = F(2*M^2+1:3*M^2,1);
F1 = col2im(F1,[M M],[M M],'distinct');
F1 = F1.';
F2 = col2im(F2,[M M],[M M],'distinct');
F2 = F2.';
F3 = col2im(F3,[M M],[M M],'distinct');
F3 = F3.';
ff1 = ifft2(F1);
ff2 = ifft2(F2);
ff3 = ifft2(F3);
ff1 = real(ff1);
ff2 = real(ff2);
ff3 = real(ff3);
imshow(ff1)
figure
imshow(ff2)
figure
imshow(ff3)
```

D.17 Inverse Filter Restoration

```
f = imread('cameraman.tif');
f = double(f)/255;
```

```
l = f(1:30,1:30);
M = 256;
nhood = [7 7];
h = ones(nhood);
h = h/49;
h(M,M) = 0;
H = fft2(h);
F = fft2(f);
G = F.*H;
g = ifft2(G);
g = real(g);
sigma = (std2(l)^2/10^5)^(1/2);
g = imnoise(g,'gaussian',0, sigma^2);
G = fft2(g);
G = G.';
G = im2col(G,[M M],'distinct');
x = 0+eps*i
Diagonals = repmat(x,M^2,2);
column = im2col(H,[M M],'distinct');
Diagonals(:,1) = column;
Diagonals(:,2) = conj(column);
I = linspace(1, M^2,M^2);
J = I;
A1 = sparse(I,J,Diagonals(:,1));
A2 = sparse(I,J,Diagonals(:,2));
A = A2*A1;
A = inv(A);
A = A*A2;
F = A*G;
F = col2im(F,[M M],[M M],'distinct');
ff = ifft2(F);
ff = real(ff);
ff = ff';
imshow(ff)
```

D.18 Adaptive Single-Channel Regularized Restoration

```
f = imread('cameraman.tif');
f = double(f)/255;
l = f(1:30,1:30);
M = 256;
nhood = [7 7];
h = ones(nhood);
h = h/49;
h(M,M) = 0;
H = fft2(h);
F = fft2(f);
```

```
G = F.*H;
g = ifft2(G);
g = real(g);
sigma = (std2(l)^2/10^2)^(1/2);
g = imnoise(g,'gaussian',0, sigma^2);
% Estimate the local mean of g.
localMean = filter2(ones(nhood), g)/prod(nhood);
% Estimate of the local variance of g.
localVar = filter2(ones(nhood), g.^2)/prod(nhood)
- localMean.^2;
BSNR = (std2(l)/sigma)^2;
sigmag = std2(g);
k = 120;
constant = (1/k)*log10(BSNR+1)*sigmag
r = constant*log10(localVar+1);
G = fft2(g);
G = G.';
G = im2col(G,[M M],'distinct');
x = 0+eps*i
Diagonals = repmat(x,M^2,4);
column = im2col(H,[M M],'distinct');
Diagonals(:,1) = column;
Diagonals(:,2) = conj(column);
q = [eps -1 eps
-1 4 -1
eps -1 eps];
q = q/4;
q(M,M) = 0;
Q = fft2(q);
column = im2col(Q,[M M],'distinct');
Diagonals(:,3) = column;
Diagonals(:,4) = conj(column);
I = linspace(1, M^2,M^2);
J = I;
A1 = sparse(I,J,Diagonals(:,1));
A2 = sparse(I,J,Diagonals(:,2));
B1 = sparse(I,J,Diagonals(:,3));
B2 = sparse(I,J,Diagonals(:,4));
localVarP = zeros(M,M);
for counter = 1:10
counter
landa = counter/10000;
A = A2*A1;
B = B2*B1;
A = A+landa*B;
A = inv(A);
A = A*A2;
F = A*G;
```

```
F = col2im(F,[M M],[M M],'distinct');
ff = ifft2(F);
ff = real(ff);
ff = ff';
if counter = =1
ft = ff;
end;
for i = 2:M
for j = 1:M
d1(i,j) = ff(i,j)-ff(i-1,j);
end;
end;
d1(1,:) = ff(1,:);
for i = 1:M
for j = 2:M
d2(i,j) = ff(i,j)-ff(i,j-1);
end;
end;
d2(:,1) = ff(:,1);
M1 = d1.^2;
M2 = d2.^2;
MM = M1+M2;
MM = MM.^(1/2);
% Estimate the local mean of g.
localMeanM = filter2(ones(nhood), MM)/prod(nhood);
% Estimate of the local variance of g.
localVarM = filter2(ones(nhood), MM.^2)/prod(nhood)
- localMean.^2;
for i = 1:M
for j = 1:M
if localVarM(i,j) < r(i,j)
if localVarM(i,j) > localVarP(i,j)
ft(i,j) = ff(i,j);
end;
end;
end;
end;
localVarP = localVarM;
end;
imshow(ft)
```

D.19 Multi-Channel Regularized Restoration

```
f = imread('cameraman.tif');
f = double(f)/255;
l = f(1:30,1:30);
M = 128;
```

```
f = f(1:128,1:128);
nhood = [5 5];
h11 = ones(nhood);
h11 = h11/25;
h11(M,M) = 0;
H11 = fft2(h11);
F = fft2(f);
G1 = F.*H11;
g1 = ifft2(G1);
sigma1 = (std2(l)^2/10^3)^(1/2);
g1 = imnoise(g1,'gaussian',0, sigma1^2);
nhood = [7 7];
h22 = ones(nhood);
h22 = h22/49;
h22(M,M) = 0;
H22 = fft2(h22);
G2 = F.*H22;
g2 = ifft2(G2);
sigma2 = (std2(l)^2/10^4)^(1/2);
g2 = imnoise(g2,'gaussian',0, sigma2^2);
nhood = [9 9];
h33 = ones(nhood);
h33 = h33/81;
h33(M,M) = 0;
H33 = fft2(h33);
G3 = F.*H33;
g3 = ifft2(G3);
sigma3 = (std2(l)^2/10^5)^(1/2);
g3 = imnoise(g3,'gaussian',0, sigma3^2);
g1 = real(g1);
g2 = real(g2);
g3 = real(g3);
G1 = fft2(g1);
G2 = fft2(g2);
G3 = fft2(g3);
G1 = G1.';
G2 = G2.';
G3 = G3.';
G1 = im2col(G1,[M M],'distinct');
G2 = im2col(G2,[M M],'distinct');
G3 = im2col(G3,[M M],'distinct');
G = [G1;G2;G3];
g1 = g1';
g2 = g2';
g3 = g3';
g1 = im2col(g1,[M M],'distinct');
g2 = im2col(g2,[M M],'distinct');
g3 = im2col(g3,[M M],'distinct');
```

```
g = [g1;g2;g3];
x = 0
HDiagonals = repmat(x,M^2,9);
QDiagonals = repmat(x,M^2,9);
Laplacianselector = 2;
switch Laplacianselector
case 1
q = [0 1 0
1 -4 1
0 1 0];
q = q/4;
case 2
q = [0 1 0
1 -6 1
0 1 0];
q = q/6;
q1 = [0 0 0
0 1/6 0
0 0 0];
q1(M,M) = 0;
Q1 = fft2(q1);
end;
q(M,M) = 0;
Q = fft2(q);
H11 = H11.';
column = im2col(H11,[M M],'distinct');
HDiagonals(:,1) = column;
H22 = H22.';
column = im2col(H22,[M M],'distinct');
HDiagonals(:,5) = column;
H33 = H33.';
column = im2col(H33,[M M],'distinct');
HDiagonals(:,9) = column;
switch Laplacianselector
case 1
Q = Q.';
column = im2col(Q,[M M],'distinct');
QDiagonals(:,1) = column;
QDiagonals(:,5) = column;
QDiagonals(:,9) = column;
case 2
Q = Q.';
column = im2col(Q,[M M],'distinct');
QDiagonals(:,1) = column;
QDiagonals(:,5) = column;
QDiagonals(:,9) = column;
Q1 = Q1.';
column = im2col(Q1,[M M],'distinct');
```

```matlab
QDiagonals(:,2) = column;
QDiagonals(:,3) = column;
QDiagonals(:,4) = column;
QDiagonals(:,6) = column;
QDiagonals(:,7) = column;
QDiagonals(:,8) = column;
end;
% Computation of the Jacobi.
I = linspace(1, M^2,M^2);
J = I;
one = ones(M^2,1);
% 5 zeros.
x1 =.0001;
x2 =.0001;
x3 =.0001;
X = [x1;x2;x3];
% Begin of computation of fifth multiplicand:
switch Laplacianselector
case 1
A1 = sparse(I,J,QDiagonals(:,1));
%A2 = sparse(I,J+M^2,QDiagonals(:,2));
%A3 = sparse(I,J+2*M^2,QDiagonals(:,3));
%A4 = sparse(I+M^2,J,QDiagonals(:,4));
A5 = sparse(I+M^2,J+M^2,QDiagonals(:,5));
%A6 = sparse(I+M^2,J+2*M^2,QDiagonals(:,6));
%A7 = sparse(I+2*M^2,J,QDiagonals(:,7));
%A8 = sparse(I+2*M^2,J+M^2,QDiagonals(:,8));
A9 = sparse(I+2*M^2,J+2*M^2,QDiagonals(:,9));
A1(3*M^2,3*M^2) = 0;
%A2(3*M^2,3*M^2) = 0;
%A3(3*M^2,3*M^2) = 0;
%A4(3*M^2,3*M^2) = 0;
A5(3*M^2,3*M^2) = 0;
%A6(3*M^2,3*M^2) = 0;
%A7(3*M^2,3*M^2) = 0;
%A8(3*M^2,3*M^2) = 0;
%Q = A1+A2+A3+A4+A5+A6+A7+A8+A9;
Q = A1+A5+A9;
A1 = sparse(I,J,conj(QDiagonals(:,1)));
%A2 = sparse(I,J+M^2,conj(QDiagonals(:,2)));
%A3 = sparse(I,J+2*M^2,conj(QDiagonals(:,3)));
%A4 = sparse(I+M^2,J,conj(QDiagonals(:,4)));
A5 = sparse(I+M^2,J+M^2,conj(QDiagonals(:,5)));
%A6 = sparse(I+M^2,J+2*M^2,conj(QDiagonals(:,6)));
%A7 = sparse(I+2*M^2,J,conj(QDiagonals(:,7)));
%A8 = sparse(I+2*M^2,J+M^2,conj(QDiagonals(:,8)));
A9 = sparse(I+2*M^2,J+2*M^2,conj(QDiagonals(:,9)));
A1(3*M^2,3*M^2) = 0;
```

```
%A2(3*M^2,3*M^2) = 0;
%A3(3*M^2,3*M^2) = 0;
%A4(3*M^2,3*M^2) = 0;
A5(3*M^2,3*M^2) = 0;
%A6(3*M^2,3*M^2) = 0;
%A7(3*M^2,3*M^2) = 0;
%A8(3*M^2,3*M^2) = 0;
%QQ = A1+A2+A3+A4+A5+A6+A7+A8+A9;
QQ = A1+A5+A9;
case 2
A1 = sparse(I,J,QDiagonals(:,1));
A2 = sparse(I,J+M^2,QDiagonals(:,4));
A3 = sparse(I,J+2*M^2,QDiagonals(:,7));
A4 = sparse(I+M^2,J,QDiagonals(:,2));
A5 = sparse(I+M^2,J+M^2,QDiagonals(:,5));
A6 = sparse(I+M^2,J+2*M^2,QDiagonals(:,8));
A7 = sparse(I+2*M^2,J,QDiagonals(:,3));
A8 = sparse(I+2*M^2,J+M^2,QDiagonals(:,6));
A9 = sparse(I+2*M^2,J+2*M^2,QDiagonals(:,9));
A1(3*M^2,3*M^2) = 0;
A2(3*M^2,3*M^2) = 0;
A3(3*M^2,3*M^2) = 0;
A4(3*M^2,3*M^2) = 0;
A5(3*M^2,3*M^2) = 0;
A6(3*M^2,3*M^2) = 0;
A7(3*M^2,3*M^2) = 0;
A8(3*M^2,3*M^2) = 0;
Q = A1+A2+A3+A4+A5+A6+A7+A8+A9;
%Q = A1+A5+A9;
A1 = sparse(I,J,conj(QDiagonals(:,1)));
A2 = sparse(I,J+M^2,conj(QDiagonals(:,4)));
A3 = sparse(I,J+2*M^2,conj(QDiagonals(:,7)));
A4 = sparse(I+M^2,J,conj(QDiagonals(:,2)));
A5 = sparse(I+M^2,J+M^2,conj(QDiagonals(:,5)));
A6 = sparse(I+M^2,J+2*M^2,conj(QDiagonals(:,8)));
A7 = sparse(I+2*M^2,J,conj(QDiagonals(:,3)));
A8 = sparse(I+2*M^2,J+M^2,conj(QDiagonals(:,6)));
A9 = sparse(I+2*M^2,J+2*M^2,conj(QDiagonals(:,9)));
A1(3*M^2,3*M^2) = 0;
A2(3*M^2,3*M^2) = 0;
A3(3*M^2,3*M^2) = 0;
A4(3*M^2,3*M^2) = 0;
A5(3*M^2,3*M^2) = 0;
A6(3*M^2,3*M^2) = 0;
A7(3*M^2,3*M^2) = 0;
A8(3*M^2,3*M^2) = 0;
QQ = A1+A2+A3+A4+A5+A6+A7+A8+A9;
%QQ = A1+A5+A9;
```

```
end;
mult5 = QQ*Q;
% End of computation of fifth multiplicand.
%Begin of computation of third multiplicand:
A1 = sparse(I,J,HDiagonals(:,1));
%A2 = sparse(I,J+M^2,HDiagonals(:,2));
%A3 = sparse(I,J+2*M^2,HDiagonals(:,3));
%A4 = sparse(I+M^2,J,HDiagonals(:,4));
A5 = sparse(I+M^2,J+M^2,HDiagonals(:,5));
%A6 = sparse(I+M^2,J+2*M^2,HDiagonals(:,6));
%A7 = sparse(I+2*M^2,J,HDiagonals(:,7));
%A8 = sparse(I+2*M^2,J+M^2,HDiagonals(:,8));
A9 = sparse(I+2*M^2,J+2*M^2,HDiagonals(:,9));
A1(3*M^2,3*M^2) = 0;
%A2(3*M^2,3*M^2) = 0;
%A3(3*M^2,3*M^2) = 0;
%A4(3*M^2,3*M^2) = 0;
A5(3*M^2,3*M^2) = 0;
%A6(3*M^2,3*M^2) = 0;
%A7(3*M^2,3*M^2) = 0;
%A8(3*M^2,3*M^2) = 0;
%H = A1+A2+A3+A4+A5+A6+A7+A8+A9;
H = A1+A5+A9;
A1 = sparse(I,J,conj(HDiagonals(:,1)));
%A2 = sparse(I,J+M^2,conj(HDiagonals(:,2)));
%A3 = sparse(I,J+2*M^2,conj(HDiagonals(:,3)));
%A4 = sparse(I+M^2,J,conj(HDiagonals(:,4)));
A5 = sparse(I+M^2,J+M^2,conj(HDiagonals(:,5)));
%A6 = sparse(I+M^2,J+2*M^2,conj(HDiagonals(:,6)));
%A7 = sparse(I+2*M^2,J,conj(HDiagonals(:,7)));
%A8 = sparse(I+2*M^2,J+M^2,conj(HDiagonals(:,8)));
A9 = sparse(I+2*M^2,J+2*M^2,conj(HDiagonals(:,9)));
A1(3*M^2,3*M^2) = 0;
%A2(3*M^2,3*M^2) = 0;
%A3(3*M^2,3*M^2) = 0;
%A4(3*M^2,3*M^2) = 0;
A5(3*M^2,3*M^2) = 0;
%A6(3*M^2,3*M^2) = 0;
%A7(3*M^2,3*M^2) = 0;
%A8(3*M^2,3*M^2) = 0;
%HH = A1+A2+A3+A4+A5+A6+A7+A8+A9;
HH = A1+A5+A9;
counter = 0;
while counter< = 10
mult3 = HH*H;
A1 = sparse(I,J,one*x1);
A2 = sparse(I+M^2,J+M^2,one*x2);
A3 = sparse(I+2*M^2,J+2*M^2,one*x3);
```

```
A1(3*M^2,3*M^2) = 0;
A2(3*M^2,3*M^2) = 0;
Landa = A1+A2+A3;
mult3 = mult3+Landa*mult5;
mult3 = inv(mult3);
% End of computation of third multiplicand.
% Begin of computation of sixth multiplicand;
mult6 = mult3*HH*G;
% End of computation of sixth multiplicand.
for i = 1:3
for j = 1:3
% Begin of computation of forth multiplicand.
switch j
case 1
mult4 = sparse(I,J,one);
mult4(3*M^2,3*M^2) = 0;
case 2
mult4 = sparse(I+M^2,J+M^2,one);
mult4(3*M^2,3*M^2) = 0;
case 3
mult4 = sparse(I+2*M^2,J+2*M^2,one);
end;
% End of computation of forth multiplicand.
% Begin of computation of second multiplicand:
switch i
case 1
mult2 = sparse(I,J,HDiagonals(:,1));
mult2(M^2,3*M^2) = 0;
case 2
mult2 = sparse(I,J+M^2,HDiagonals(:,5));
mult2(M^2,3*M^2) = 0;
case 3
mult2 = sparse(I,J+2*M^2,HDiagonals(:,9));
end;
% End of computation of the second multiplicand.
mult = mult2*mult3;
mult = mult*mult4;
mult = mult*mult5;
mult = mult*mult6;
mult = col2im(mult,[M M],[M M],'distinct');
mult = mult.';
mult = ifft2(mult);
mult = real(mult);
mult = mult.';
mult = im2col(mult,[M M],'distinct');
% Begin of computation of first multiplicand:
switch i
case 1
```

```
F1 = mult6(1:M^2);
case 2
F1 = mult6(M^2+1:2*M^2);
case 3
F1 = mult6(2*M^2+1:3*M^2);
end;
F1 = col2im(F1,[M M],[M M],'distinct');
F1 = F1.';
switch i
case 1
Gii = F1.*H11;
case 2
Gii = F1.*H22;
case 3
Gii = F1.*H33;
end;
gii = ifft2(Gii);
switch i
case 1
gi = col2im(g1,[M M],[M M],'distinct');
case 2
gi = col2im(g2,[M M],[M M],'distinct');
case 3
gi = col2im(g3,[M M],[M M],'distinct');
end;
gi = gi.';
mult1 = gii-gi;
mult1 = mult1';
mult1 = im2col(mult1,[M M],'distinct');
mult1 = real(mult1);
mult1 = mult1.';
switch i
case 1
i;
Z1 = (norm(mult1,'fro'))^2-M^2*sigma1^2;
case 2
i;
Z2 = (norm(mult1,'fro'))^2-M^2*sigma2^2;
case 3
i;
Z3 = (norm(mult1,'fro'))^2-M^2*sigma3^2;
end;
% End of computation of first multiplicand:
mult = -2*mult1*mult;
Jacobian(i,j) = mult;
end;
end;
Jacobian = inv(Jacobian);
```

```
Z = [Z1;Z2;Z3];
Y = -Jacobian*Z;
X = X+Y;
x1 = X(1);
x2 = X(2);
x3 = X(3);
counter = counter+1;
end;
F1 = mult6(1:M^2);
F1 = col2im(F1,[M M],[M M],'distinct');
F1 = F1.';
f1 = ifft2(F1);
F2 = mult6(M^2+1:2*M^2);
F2 = col2im(F2,[M M],[M M],'distinct');
F2 = F2.';
f2 = ifft2(F2);
F3 = mult6(2*M^2+1:3*M^2);
F3 = col2im(F3,[M M],[M M],'distinct');
F3 = F3.';
f3 = ifft2(F3);
f1 = real(f1);
f2 = real(f2);
f3 = real(f3);
imshow(f1)
figure
imshow(f2)
figure
imshow(f3)
```

D.20 Iterative Restoration of Color Image

```
load trees
[r g b] = ind2rgb(X,map);
[N L] = size(r);
r = r(1:N,1:N);
nhood = [3 3];
h = ones(nhood);
h = h/9;
rr = filter2(h,r);
r = rr;
i = zeros(N,1,N);
for l = 1:N
i(:,:,l) = r(:,l);
end;
PSFlength = 3;
H = zeros(N,N,PSFlength);
for ll = 1:PSFlength
```

```
for l = 1 :N-2
H(l:l+2,l,ll) = h(1:3,ll);
end
end;
o = i;
alpha = 0.05
D = zeros(N,N);
for k = 1:3
for l = 1:N
D(l,l) = H(1,1,1);
end;
D = inv(D);
o(:,:,1) = o(:,:,1)+alpha*D*(i(:,:,1)-H(:,:,1)*o(:,:,1));
o(:,:,2) = o(:,:,2)+alpha*D*(i(:,:,2)-H(:,:,1)*o(:,:,2)-
H(:,:,2)*o(:,:,1));
for l = PSFlength:N
o(:,:,l) = o(:,:,l)+alpha*D*(i(:,:,l)-H(:,:,1)*o(:,:,l)-
H(:,:,2)*o(:,:,l-1)- H(:,:,2)*o(:,:,l-2));
end;
end;
oo = zeros(N,N);
for l = 1:N
oo(:,l) = o(:,1,l);
end;
imshow(oo)
```

D.21 Evaluation of Regularization Parameter

```
%Method to choose the global regularization parameter for
Gaussian noise.
% Single Channel regularized restoration.
f = imread('cameraman.tif');
f = double(f)/255;
l = f(1:30,1:30);
M = 256;
nhood = [7 7];
h = ones(nhood);
h = h/49;
h(M,M) = 0;
H = fft2(h);
F = fft2(f);
G = F.*H;
g = ifft2(G);
sigma = (std2(l)^2/10^5)^(1/2);
g = imnoise(g,'gaussian',0, sigma^2);
G = fft2(g);
HH = conj(H);
```

```
l = [0 1 0
1 -4 1
0 1 0];
l = l/4;
l(M,M) = 0;
L = fft2(l);
LL = conj(L);
Num = G.*HH;
Denum1 = HH.*H;
Denum2 = LL.*L;
Alpha = 0.0000001;
for counter = 0:19
Alpha = Alpha+counter*0.0000001;
Denum = Denum1+Alpha*Denum2;
F = Num./Denum;
ff = ifft2(F);
ff = real(ff);
% End of single-channel regularized restoration.
wlength = 3;
ff(M+wlength,M+wlength) = 0;
for j = 0:M-1
for k = 0:M-1
sum = 0;
for n = 1:wlength
for m = 1:wlength
sum = sum+ff(n,m)*ff(n+j,m+k);
end;
end;
R (j+1,k+1) = 1/((wlength)^2)*sum;
end;
end;
Pf = fft2(R);
Pf = abs(Pf);
Pf = Pf.^2;
summation = 0;
for counter1 = 1:M
for counter2 = 1:M
summation = summation+Pf(counter1,counter2);
end;
end;
summation
MSE = 0;
for m = 1:M
for n = 1:M
MSE = MSE+(f(m,n)-ff(m,n))^2;
end;
end;
MSE = MSE/M^2;
```

```
MSE = MSE^.5
A(counter+1) = Alpha;
B(counter+1) = summation;
C(counter+1) = MSE;
end;
% End of the method to choose the global regularization
parameter.
```

D.22 Greatest Common Divisor Restoration

```
%GCD for Sandy image.
% GCD between 2-observations:
f = imread('c33.bmp');
f = f(:,:,1);
f = double(f)/255;
[M1 N1] = size(f);
h1 = ones([5 5]);
h1 = h1/25;
h2 = [0.041 0.033 0.038 0.042 0.04
0.04 0.038 0.035 0.041 0.046
0.039 0.04 0.035 0.043 0.049
0.037 0.038 0.032 0.042 0.043
0.04 0.05 0.035 0.035 0.04];
[M2 N2] = size(h1)
M = M1+M2-1;
N = N1+N2-1;
f(M,N) = 0;
h1 (M,N) = 0;
h2(M,N) = 0;
F = fft2(f);
H1 = fft2(h1);
F1 = H1.*F;
f1 = ifft2(F1);
f1 = real(f1);
H2 = fft2(h2);
F2 = H2.*F;
f2 = ifft2(F2);
f2 = real(f2);
% Noise addition statements.
l = f(1:20,70:90);
sigma = std2(l)/(100000)^(1/2);
f1 = imnoise(f1,'gaussian',0,sigma^2);
f2 = imnoise(f2,'gaussian',0,sigma^2);
% End of noise addition statements.
r = M1-1;
A = zeros(M,N);
ff1 = fft(f1);
```

```
ff2 = fft(f2);
ff1 = ff1.';
ff2 = ff2.';
ff1 = flipud(ff1);
ff2 = flipud(ff2);
ff1 = ff1.';
ff2 = ff2.';
for row = 1:M
row
R = ff1(row,:);
R(1,2*M-r-1) = 0;
C = R(1,1);
C(M-r,1) = 0;
S01 = toeplitz(C,R);
R = ff2(row,:);
R(1,2*M-r-1) = 0;
C = R(1,1);
C(M-r,1) = 0;
S02 = toeplitz(C,R);
S02 = flipud(S02);
S0 = [S01;S02];
[U,X,V] = svd (S0,0);
[u1 u2] = size(U);
Dividend = ff2(row,:);
Divisor = U(1:N-r,u2);
Divisor = Divisor';
Divisor(1,N) = 0;
Z1 = fft(Dividend);
Z2 = fft(Divisor);
Q = Z1./Z2;
Q = ifft(Q);
Q = Q(1:r+1);
Q = rot90(Q,2);
A(row,1:r+1) = Q;
end;
A = A.';
A = fft(A);
A = A.';
f1 = f1';
f2 = f2';
r = M1-1;
B = zeros(M,N);
ff1 = fft(f1);
ff2 = fft(f2);
ff1 = ff1.';
ff2 = ff2.';
ff1 = flipud(ff1);
ff2 = flipud(ff2);
```

```
ff1 = ff1.';
ff2 = ff2.';
for row = 1:M
row
R = ff1(row,:);
R(1,2*M-r-1) = 0;
C = R(1,1);
C(M-r,1) = 0;
S01 = toeplitz(C,R);
R = ff2(row,:);
R(1,2*M-r-1) = 0;
C = R(1,1);
C(M-r,1) = 0;
S02 = toeplitz(C,R);
S02 = flipud(S02);
S0 = [S01;S02];
[U,X,V] = svd (S0,0);
[u1 u2] = size(U);
Dividend = ff2(row,:);
Divisor = U(1:N-r,u2);
Divisor = Divisor';
Divisor(1,N) = 0;
Z1 = fft(Dividend);
Z2 = fft(Divisor);
Q = Z1./Z2;
Q = ifft(Q);
Q = Q(1:r+1);
Q = rot90(Q,2);
B(row,1:r+1) = Q;
end;
B = B.';
B = fft(B);
Gamma = spalloc(N*M,M+N,2*N*M);
for I = 0:M-1
Gamma(I*N+1:I*N+N,I+1) = A(I+1,:).';
I
end;
for I = 0:M-1
k = 0;
for j = M:M+N-1
Gamma(I*N+1+k,j+1) = -B (I+1,j-M+1).';
k = k+1;
end;
I
end;
Gamma1 = Gamma.'*Gamma;
[V,D,FLAG] = EIGS(Gamma1,1,'SM');
a = V(1:M,1);
```

```
b = V(M+1:N+M,1);
P1 = zeros(M,M);
P2 = zeros(M,M);
for I = 1:M
P1(I,:) = A(I,:)*a(I);
end;
for j = 1:N
P2(:,j) = B(:,j)*b(j);
end;
P = (1/2)*(P1+P2);
PP = ifft2(P);
t2 = max(f);
t2 = max(t2);
t1 = max(PP);
t1 = max(t1);
PP = t2*PP/t1;
PP = real(PP);
% End of GCD between 2- observations.
% GCD between 3-observations:
f = imread('c33.bmp');
f = f(:,:,1);
f = double(f)/255;
[M1 N1] = size(f);
h1 = ones([5 5]);
h1 = h1/25;
h2 = [0.038 0.033 0.034 0.042 0.04
0.04 0.038 0.035 0.045 0.046
0.039 0.04 0.035 0.043 0.049
0.037 0.038 0.032 0.042 0.043
0.04 0.05 0.035 0.035 0.04];
h3 = [0.044 0.043 0.035 0.0392 0.042
0.042 0.037 0.037 0.042 0.043
0.041 0.029 0.037 0.049 0.045
0.039 0.037 0.036 0.041 0.041
0.041 0.044 0.039 0.032 0.038];
[M2 N2] = size(h1)
M = M1+M2-1;
N = N1+N2-1;
f(M,N) = 0;
h1 (M,N) = 0;
h2(M,N) = 0;
h3(M,N) = 0;
F = fft2(f);
H1 = fft2(h1);
F1 = H1.*F;
f1 = ifft2(F1);
f1 = real(f1);
H2 = fft2(h2);
```

```
F2 = H2.*F;
f2 = ifft2(F2);
f2 = real(f2);
H3 = fft2(h3);
F3 = H3.*F;
f3 = ifft2(F3);
f3 = real(f3);
% Noise addition statements.
l = f(1:20,70:90);
sigma = std2(l)/(100000)^(1/2);
f1 = imnoise(f1,'gaussian',0,sigma^2);
f2 = imnoise(f2,'gaussian',0,sigma^2);
f3 = imnoise(f3,'gaussian',0,sigma^2);
im1 = f1;
im2 = f2;
im3 = f3;
% End of noise addition statements.
for counter = 1:3
switch counter
case 1
f1 = im1;
f2 = im2;
case 2
f1 = im1;
f2 = im3;
case 3
f1 = im2;
f2 = im3;
end;
r = M1-1;
A = zeros(M,N);
ff1 = fft(f1);
ff2 = fft(f2);
ff1 = ff1.';
ff2 = ff2.';
ff1 = flipud(ff1);
ff2 = flipud(ff2);
ff1 = ff1.';
ff2 = ff2.';
for row = 1:M
row
R = ff1(row,:);
R(1,2*M-r-1) = 0;
C = R(1,1);
C(M-r,1) = 0;
S01 = toeplitz(C,R);
R = ff2(row,:);
R(1,2*M-r-1) = 0;
```

```
C = R(1,1);
C(M-r,1) = 0;
S02 = toeplitz(C,R);
S02 = flipud(S02);
S0 = [S01;S02];
[U,X,V] = svd (S0,0);
[u1 u2] = size(U);
Dividend = ff2(row,:);
Divisor = U(1:N-r,u2);
Divisor = Divisor';
Divisor(1,N) = 0;
Z1 = fft(Dividend);
Z2 = fft(Divisor);
Q = Z1./Z2;
Q = ifft(Q);
Q = Q(1:r+1);
Q = rot90(Q,2);
A(row,1:r+1) = Q;
end;
A = A.';
A = fft(A);
A = A.';
f1 = f1';
f2 = f2';
r = M1-1;
B = zeros(M,N);
ff1 = fft(f1);
ff2 = fft(f2);
ff1 = ff1.';
ff2 = ff2.';
ff1 = flipud(ff1);
ff2 = flipud(ff2);
ff1 = ff1.';
ff2 = ff2.';
for row = 1:M
row
R = ff1(row,:);
R(1,2*M-r-1) = 0;
C = R(1,1);
C(M-r,1) = 0;
S01 = toeplitz(C,R);
R = ff2(row,:);
R(1,2*M-r-1) = 0;
C = R(1,1);
C(M-r,1) = 0;
S02 = toeplitz(C,R);
S02 = flipud(S02);
S0 = [S01;S02];
```

```
[U,X,V] = svd (S0,0);
[u1 u2] = size(U);
Dividend = ff2(row,:);
Divisor = U(1:N-r,u2);
Divisor = Divisor';
Divisor(1,N) = 0;
Z1 = fft(Dividend);
Z2 = fft(Divisor);
Q = Z1./Z2;
Q = ifft(Q);
Q = Q(1:r+1);
Q = rot90(Q,2);
B(row,1:r+1) = Q;
end;
B = B.';
B = fft(B);
Gamma = spalloc(N*M,M+N,2*N*M);
for I = 0:M-1
Gamma(I*N+1:I*N+N,I+1) = A(I+1,:).';
I
end;
for I = 0:M-1
k = 0;
for j = M:M+N-1
Gamma(I*N+1+k,j+1) = -B (I+1,j-M+1).';
k = k+1;
end;
I
end;
Gamma1 = Gamma.'*Gamma;
[V,D,FLAG] = EIGS(Gamma1,1,'SM');
a = V(1:M,1);
b = V(M+1:N+M,1);
P1 = zeros(M,M);
P2 = zeros(M,M);
for I = 1:M
P1(I,:) = A(I,:)*a(I);
end;
for j = 1:N
P2(:,j) = B(:,j)*b(j);
end;
P = (1/2)*(P1+P2);
PP = ifft2(P);
t2 = max(f);
t2 = max(t2);
t1 = max(PP);
t1 = max(t1);
PP = t2*PP/t1;
```

```
PP = real(PP);
switch counter
case 1
PP1 = PP;
case 2
PP2 = PP;
case 3
PP3 = PP;
end;
end;
% End of GCD between 3-observations.
% GCD between 3-observations using the suggested algorithm.:
f = imread('c33.bmp');
f = f(:,:,1);
f = double(f)/255;
[M1 N1] = size(f);
h1 = ones([5 5]);
h1 = h1/25;
h2 = [0.038 0.033 0.034 0.042 0.04
0.04 0.038 0.035 0.045 0.046
0.039 0.04 0.035 0.043 0.049
0.037 0.038 0.032 0.042 0.043
0.04 0.05 0.035 0.035 0.04];
h3 = [0.044 0.043 0.035 0.0392 0.042
0.042 0.037 0.037 0.042 0.043
0.041 0.029 0.037 0.049 0.045
0.039 0.037 0.036 0.041 0.041
0.041 0.044 0.039 0.032 0.038];
[M2 N2] = size(h1)
M = M1+M2-1;
N = N1+N2-1;
f(M,N) = 0;
h1 (M,N) = 0;
h2(M,N) = 0;
h3(M,N) = 0;
F = fft2(f);
H1 = fft2(h1);
F1 = H1.*F;
f1 = ifft2(F1);
f1 = real(f1);
H2 = fft2(h2);
F2 = H2.*F;
f2 = ifft2(F2);
f2 = real(f2);
H3 = fft2(h3);
F3 = H3.*F;
f3 = ifft2(F3);
f3 = real(f3);
```

```
% Noise addition statements.
l = f(1:20,70:90);
sigma = std2(l)/(500000)^(1/2);
f1 = imnoise(f1,'gaussian',0,sigma^2);
f2 = imnoise(f2,'gaussian',0,sigma^2);
f3 = imnoise(f3,'gaussian',0,sigma^2);
im1 = f1;
im2 = f2;
im3 = f3;
% End of noise addition statements.
for counter = 1:3
switch counter
case 1
f1 = im1;
f2 = (im1+im2+im3)/3;
case 2
f1 = im2;
f2 = (im1+im2+im3)/3;
case 3
f1 = im3;
f2 = (im1+im2+im3)/3;
end;
r = M1-1;
A = zeros(M,N);
ff1 = fft(f1);
ff2 = fft(f2);
ff1 = ff1.';
ff2 = ff2.';
ff1 = flipud(ff1);
ff2 = flipud(ff2);
ff1 = ff1.';
ff2 = ff2.';
for row = 1:M
row
R = ff1(row,:);
R(1,2*M-r-1) = 0;
C = R(1,1);
C(M-r,1) = 0;
S01 = toeplitz(C,R);
R = ff2(row,:);
R(1,2*M-r-1) = 0;
C = R(1,1);
C(M-r,1) = 0;
S02 = toeplitz(C,R);
S02 = flipud(S02);
S0 = [S01;S02];
[U,X,V] = svd (S0,0);
[u1 u2] = size(U);
```

```
Dividend = ff2(row,:);
Divisor = U(1:N-r,u2);
Divisor = Divisor';
Divisor(1,N) = 0;
Z1 = fft(Dividend);
Z2 = fft(Divisor);
Q = Z1./Z2;
Q = ifft(Q);
Q = Q(1:r+1);
Q = rot90(Q,2);
A(row,1:r+1) = Q;
end;
A = A.';
A = fft(A);
A = A.';
f1 = f1';
f2 = f2';
r = M1-1;
B = zeros(M,N);
ff1 = fft(f1);
ff2 = fft(f2);
ff1 = ff1.';
ff2 = ff2.';
ff1 = flipud(ff1);
ff2 = flipud(ff2);
ff1 = ff1.';
ff2 = ff2.';
for row = 1:M
row
R = ff1(row,:);
R(1,2*M-r-1) = 0;
C = R(1,1);
C(M-r,1) = 0;
S01 = toeplitz(C,R);
R = ff2(row,:);
R(1,2*M-r-1) = 0;
C = R(1,1);
C(M-r,1) = 0;
S02 = toeplitz(C,R);
S02 = flipud(S02);
S0 = [S01;S02];
[U,X,V] = svd (S0,0);
[u1 u2] = size(U);
Dividend = ff2(row,:);
Divisor = U(1:N-r,u2);
Divisor = Divisor';
Divisor(1,N) = 0;
Z1 = fft(Dividend);
```

```
Z2 = fft(Divisor);
Q = Z1./Z2;
Q = ifft(Q);
Q = Q(1:r+1);
Q = rot90(Q,2);
B(row,1:r+1) = Q;
end;
B = B.';
B = fft(B);
Gamma = spalloc(N*M,M+N,2*N*M);
for I = 0:M-1
Gamma(I*N+1:I*N+N,I+1) = A(I+1,:).';
I
end;
for I = 0:M-1
k = 0;
for j = M:M+N-1
Gamma(I*N+1+k,j+1) = -B (I+1,j-M+1).';
k = k+1;
end;
I
end;
Gamma1 = Gamma.'*Gamma;
[V,D,FLAG] = EIGS(Gamma1,1,'SM');
a = V(1:M,1);
b = V(M+1:N+M,1);
P1 = zeros(M,M);
P2 = zeros(M,M);
for I = 1:M
P1(I,:) = A(I,:)*a(I);
end;
for j = 1:N
P2(:,j) = B(:,j)*b(j);
end;
P = (1/2)*(P1+P2);
PP = ifft2(P);
t2 = max(f);
t2 = max(t2);
t1 = max(PP);
t1 = max(t1);
PP = t2*PP/t1;
PP = real(PP);
switch counter
case 1
PP1 = PP;
case 2
PP2 = PP;
case 3
```

```
PP3 = PP;
end;
end;
% End of GCD between 3-observations.
```

D.23 LMMSE Super-Resolution

```
tic
f = imread('lenna.bmp');
f = f(:,:,1);
f = double(f)/255;
imwrite(f,'orig.tif','tif')
l = f(1:30,1:30);
[M,M] = size(f);
nhood = [5 5];
h1 = ones(nhood);
h1 = h1/25;
h1(M,M) = 0;
H1 = fft2(h1);
F = fft2(f);
G1 = F.*H1;
g1 = ifft2(G1);
g1 = real(g1);
SNR = input('Enter the value of SNR1 in dB');
gg = im2col(g1,[M,M],'distinct');
n_var = var(gg)/10^(SNR/10)
sigma1 = sqrt(n_var)
g1 = imnoise(g1,'gaussian',0, n_var);
g1 = max(g1,0);
g1 = min(g1,1);
imwrite(g1,'ob1.tif','tif')
nhood = [7 7];
h2 = ones(nhood);
h2 = h2/49;
h2(M,M) = 0;
H2 = fft2(h2);
G2 = F.*H2;
g2 = ifft2(G2);
g2 = real(g2);
SNR = input('Enter the value of SNR2 in dB');
gg = im2col(g2,[M,M],'distinct');
n_var = var(gg)/10^(SNR/10)
sigma2 = sqrt(n_var)
g2 = imnoise(g2,'gaussian',0, n_var);
g2 = max(g2,0);
g2 = min(g2,1);
imwrite(g2,'ob2.tif','tif')
```

```
nhood = [9 9];
h3 = ones(nhood);
h3 = h3/81;
h3(M,M) = 0;
H3 = fft2(h3);
G3 = F.*H3;
g3 = ifft2(G3);
g3 = real(g3);
SNR = input('Enter the value of SNR1 in dB');
gg = im2col(g3,[M,M],'distinct');
n_var = var(gg)/10^(SNR/10)
sigma3 = sqrt(n_var)
g3 = imnoise(g3,'gaussian',0, n_var);
g3 = max(g3,0);
g3 = min(g3,1);
imwrite(g3,'ob3.tif','tif')
G1 = fft2(g1);
G2 = fft2(g2);
G3 = fft2(g3);
G1 = G1.';
G2 = G2.';
G3 = G3.';
G1 = im2col(G1,[M M],'distinct');
G2 = im2col(G2,[M M],'distinct');
G3 = im2col(G3,[M M],'distinct');
G = [G1;G2;G3];
wlength = 3;
g1(M+wlength,M+wlength) = 0;
g2(M+wlength,M+wlength) = 0;
g3(M+wlength,M+wlength) = 0;
x = 0+eps*i
Diagonals = repmat(x,M^2,18);
for counter = 1:9
% Begin of formation of correlation matrices.
switch counter
case 1
for j = 0:M-1
for k = 0:M-1
sum = 0;
for n = 1:wlength
for m = 1:wlength
sum = sum+g1(n,m)*g1(n+j,m+k);
end;
end;
R (j+1,k+1) = 1/((wlength)^2)*sum;
end;
end;
case 2
```

```
for j = 0:M-1
for k = 0:M-1
sum = 0;
for n = 1:wlength
for m = 1:wlength
sum = sum+g2(n,m)*g1(n+j,m+k);
end;
end;
R (j+1,k+1) = 1/((wlength)^2)*sum;
end;
end;
case 3
for j = 0:M-1
for k = 0:M-1
sum = 0;
for n = 1:wlength
for m = 1:wlength
sum = sum+g3(n,m)*g1(n+j,m+k);
end;
end;
R (j+1,k+1) = 1/((wlength)^2)*sum;
end;
end;
case 4
for j = 0:M-1
for k = 0:M-1
sum = 0;
for n = 1:wlength
for m = 1:wlength
sum = sum+g1(n,m)*g2(n+j,m+k);
end;
end;
R (j+1,k+1) = 1/((wlength)^2)*sum;
end;
end;
case 5
for j = 0:M-1
for k = 0:M-1
sum = 0;
for n = 1:wlength
for m = 1:wlength
sum = sum+g2(n,m)*g2(n+j,m+k);
end;
end;
R (j+1,k+1) = 1/((wlength)^2)*sum;
end;
end;
```

```
case 6
for j = 0:M-1
for k = 0:M-1
sum = 0;
for n = 1:wlength
for m = 1:wlength
sum = sum+g3(n,m)*g2(n+j,m+k);
end;
end;
R (j+1,k+1) = 1/((wlength)^2)*sum;
end;
end;
case 7
for j = 0:M-1
for k = 0:M-1
sum = 0;
for n = 1:wlength
for m = 1:wlength
sum = sum+g1(n,m)*g3(n+j,m+k);
end;
end;
R (j+1,k+1) = 1/((wlength)^2)*sum;
end;
end;
case 8
for j = 0:M-1
for k = 0:M-1
sum = 0;
for n = 1:wlength
for m = 1:wlength
sum = sum+g2(n,m)*g3(n+j,m+k);
end;
end;
R (j+1,k+1) = 1/((wlength)^2)*sum;
end;
end;
case 9
for j = 0:M-1
for k = 0:M-1
sum = 0;
for n = 1:wlength
for m = 1:wlength
sum = sum+g3(n,m)*g3(n+j,m+k);
end;
end;
R (j+1,k+1) = 1/((wlength)^2)*sum;
end;
end;
```

```
end;% End of switch.
Pf = fft2(R);
Pf = Pf.';
column = im2col(Pf,[M M],'distinct');
Diagonals(:,counter) = column;
end;
for counter = 1:3
switch counter
case 1
H = H1;
case 2
H = H2;
case 3
H = H3;
end;
H = H.';
column = im2col(H,[M M],'distinct');
Diagonals(:,9+counter) = column;
Diagonals(:,12+counter) = conj(column);
end;
Diagonals(:,16) = sigma1^2;
Diagonals(:,17) = sigma2^2;
Diagonals(:,18) = sigma3^2;
I = linspace(1, M^2,M^2);
J = I;
A1 = sparse(I,J,Diagonals(:,1));
A2 = sparse(I,J+M^2,Diagonals(:,2));
A3 = sparse(I,J+2*M^2,Diagonals(:,3));
A4 = sparse(I+M^2,J,Diagonals(:,4));
A5 = sparse(I+M^2,J+M^2,Diagonals(:,5));
A6 = sparse(I+M^2,J+2*M^2,Diagonals(:,6));
A7 = sparse(I+2*M^2,J,Diagonals(:,7));
A8 = sparse(I+2*M^2,J+M^2,Diagonals(:,8)) ;
A9 = sparse(I+2*M^2,J+2*M^2,Diagonals(:,9));
A1(3*M^2,3*M^2) = 0;
A2(3*M^2,3*M^2) = 0;
A3(3*M^2,3*M^2) = 0;
A4(3*M^2,3*M^2) = 0;
A5(3*M^2,3*M^2) = 0;
A6(3*M^2,3*M^2) = 0;
A7(3*M^2,3*M^2) = 0;
A8(3*M^2,3*M^2) = 0;
A = A1+A2+A3+A4+A5+A6+A7+A8+A9;
A1 = sparse(I,J,Diagonals(:,13));
A2 = sparse(I+M^2,J+M^2,Diagonals(:,14));
A3 = sparse(I+2*M^2,J+2*M^2,Diagonals(:,15));
A1(3*M^2,3*M^2) = 0;
A2(3*M^2,3*M^2) = 0;
```

```
B = A1+A2+A3;
column = Diagonals(:,10).* Diagonals(:,1);
column = column.* Diagonals(:,13);
column = column+ Diagonals(:,16);
Diagonals(:,1) = column;
column = Diagonals(:,10).* Diagonals(:,2);
column = column.* Diagonals(:,14);
Diagonals(:,2) = column;
column = Diagonals(:,10).* Diagonals(:,3);
column = column.* Diagonals(:,15);
Diagonals(:,3) = column;
column = Diagonals(:,11).* Diagonals(:,4);
column = column.* Diagonals(:,13);
Diagonals(:,4) = column;
column = Diagonals(:,11).* Diagonals(:,5);
column = column.* Diagonals(:,14);
column = column+ Diagonals(:,17);
Diagonals(:,5) = column;
column = Diagonals(:,11).* Diagonals(:,6);
column = column.* Diagonals(:,15);
Diagonals(:,6) = column;
column = Diagonals(:,12).* Diagonals(:,7);
column = column.* Diagonals(:,13);
Diagonals(:,7) = column;
column = Diagonals(:,12).* Diagonals(:,8);
column = column.* Diagonals(:,14);
Diagonals(:,8) = column;
column = Diagonals(:,12).* Diagonals(:,9);
column = column.* Diagonals(:,15);
column = column+ Diagonals(:,18);
Diagonals(:,9) = column;
A1 = sparse(I,J,Diagonals(:,1));
A2 = sparse(I,J+M^2,Diagonals(:,2));
A3 = sparse(I,J+2*M^2,Diagonals(:,3));
A4 = sparse(I+M^2,J,Diagonals(:,4));
A5 = sparse(I+M^2,J+M^2,Diagonals(:,5));
A6 = sparse(I+M^2,J+2*M^2,Diagonals(:,6));
A7 = sparse(I+2*M^2,J,Diagonals(:,7));
A8 = sparse(I+2*M^2,J+M^2,Diagonals(:,8)) ;
A9 = sparse(I+2*M^2,J+2*M^2,Diagonals(:,9));
A1(3*M^2,3*M^2) = 0;
A2(3*M^2,3*M^2) = 0;
A3(3*M^2,3*M^2) = 0;
A4(3*M^2,3*M^2) = 0;
A5(3*M^2,3*M^2) = 0;
A6(3*M^2,3*M^2) = 0;
A7(3*M^2,3*M^2) = 0;
A8(3*M^2,3*M^2) = 0;
```

```
C = A1+A2+A3+A4+A5+A6+A7+A8+A9;
C = inv(C);
A = A*B;
C = C*G;
F = A*C;
F1 = F(1:M^2,1);
F2 = F(M^2+1:2*M^2,1);
F3 = F(2*M^2+1:3*M^2,1);
F1 = col2im(F1,[M M],[M M],'distinct');
F1 = F1.';
F2 = col2im(F2,[M M],[M M],'distinct');
F2 = F2.';
F3 = col2im(F3,[M M],[M M],'distinct');
F3 = F3.';
ff1 = ifft2(F1);
ff2 = ifft2(F2);
ff3 = ifft2(F3);
ff1 = real(ff1);
ff2 = real(ff2);
ff3 = real(ff3);
fff = wavelet_fusion(ff1,ff2);
fff = wavelet_fusion(fff,ff3);
fff = max(fff,0);
fff = min(fff,1);
clear sum;
fff = max(fff,0);
fff = min(fff,1);
MSE = sum(sum((f-fff).^2))/(M^2);
PSNR = 10*log(1/MSE)/log(10)
f = fff;
imwrite(f,'fused.tif','tif')
save data f
toc
% Image interpolation using the LMMSE algorithm
clear all;
tic
load data f
g = f;
n_var = 0.0001
[M,N] = size(f);
M = 2*M;
N = 2*N;
key0 = 4;
a = -1/2;
s = 0.5;
[M,N] = size(f);
ff = zeros(M,N);
x = f(:,N-1:N);
```

```
x = rot90(x,2);
y = f(:,1);
f = [y,f,x];
for i = 1:M
for j = 2:N+1
switch key0
case 1
ff(i,j-1) = f(i,j)*(1-s)+f(i,j+1)*s;%bilinear
case 2
ff(i,j-1) = f(i,j-1)*(a*s^3-2*a*s^2+a*s)+f(i,j)*((a+2)*s^3
-(3+a)*s^2+1)+f(i,j+1)*(-(a+2)*s^3+(2*a+3)*s^2-a*s)+f(i,j+2)
*(-a*s^3+a*s^2); % Bicubic
case 3
ff(i,j-1) = f(i,j-1)*((3+s)^3-4*(2+s)^3+6*(1+s)^3-4*s^3)/6+f
(i,j)*((2+s)^3-4*(1+s)^3+6*s^3)/6+f(i,j+1)*((1+s)^3-4*s^3)/6+f
(i,j+2)*s^3/6;% Cubic-Spline
case 4
ff(i,j-1) = f(i,j-1)*((-1/6)*(1+s)^3+(1+s)^2+(-85/42)*(1+s)+
(29/21))+f(i,j)*(0.5*s^3-s^2+(1/14)*s+13/21)+f(i,j+1)*(0.5*
(1-s)^3-(1-s)^2+(1/14)*s+13/21)+f(i,j+2)*((-1/6)*(2-s)^
3+(2-s)^2-(85/42)*(2-s)+29/21);% Cubic o- Moms
end;
end;
end;
ff = ff(:,1:N);
fff(1:M,1:2:2*N) = f(1:M,2:N+1);
fff(1:M,2:2:2*N) = ff(1:M,1:N);
f = fff';
clear ff,fff;
a = -1/2;
s = 0.5;
[M,N] = size(f);
ff = zeros(M,N);
x = f(:,N-1:N);
x = rot90(x,2);
y = f(:,1);
f = [y,f,x];
for i = 1:M
for j = 2:N+1
switch key0
case 1
ff(i,j-1) = f(i,j)*(1-s)+f(i,j+1)*s;%bilinear
case 2
ff(i,j-1) = f(i,j-1)*(a*s^3-2*a*s^2+a*s)+f(i,j)*((a+2)*s^3-
(3+a)*s^2+1)+f(i,j+1)*(-(a+2)*s^3+(2*a+3)*s^2-a*s)+f(i,j+2)*
(-a*s^3+a*s^2); % Bicubic
case 3
```

```
ff(i,j-1) = f(i,j-1)*((3+s)^3-4*(2+s)^3+6*(1+s)^3-4*s^3)/6+f
(i,j)*((2+s)^3-4*(1+s)^3+6*s^3)/6+f(i,j+1)*((1+s)^3-4*s^3)/6+f
(i,j+2)*s^3/6;% Cubic-Spline
case 4
ff(i,j-1) = f(i,j-1)*((-1/6)*(1+s)^3+(1+s)^2+(-85/42)*(1+s)+
(29/21))+f(i,j)*(0.5*s^3-s^2+(1/14)*s+13/21)+f(i,j+1)*(0.5*
(1-s)^3-(1-s)^2+(1/14)*s+13/21)+f(i,j+2)*((-1/6)*(2-s)^3+(2-s)
^2-(85/42)*(2-s)+29/21);% Cubic o- Moms
end;
end;
end;
ff = ff(:,1:N);
fff(1:M,1:2:2*N) = f(1:M,2:N+1);
fff(1:M,2:2:2*N) = ff(1:M,1:N);
fff = fff';
wlength = 3;
[M1,M1] = size(fff);
fff(M1+wlength,M1+wlength) = 0;
for j = 0:M1-1
for k = 0:M1-1
sum = 0;
for n = 1:wlength
for m = 1:wlength
sum = sum+fff(n,m)*fff(n+j,m+k);
end;
end;
RRR(j+1,k+1) = 1/((wlength)^2)*sum;
end;
end;
[M M] = size(g);
N = 2*M;
R = 2;
kff = zeros(N,N);
kff = RRR';
kff = im2col(kff,[N N],'distinct');
kff = sparse(1:N^2,1:N^2,kff);
g = g';
g = im2col(g,[M M],'distinct');
I = speye(M^2)/12;
H1 = sparse(M,N);
counter = 1;
for i = 1:M
H1(i,counter) = 1;
H1(i,counter+1) = 1;
counter = counter+2;
end;
H1 = H1/2;
H = kron(H1,H1);
```

```
Hz1 = speye(M^2)*n_var;
HH = H*kff*H';
I = speye(size(HH));
Hopt = kff*H'*inv(HH+Hz1);
f = Hopt*g;
f = col2im(f,[N N],[N N],'distinct');
f = max(f,0);
f = min(f,1);
imshow(f');
imwrite(f','sup.tif','tif')
fg = f';
save dat fg
toc
figure
clear all;
f = imread('lenna.bmp');
f = f(:,:,1);
f = double(f)/255;
g = f;
n_var = 0.0001
[M,N] = size(f);
M = 2*M;
N = 2*N;
key0 = 4;
a = -1/2;
s = 0.5;
[M,N] = size(f);
ff = zeros(M,N);
x = f(:,N-1:N);
x = rot90(x,2);
y = f(:,1);
f = [y,f,x];
for i = 1:M
for j = 2:N+1
switch key0
case 1
ff(i,j-1) = f(i,j)*(1-s)+f(i,j+1)*s;%bilinear
case 2
ff(i,j-1) = f(i,j-1)*(a*s^3-2*a*s^2+a*s)+f(i,j)*((a+2)*s^3-
(3+a)*s^2+1)+f(i,j+1)*(-(a+2)*s^3+(2*a+3)*s^2-a*s)+f(i,j+2)*
(-a*s^3+a*s^2); % Bicubic
case 3
ff(i,j-1) = f(i,j-1)*((3+s)^3-4*(2+s)^3+6*(1+s)^3-4*s^3)/6+f
(i,j)*((2+s)^3-4*(1+s)^3+6*s^3)/6+f(i,j+1)*((1+s)^3-4*s^3)/6+f
(i,j+2)*s^3/6;% Cubic-Spline
case 4
ff(i,j-1) = f(i,j-1)*((-1/6)*(1+s)^3+(1+s)^2+(-85/42)*(1+s)+
(29/21))+f(i,j)*(0.5*s^3-s^2+(1/14)*s+13/21)+f(i,j+1)*(0.5*
```

```
(1-s)^3-(1-s)^2+(1/14)*s+13/21)+f(i,j+2)*((-1/6)*(2-s)^3+(2-s)
^2-(85/42)*(2-s)+29/21);% Cubic o- Moms
end;
end;
end;
ff = ff(:,1:N);
fff(1:M,1:2:2*N) = f(1:M,2:N+1);
fff(1:M,2:2:2*N) = ff(1:M,1:N);
f = fff';
clear ff,fff;
a = -1/2;
s = 0.5;
[M,N] = size(f);
ff = zeros(M,N);
x = f(:,N-1:N);
x = rot90(x,2);
y = f(:,1);
f = [y,f,x];
for i = 1:M
for j = 2:N+1
switch key0
case 1
ff(i,j-1) = f(i,j)*(1-s)+f(i,j+1)*s;%bilinear
case 2
ff(i,j-1) = f(i,j-1)*(a*s^3-2*a*s^2+a*s)+f(i,j)*((a+2)*s^3-
(3+a)*s^2+1)+f(i,j+1)*(-(a+2)*s^3+(2*a+3)*s^2-a*s)+f(i,j+2)*
(-a*s^3+a*s^2); % Bicubic
case 3
ff(i,j-1) = f(i,j-1)*((3+s)^3-4*(2+s)^3+6*(1+s)^3-4*s^3)/6+f
(i,j)*((2+s)^3-4*(1+s)^3+6*s^3)/6+f(i,j+1)*((1+s)^3-4*s^3)/6+f
(i,j+2)*s^3/6;% Cubic-Spline
case 4
ff(i,j-1) = f(i,j-1)*((-1/6)*(1+s)^3+(1+s)^2+(-85/42)*(1+s)+
(29/21))+f(i,j)*(0.5*s^3-s^2+(1/14)*s+13/21)+f(i,j+1)*(0.5*
(1-s)^3-(1-s)^2+(1/14)*s+13/21)+f(i,j+2)*((-1/6)*(2-s)^
3+(2-s)^2-(85/42)*(2-s)+29/21);% Cubic o- Moms
end;
end;
end;
ff = ff(:,1:N);
fff(1:M,1:2:2*N) = f(1:M,2:N+1);
fff(1:M,2:2:2*N) = ff(1:M,1:N);
fff = fff';
wlength = 3;
[M1,M1] = size(fff);
fff(M1+wlength,M1+wlength) = 0;
for j = 0:M1-1
for k = 0:M1-1
sum = 0;
```

```
for n = 1:wlength
for m = 1:wlength
sum = sum+fff(n,m)*fff(n+j,m+k);
end;
end;
RRR(j+1,k+1) = 1/((wlength)^2)*sum;
end;
end;
[M M] = size(g);
N = 2*M;
R = 2;
kff = zeros(N,N);
kff = RRR';
kff = im2col(kff,[N N],'distinct');
kff = sparse(1:N^2,1:N^2,kff);
g = g';
g = im2col(g,[M M],'distinct');
I = speye(M^2)/12;
H1 = sparse(M,N);
counter = 1;
for i = 1:M
H1(i,counter) = 1;
H1(i,counter+1) = 1;
counter = counter+2;
end;
H1 = H1/2;
H = kron(H1,H1);
Hz1 = speye(M^2)*n_var;
HH = H*kff*H';
I = speye(size(HH));
Hopt = kff*H'*inv(HH+Hz1);
f = Hopt*g;
f = col2im(f,[N N],[N N],'distinct');
f = max(f,0);
f = min(f,1);
imshow(f');
imwrite(f','sup_orig.tif','tif')
fg1 = f';
load dat fg
clear sum
MSE = sum(sum((fg-fg1).^2))/prod(size(fg))
PSNR = 10*log(1/MSE)/log(10)
```

D.24 Maximum Entropy Super-Resolution

```
clear all;
tic
```

```
f = imread('lenna.bmp');
f = f(:,:,1);
f = double(f)/255;
imwrite(f,'orig.tif','tif')
save datax f
l = f(1:30,1:30);
[M,M] = size(f);
nhood = [5 5];
h1 = ones(nhood);
h1 = h1/25;
h1(M,M) = 0;
H1 = fft2(h1);
F = fft2(f);
G1 = F.*H1;
g1 = ifft2(G1);
g1 = real(g1);
SNR = input('Enter the value of SNR1 in dB');
gg = im2col(g1,[M,M],'distinct');
n_var = var(gg)/10^(SNR/10)
sigma1 = sqrt(n_var)
g1 = imnoise(g1,'gaussian',0, n_var);
g1 = max(g1,0);
g1 = min(g1,1);
imwrite(g1,'ob1.tif','tif')
nhood = [7 7];
h2 = ones(nhood);
h2 = h2/49;
h2(M,M) = 0;
H2 = fft2(h2);
G2 = F.*H2;
g2 = ifft2(G2);
g2 = real(g2);
SNR = input('Enter the value of SNR2 in dB');
gg = im2col(g2,[M,M],'distinct');
n_var = var(gg)/10^(SNR/10)
sigma2 = sqrt(n_var)
g2 = imnoise(g2,'gaussian',0, n_var);
g2 = max(g2,0);
g2 = min(g2,1);
imwrite(g2,'ob2.tif','tif')
nhood = [9 9];
h3 = ones(nhood);
h3 = h3/81;
h3(M,M) = 0;
H3 = fft2(h3);
G3 = F.*H3;
g3 = ifft2(G3);
g3 = real(g3);
```

```
SNR = input('Enter the value of SNR1 in dB');
gg = im2col(g3,[M,M],'distinct');
n_var = var(gg)/10^(SNR/10)
sigma3 = sqrt(n_var)
g3 = imnoise(g3,'gaussian',0, n_var);
g3 = max(g3,0);
g3 = min(g3,1);
G1 = fft2(g1);
G2 = fft2(g2);
G3 = fft2(g3);
imwrite(g3,'ob3.tif','tif')
eta = input('Enter the value of eta');
G1 = G1.';
G2 = G2.';
G3 = G3.';
G1 = im2col(G1,[M M],'distinct');
G2 = im2col(G2,[M M],'distinct');
G3 = im2col(G3,[M M],'distinct');
G = [G1;G2;G3];
x = 0+eps*i
Diagonals = repmat(x,M^2,6);
for counter = 1:3
switch counter
case 1
H = H1;
case 2
H = H2;
case 3
H = H3;
end;
H = H.';
column = im2col(H,[M M],'distinct');
Diagonals(:,counter) = column;
Diagonals(:,3+counter) = conj(column);
end;
I = linspace(1, M^2,M^2);
J = I;
column = Diagonals(:,1).*Diagonals(:,4)+eta*ones(size
(Diagonals(:,1)));
A1 = sparse(I,J,column);
column = Diagonals(:,2).*Diagonals(:,5)+eta*ones(size
(Diagonals(:,2)));
A5 = sparse(I+M^2,J+M^2,column);
column = Diagonals(:,3).*Diagonals(:,6)+eta*ones(size
(Diagonals(:,3)));
A9 = sparse(I+2*M^2,J+2*M^2,column);
A1(3*M^2,3*M^2) = 0;
A5(3*M^2,3*M^2) = 0;
```

```
C = A1+A5+A9;
C = inv(C);
column = Diagonals(:,4);
A1 = sparse(I,J,column);
column = Diagonals(:,5);
A5 = sparse(I+M^2,J+M^2,column);
column = Diagonals(:,6);
A9 = sparse(I+2*M^2,J+2*M^2,column);
A1(3*M^2,3*M^2) = 0;
A5(3*M^2,3*M^2) = 0;
A = A1+A5+A9;
A = C*A;
F = A*G;
F1 = F(1:M^2,1);
F2 = F(M^2+1:2*M^2,1);
F3 = F(2*M^2+1:3*M^2,1);
F1 = col2im(F1,[M M],[M M],'distinct');
F1 = F1.';
F2 = col2im(F2,[M M],[M M],'distinct');
F2 = F2.';
F3 = col2im(F3,[M M],[M M],'distinct');
F3 = F3.';
ff1 = ifft2(F1);
ff2 = ifft2(F2);
ff3 = ifft2(F3);
ff1 = real(ff1);
ff2 = real(ff2);
ff3 = real(ff3);
fff = wavelet_fusion(ff1,ff2);
fff = wavelet_fusion(fff,ff3);
fff = max(fff,0);
fff = min(fff,1);
clear sum;
fff = max(fff,0);
fff = min(fff,1);
MSE = sum(sum((f-fff).^2))/(M^2);
PSNR = 10*log(1/MSE)/log(10)
f = fff;
imwrite(f,'fused.tif','tif')
save data f
toc
% Image interpolation using the Maximum Entropy algorithm
clear all;
tic
load data f
g = f;
[M M] = size(g);
N = 2*M;
```

```
g = g';
g = im2col(g,[M M],'distinct');
I = speye(N^2);
H1 = sparse(M,N);
counter = 1;
for i = 1:M
H1(i,counter) = 1;
H1(i,counter+1) = 1;
counter = counter+2;
end;
H1 = H1/2;
H = kron(H1,H1);
HH = H'*H;
gama = input('Enter the value of gama');
Hopt = inv(HH+gama*I)*H';
f = Hopt*g;
f = col2im(f,[N N],[N N],'distinct');
imwrite(f','sup.tif','tif')
fg = f';
save dat fg
toc
load datax f
g = f;
[M M] = size(g);
N = 2*M;
g = g';
g = im2col(g,[M M],'distinct');
I = speye(N^2);
H1 = sparse(M,N);
counter = 1;
for i = 1:M
H1(i,counter) = 1;
H1(i,counter+1) = 1;
counter = counter+2;
end;
H1 = H1/2;
H = kron(H1,H1);
HH = H'*H;
gama = input('Enter the value of gama');
Hopt = inv(HH+gama*I)*H';
f = Hopt*g;
f = col2im(f,[N N],[N N],'distinct');
imwrite(f','sup_orig.tif','tif')
fg1 = f';
load dat fg
clear sum
MSE = sum(sum((fg-fg1).^2))/prod(size(fg))
PSNR = 10*log(1/MSE)/log(10)
```

D.25 Regularized Super-Resolution

```
clear all;
tic
f = imread('lenna.bmp');
f = f(:,:,1);
f = double(f)/255;
imwrite(f,'orig.tif','tif')
save datax f
l = f(1:30,1:30);
[M,M] = size(f);
nhood = [5 5];
h11 = ones(nhood);
h11 = h11/25;
h11(M,M) = 0;
H11 = fft2(h11);
F = fft2(f);
G1 = F.*H11;
g1 = ifft2(G1);
g1 = real(g1);
SNR = input('Enter the value of SNR1 in dB');
gg = im2col(g1,[M,M],'distinct');
n_var = var(gg)/10^(SNR/10)
sigma1 = sqrt(n_var)
g1 = imnoise(g1,'gaussian',0, n_var);
g1 = max(g1,0);
g1 = min(g1,1);
imwrite(g1,'ob1.tif','tif')
nhood = [7 7];
h22 = ones(nhood);
h22 = h22/49;
h22(M,M) = 0;
H22 = fft2(h22);
G2 = F.*H22;
g2 = ifft2(G2);
g2 = real(g2);
SNR = input('Enter the value of SNR2 in dB');
gg = im2col(g2,[M,M],'distinct');
n_var = var(gg)/10^(SNR/10)
sigma2 = sqrt(n_var)
g2 = imnoise(g2,'gaussian',0, n_var);
g2 = max(g2,0);
g2 = min(g2,1);
imwrite(g2,'ob2.tif','tif')
nhood = [9 9];
h33 = ones(nhood);
h33 = h33/81;
h33(M,M) = 0;
```

```
H33 = fft2(h33);
G3 = F.*H33;
g3 = ifft2(G3);
g3 = real(g3);
SNR = input('Enter the value of SNR1 in dB');
gg = im2col(g3,[M,M],'distinct');
n_var = var(gg)/10^(SNR/10)
sigma3 = sqrt(n_var)
g3 = imnoise(g3,'gaussian',0, n_var);
g3 = max(g3,0);
g3 = min(g3,1);
G1 = fft2(g1);
G2 = fft2(g2);
G3 = fft2(g3);
imwrite(g3,'ob3.tif','tif')
G1 = G1.';
G2 = G2.';
G3 = G3.';
G1 = im2col(G1,[M M],'distinct');
G2 = im2col(G2,[M M],'distinct');
G3 = im2col(G3,[M M],'distinct');
g1 = g1';
g2 = g2';
g3 = g3';
g1 = im2col(g1,[M M],'distinct');
g2 = im2col(g2,[M M],'distinct');
g3 = im2col(g3,[M M],'distinct');
G = [G1;G2;G3];
x = 0;
HDiagonals = repmat(x,M^2,9);
QDiagonals = repmat(x,M^2,9);
Laplacianselector = 2;
switch Laplacianselector
case 1
q = [0 1 0
1 -4 1
0 1 0];
q = q/4;
case 2
q = [0 1 0
1 -6 1
0 1 0];
q = q/6;
q1 = [0 0 0
0 1/6 0
0 0 0];
q1(M,M) = 0;
Q1 = fft2(q1);
```

```
end;
q(M,M) = 0;
Q = fft2(q);
H11 = H11.';
column = im2col(H11,[M M],'distinct');
HDiagonals(:,1) = column;
H22 = H22.';
column = im2col(H22,[M M],'distinct');
HDiagonals(:,5) = column;
H33 = H33.';
column = im2col(H33,[M M],'distinct');
HDiagonals(:,9) = column;
switch Laplacianselector
case 1
Q = Q.';
column = im2col(Q,[M M],'distinct');
QDiagonals(:,1) = column;
QDiagonals(:,5) = column;
QDiagonals(:,9) = column;
case 2
Q = Q.';
column = im2col(Q,[M M],'distinct');
QDiagonals(:,1) = column;
QDiagonals(:,5) = column;
QDiagonals(:,9) = column;
Q1 = Q1.';
column = im2col(Q1,[M M],'distinct');
QDiagonals(:,2) = column;
QDiagonals(:,3) = column;
QDiagonals(:,4) = column;
QDiagonals(:,6) = column;
QDiagonals(:,7) = column;
QDiagonals(:,8) = column;
end;
% Computation of the Jacobi.
I = linspace(1, M^2,M^2);
J = I;
one = ones(M^2,1);
% 5 zeros.
x1 =.0001;
x2 =.0001;
x3 =.0001;
X = [x1;x2;x3];
% Begin of computation of fifth multiplicand:
switch Laplacianselector
case 1
A1 = sparse(I,J,QDiagonals(:,1));
A5 = sparse(I+M^2,J+M^2,QDiagonals(:,5));
```

```
A9 = sparse(I+2*M^2,J+2*M^2,QDiagonals(:,9));
A1(3*M^2,3*M^2) = 0;
A5(3*M^2,3*M^2) = 0;
Q = A1+A5+A9;
A1 = sparse(I,J,conj(QDiagonals(:,1)));
A5 = sparse(I+M^2,J+M^2,conj(QDiagonals(:,5)));
A9 = sparse(I+2*M^2,J+2*M^2,conj(QDiagonals(:,9)));
A1(3*M^2,3*M^2) = 0;
A5(3*M^2,3*M^2) = 0;
QQ = A1+A5+A9;
case 2
A1 = sparse(I,J,QDiagonals(:,1));
A2 = sparse(I,J+M^2,QDiagonals(:,4));
A3 = sparse(I,J+2*M^2,QDiagonals(:,7));
A4 = sparse(I+M^2,J,QDiagonals(:,2));
A5 = sparse(I+M^2,J+M^2,QDiagonals(:,5));
A6 = sparse(I+M^2,J+2*M^2,QDiagonals(:,8));
A7 = sparse(I+2*M^2,J,QDiagonals(:,3));
A8 = sparse(I+2*M^2,J+M^2,QDiagonals(:,6)) ;
A9 = sparse(I+2*M^2,J+2*M^2,QDiagonals(:,9));
A1(3*M^2,3*M^2) = 0;
A2(3*M^2,3*M^2) = 0;
A3(3*M^2,3*M^2) = 0;
A4(3*M^2,3*M^2) = 0;
A5(3*M^2,3*M^2) = 0;
A6(3*M^2,3*M^2) = 0;
A7(3*M^2,3*M^2) = 0;
A8(3*M^2,3*M^2) = 0;
Q = A1+A2+A3+A4+A5+A6+A7+A8+A9;
A1 = sparse(I,J,conj(QDiagonals(:,1)));
A2 = sparse(I,J+M^2,conj(QDiagonals(:,4)));
A3 = sparse(I,J+2*M^2,conj(QDiagonals(:,7)));
A4 = sparse(I+M^2,J,conj(QDiagonals(:,2)));
A5 = sparse(I+M^2,J+M^2,conj(QDiagonals(:,5)));
A6 = sparse(I+M^2,J+2*M^2,conj(QDiagonals(:,8)));
A7 = sparse(I+2*M^2,J,conj(QDiagonals(:,3)));
A8 = sparse(I+2*M^2,J+M^2,conj(QDiagonals(:,6))) ;
A9 = sparse(I+2*M^2,J+2*M^2,conj(QDiagonals(:,9)));
A1(3*M^2,3*M^2) = 0;
A2(3*M^2,3*M^2) = 0;
A3(3*M^2,3*M^2) = 0;
A4(3*M^2,3*M^2) = 0;
A5(3*M^2,3*M^2) = 0;
A6(3*M^2,3*M^2) = 0;
A7(3*M^2,3*M^2) = 0;
A8(3*M^2,3*M^2) = 0;
QQ = A1+A2+A3+A4+A5+A6+A7+A8+A9;
end;
```

```
mult5 = QQ*Q;
% End of computation of fifth multiplicand.
%Begin of computation of third multiplicand:
A1 = sparse(I,J,HDiagonals(:,1));
A5 = sparse(I+M^2,J+M^2,HDiagonals(:,5));
A9 = sparse(I+2*M^2,J+2*M^2,HDiagonals(:,9));
A1(3*M^2,3*M^2) = 0;
A5(3*M^2,3*M^2) = 0;
H = A1+A5+A9;
A1 = sparse(I,J,conj(HDiagonals(:,1)));
A5 = sparse(I+M^2,J+M^2,conj(HDiagonals(:,5)));
A9 = sparse(I+2*M^2,J+2*M^2,conj(HDiagonals(:,9)));
A1(3*M^2,3*M^2) = 0;
A5(3*M^2,3*M^2) = 0;
HH = A1+A5+A9;
counter = 0;
while counter< = 10
mult3 = HH*H;
A1 = sparse(I,J,one*x1);
A2 = sparse(I+M^2,J+M^2,one*x2);
A3 = sparse(I+2*M^2,J+2*M^2,one*x3);
A1(3*M^2,3*M^2) = 0;
A2(3*M^2,3*M^2) = 0;
Landa = A1+A2+A3;
mult3 = mult3+Landa*mult5;
mult3 = inv(mult3);
% End of computation of third multiplicand.
% Begin of computation of sixth multiplicand;
mult6 = mult3*HH*G;
% End of computation of sixth multiplicand.
for i = 1:3
for j = 1:3
% Begin of computation of forth multiplicand.
switch j
case 1
mult4 = sparse(I,J,one);
mult4(3*M^2,3*M^2) = 0;
case 2
mult4 = sparse(I+M^2,J+M^2,one);
mult4(3*M^2,3*M^2) = 0;
case 3
mult4 = sparse(I+2*M^2,J+2*M^2,one);
end;
% End of computation of forth multiplicand.
% Begin of computation of second multiplicand:
switch i
case 1
mult2 = sparse(I,J,HDiagonals(:,1));
```

```
mult2(M^2,3*M^2) = 0;
case 2
mult2 = sparse(I,J+M^2,HDiagonals(:,5));
mult2(M^2,3*M^2) = 0;
case 3
mult2 = sparse(I,J+2*M^2,HDiagonals(:,9));
end;
% End of computation of the second multiplicand.
mult = mult2*mult3;
mult = mult*mult4;
mult = mult*mult5;
mult = mult*mult6;
mult = col2im(mult,[M M],[M M],'distinct');
mult = mult.';
mult = ifft2(mult);
mult = real(mult);
mult = mult.';
mult = im2col(mult,[M M],'distinct');
% Begin of computation of first multiplicand:
switch i
case 1
F1 = mult6(1:M^2);
case 2
F1 = mult6(M^2+1:2*M^2);
case 3
F1 = mult6(2*M^2+1:3*M^2);
end;
F1 = col2im(F1,[M M],[M M],'distinct');
F1 = F1.';
switch i
case 1
Gii = F1.*H11;
case 2
Gii = F1.*H22;
case 3
Gii = F1.*H33;
end;
gii = ifft2(Gii);
switch i
case 1
gi = col2im(g1,[M M],[M M],'distinct');
case 2
gi = col2im(g2,[M M],[M M],'distinct');
case 3
gi = col2im(g3,[M M],[M M],'distinct');
end;
gi = gi.';
mult1 = gii-gi;
```

```
mult1 = mult1';
mult1 = im2col(mult1,[M M],'distinct');
mult1 = real(mult1);
mult1 = mult1.';
switch i
case 1
i;
Z1 = (norm(mult1,'fro'))^2-M^2*sigma1^2;
case 2
i;
Z2 = (norm(mult1,'fro'))^2-M^2*sigma2^2;
case 3
i;
Z3 = (norm(mult1,'fro'))^2-M^2*sigma3^2;
end;
% End of computation of first multiplicand:
mult = -2*mult1*mult;
Jacobian(i,j) = mult;
end;
end;
Jacobian = inv(Jacobian);
Z = [Z1;Z2;Z3];
Y = -Jacobian*Z;
X = X+Y;
x1 = X(1);
x2 = X(2);
x3 = X(3);
counter = counter+1;
end;
F1 = mult6(1:M^2);
F1 = col2im(F1,[M M],[M M],'distinct');
F1 = F1.';
f1 = ifft2(F1);
F2 = mult6(M^2+1:2*M^2);
F2 = col2im(F2,[M M],[M M],'distinct');
F2 = F2.';
f2 = ifft2(F2);
F3 = mult6(2*M^2+1:3*M^2);
F3 = col2im(F3,[M M],[M M],'distinct');
F3 = F3.';
f3 = ifft2(F3);
ff1 = real(f1);
ff2 = real(f2);
ff3 = real(f3);
fff = wavelet_fusion(ff1,ff2);
fff = wavelet_fusion(fff,ff3);
fff = max(fff,0);
fff = min(fff,1);
```

```
clear sum;
fff = max(fff,0);
fff = min(fff,1);
MSE = sum(sum((f-fff).^2))/(M^2);
PSNR = 10*log(1/MSE)/log(10)
f = fff;
imwrite(f,'fused.tif','tif')
save data f
toc
% Image interpolation using the regularized algorithm
clear all;
tic
load data f
g = f;
%clear f
[M,N] = size(g);
M = M*2;
%landa = input('Enter the value of regularization parameter');
landa = 0.001;
g = [rot90(g(:,1:4),2),g,rot90(g(:,M/2-8:M/2),2)];
g = [rot90(g(1:4,:),2);g;rot90(g(M/2-8:M/2,:),2)];
[L1,L2] = size(f);
L1 = L1*2;
L2 = L2*2;
M = 24;
N = 12;
I = speye(M^2);
H1 = sparse(M/2,M);
counter = 1;
for i = 1:M/2
H1(i,counter) = 1;
H1(i,counter+1) = 1;
counter = counter+2;
end;
H1 = H1/2;
H = kron(H1,H1);
HH = H'*H;
beta = 0.125;
Q1 = sparse(M,M);
for i = 1:M
Q1(i,i) = -2;
end;
for i = 1:M-1
Q1(i,i+1) = 1;
end;
for i = 2:M
Q1(i,i-1) = 1;
end;
```

```
Q = kron(Q1,Q1);
QQ = Q'*Q;
L = inv(HH+landa*QQ);
for ii = 1:L1/8
for jj = 1:L2/8
ii;
f = g(4*ii+1-4:4*(ii+1)+4,4*jj+1-4:4*(jj+1)+4);
z = f;
y = im2col(z,[N N],'distinct');
x1 = L*H'*y;
x1 = col2im(x1,[M M],[M M],'distinct');
xx1(8*(ii-1)+1:8*(ii),8*(jj-1)+1:8*(jj)) = x1(9:16,9:16);
end;
end;
[a,b] = size(xx1)
xx1 = max(xx1,0);
xx1 = min(xx1,1);
imshow(xx1)
imwrite(xx1,'sup.tif','tif')
fg = xx1;
save dat fg
toc
clear all;
tic
load datax f
g = f;
%landa = input('Enter the value of regularization parameter');
landa = 0.001;
[M,N] = size(g);
M = M*2;
g = [rot90(g(:,1:4),2),g,rot90(g(:,M/2-8:M/2),2)];
g = [rot90(g(1:4,:),2);g;rot90(g(M/2-8:M/2,:),2)];
[L1,L2] = size(f);
L1 = L1*2;
L2 = L2*2;
M = 24;
N = 12;
I = speye(M^2);
H1 = sparse(M/2,M);
counter = 1;
for i = 1:M/2
H1(i,counter) = 1;
H1(i,counter+1) = 1;
counter = counter+2;
end;
H1 = H1/2;
H = kron(H1,H1);
HH = H'*H;
```

```
beta = 0.125;
Q1 = sparse(M,M);
for i = 1:M
Q1(i,i) = -2;
end;
for i = 1:M-1
Q1(i,i+1) = 1;
end;
for i = 2:M
Q1(i,i-1) = 1;
end;
Q = kron(Q1,Q1);
QQ = Q'*Q;
L = inv(HH+landa*QQ);
for ii = 1:L1/8
for jj = 1:L2/8
f = g(4*ii+1-4:4*(ii+1)+4,4*jj+1-4:4*(jj+1)+4);
z = f;
y = im2col(z,[N N],'distinct');
x1 = L*H'*y;
x1 = col2im(x1,[M M],[M M],'distinct');
xx1(8*(ii-1)+1:8*(ii),8*(jj-1)+1:8*(jj)) = x1(9:16,9:16);
end;
end;
[a,b] = size(xx1)
xx1 = max(xx1,0);
xx1 = min(xx1,1);
imshow(xx1)
imwrite(xx1,'sup_orig.tif','tif')
fg1 = xx1;
toc
load dat fg
clear sum
MSE = sum(sum((fg-fg1).^2))/prod(size(fg))
PSNR = 10*log(1/MSE)/log(10)
```

D.26 Blind Super-Resolution

```
f = imread('lenna.bmp');
f = f(:,:,1);
f = f(1:128,1:128);
f = double(f)/255;
imwrite(f,'orig.tif','tif')
[M1 N1] = size(f);
M0 = [M1,N1];
save sdata M0
h1 = ones([5 5]);
```

```
h1 = h1/25;
h1 = [0.03,0.035,0.04,0.045,0.05;0.05,0.025,0.055,0.03,0.04;
0.033,0.043,0.053,0.025,0.055;0.04,0.04,0.038,0.042,0.04;
0.045,0.05,0.065,0.025,0.015];
h2 = h1';
h3 = rot90(h1,2);
[M2 N2] = size(h1)
M = M1+M2-1;
N = N1+N2-1;
f(M,N) = 0;
save data00 f
h1 (M,N) = 0;
h2(M,N) = 0;
h3(M,N) = 0;
F = fft2(f);
H1 = fft2(h1);
F1 = H1.*F;
f1 = ifft2(F1);
f1 = real(f1);
H2 = fft2(h2);
F2 = H2.*F;
f2 = ifft2(F2);
f2 = real(f2);
H3 = fft2(h3);
F3 = H3.*F;
f3 = ifft2(F3);
f3 = real(f3);
% Noise addition statements.
SNR = input('Enter the value of SNR in for image 1 in dB');
gg = im2col(f1,[M,N],'distinct');
n_var = var(gg)/10^(SNR/10)
f1 = imnoise(f1,'gaussian',0,n_var);
SNR = input('Enter the value of SNR in for image 2 in dB');
gg = im2col(f2,[M,N],'distinct');
n_var = var(gg)/10^(SNR/10)
f2 = imnoise(f2,'gaussian',0,n_var);
SNR = input('Enter the value of SNR in for image 3 in dB');
gg = im2col(f3,[M,N],'distinct');
n_var = var(gg)/10^(SNR/10)
f3 = imnoise(f3,'gaussian',0,n_var);
imwrite(f1,'observ1.tif','tif');
imwrite(f2,'observ2.tif','tif');
imwrite(f3,'observ3.tif','tif');
save zdata1 f1
save zdata2 f2
save zdata3 f3
im1 = f1;
im2 = f2;
```

```
im3 = f3;
selector = input('Enter 1 for separate operation and 2 for
combinational operation');
tic
% End of noise addition statements.
for counter = 1:3
switch counter
case 1
if selector = = 1
f1 = im1;
f2 = im2;
else
f1 = im1;
f2 = (im1+im2+im3)/3;
end;
case 2
if selector = = 1
f1 = im2;
f2 = im3;
else
f1 = im2;
f2 = (im1+im2+im3)/3;
end;
case 3
if selector = = 1
f1 = im1;
f2 = im3;
else
f1 = im3;
f2 = (im1+im2+im3)/3;
imwrite(f2,'comb.tif','tif');
end;
end;
r = M1-1;
A = zeros(M,N);
ff1 = fft(f1);
ff2 = fft(f2);
ff1 = ff1.';
ff2 = ff2.';
ff1 = flipud(ff1);
ff2 = flipud(ff2);
ff1 = ff1.';
ff2 = ff2.';
for row = 1:M
row;
R = ff1(row,:);
R(1,2*M-r-1) = 0;
C = R(1,1);
```

```
C(M-r,1) = 0;
S01 = toeplitz(C,R);
R = ff2(row,:);
R(1,2*M-r-1) = 0;
C = R(1,1);
C(M-r,1) = 0;
S02 = toeplitz(C,R);
S02 = flipud(S02);
S0 = [S01;S02];
[U,X,V] = svd (S0,0);
[u1 u2] = size(U);
Dividend = ff2(row,:);
Divisor = U(1:N-r,u2);
Divisor = Divisor';
Divisor(1,N) = 0;
Z1 = fft(Dividend);
Z2 = fft(Divisor);
Q = Z1./Z2;
Q = ifft(Q);
Q = Q(1:r+1);
Q = rot90(Q,2);
A(row,1:r+1) = Q;
end;
A = A.';
A = fft(A);
A = A.';
f1 = f1';
f2 = f2';
r = M1-1;
B = zeros(M,N);
ff1 = fft(f1);
ff2 = fft(f2);
ff1 = ff1.';
ff2 = ff2.';
ff1 = flipud(ff1);
ff2 = flipud(ff2);
ff1 = ff1.';
ff2 = ff2.';
for row = 1:M
row;
R = ff1(row,:);
R(1,2*M-r-1) = 0;
C = R(1,1);
C(M-r,1) = 0;
S01 = toeplitz(C,R);
R = ff2(row,:);
R(1,2*M-r-1) = 0;
C = R(1,1);
```

```
C(M-r,1) = 0;
S02 = toeplitz(C,R);
S02 = flipud(S02);
S0 = [S01;S02];
[U,X,V] = svd (S0,0);
[u1 u2] = size(U);
Dividend = ff2(row,:);
Divisor = U(1:N-r,u2);
Divisor = Divisor';
Divisor(1,N) = 0;
Z1 = fft(Dividend);
Z2 = fft(Divisor);
Q = Z1./Z2;
Q = ifft(Q);
Q = Q(1:r+1);
Q = rot90(Q,2);
B(row,1:r+1) = Q;
end;
B = B.';
B = fft(B);
Gamma = spalloc(N*M,M+N,2*N*M);
for I = 0:M-1
Gamma(I*N+1:I*N+N,I+1) = A(I+1,:).';
I;
end;
for I = 0:M-1
k = 0;
for j = M:M+N-1
Gamma(I*N+1+k,j+1) = -B (I+1,j-M+1).';
k = k+1;
end;
I;
end;
Gamma1 = Gamma.'*Gamma;
[V,D,FLAG] = EIGS(Gamma1,1,'SM');
a = V(1:M,1);
b = V(M+1:N+M,1);
P1 = zeros(M,M);
P2 = zeros(M,M);
for I = 1:M
P1(I,:) = A(I,:)*a(I);
end;
for j = 1:N
P2(:,j) = B(:,j)*b(j);
end;
P = (1/2)*(P1+P2);
PP = ifft2(P);
t2 = max(f);
```

```
t2 = max(t2);
t1 = max(PP);
t1 = max(t1);
PP = t2*PP/t1;
PP = real(PP);
gg = im2col(PP,[M,N],'distinct');
imshow(PP)
figure
switch counter
case 1
PP1 = PP;
var1 = var(gg);
PP1 = max(PP1,0);
PP1 = min(PP1,1);
imwrite(PP1,'R1.tif','tif');
case 2
PP2 = PP;
var2 = var(gg);
PP2 = max(PP2,0);
PP2 = min(PP2,1);
imwrite(PP2,'R2.tif','tif');
case 3
PP3 = PP;
var3 = var(gg);
PP3 = max(PP3,0);
PP3 = min(PP3,1);
imwrite(PP3,'R3.tif','tif');
end;
end;
% End of GCD between 3- observations.
if selector = =2
PP = wavelet_fusion(PP1,PP2);
PP = wavelet_fusion(PP,PP3);
else
select = input('Enter the number of the result to be
interpolated');
switch select
case 1
PP = PP1;
case 2
PP = PP2;
case 3
PP = PP3;
end;
end;
load sdata M0
PP = max(PP,0);
PP = min(PP,1);
```

```
imwrite(PP(1:M0(1),1:M0(2)),'combob.tif','tif');
save data PP;
toc
clear all
tic
load data PP
g = PP;
f = g;
[M M] = size(g);
N = 2*M;
g = g';
g = im2col(g,[M M],'distinct');
I = speye(N^2);
H1 = sparse(M,N);
counter = 1;
for i = 1:M
H1(i,counter) = 1;
H1(i,counter+1) = 1;
counter = counter+2;
end;
H1 = H1/2;
H = kron(H1,H1);
HH = H'*H;
gama = 0.001;
Hopt = inv(HH+gama*I)*H';
f = Hopt*g;
f = col2im(f,[N N],[N N],'distinct');
ff = f';
ff = max(ff,0);
ff = min(ff,1);
load sdata M0
imwrite(ff(1:M0(1)*2,1:M0(2)*2),'sup.tif','tif')
fg = f';
save dat fg
toc
clear all
load data00 f
g = f;
[M M] = size(g);
N = 2*M;
g = g';
g = im2col(g,[M M],'distinct');
I = speye(N^2);
H1 = sparse(M,N);
counter = 1;
for i = 1:M
H1(i,counter) = 1;
H1(i,counter+1) = 1;
```

```
counter = counter+2;
end;
H1 = H1/2;
H = kron(H1,H1);
HH = H'*H;
gama = 0.001;
Hopt = inv(HH+gama*I)*H';
f = Hopt*g;
f = col2im(f,[N N],[N N],'distinct');
ff = f';
ff = max(ff,0);
ff = min(ff,1);
load sdata M0
imwrite(ff(1:M0(1)*2,1:M0(2)*2),'sup_orig.tif','tif')
fg = f';
f = fg;
load dat fg
imshow(f);
figure
imshow(fg);
MSE = sum(sum((f-fg).^2))/prod(size(fg))
PSNR = 10*log(1/MSE)/log(10)
```

References

1. T. Sigitani, Y. Iiguni and H. Maeda. 1999. Image Interpolation for Progressive Transmission by Using Radial Basis Function Networks, *IEEE Trans. Neural Networks*, 10, 381–390.
2. N. Plaziac. 1999. Image Interpolation Using Neural Networks, *IEEE Trans. Image Proc.*, 8, 1647–1651.
3. W. K. Carey, D. B. Chuang and S. S. Hemami. 1999. Regularity Preserving Image Interpolation, *IEEE Trans. Image Proc.*, 8, 1293–1297.
4. G. Chen and R. J. P. de Figueiredo. 1993. A Unified Approach to Optimal Image Interpolation Problems Based on Linear Partial Differential Equation Models, *IEEE Trans. Image Proc.*, 2, 41–49.
5. C. Lee, M. Eden and M. Unser. 1998. High-Quality Image Resizing Using Oblique Projection Operators, *IEEE Trans. Image Proc.*, 7, 679–692.
6. D. Ramanan and K. E. Barner. 2002. Nonlinear Image Interpolation through Extended Permutation Filters. In *Proceedings of International Conference on Image Processing*.
7. A. Munoz, T. Blu and M. Unser. 2001. Least-Squares Image Resizing Using Finite Differences, *IEEE Trans. Image Proc.*, 10, 1365–1378.
8. D. Darian and T. W. Parks. 2004. Adaptively Quadratic (Aqua) Image Interpolation, *IEEE Trans. Image Proc.*, 13, 690–698.
9. J. Vesma. 2000. A Frequency-Domain Approach to Polynomial-Based Interpolation and the Farrow Structure, *IEEE Trans. Circuits Syst* II, 47, 206–209.
10. X. Pan. 1999. A Novel Approach for Multi-Dimensional Interpolation, *IEEE Trans. Signal Proc.*, 6, 38–40.
11. T. M. Lehman, C. Conner and K. Spitzer. 2001. Addendum: B-Spline Interpolation in Medical Image Processing, *IEEE Trans. Med. Imaging*, 20, 660–665.
12. T. Bretschneider, C. Miller and O. Kao. 2000. Interpolation of Scratches in Motion Picture Films. In *Proceedings of ICASSP*.
13. V. Caselles, J. M. Morel and C. Sbert. 2004. An Axiomatic Approach to Image Interpolation, *IEEE Trans. Image Proc.*, 7, 376–386.
14. H. S. Hou and H. C. Andrews. 1978. Cubic Spline for Image Interpolation and Digital Filtering, *IEEE Trans. Acoustic Speech Signal Proc.*, 26, 508–517.
15. M. Unser, A. Aldroubi and M. Eden. 1993. B-Spline Signal Processing. I: Theory, *IEEE Trans. Signal Proc.*, 41, 821–833.
16. M. Unser, A. Aldroubi and M. Eden. 1993. B-Spline Signal Processing. II: Efficient Design and Applications, *IEEE Trans. Signal Proc.*, 41, 834–848.

17. B. Vrcelj and P. P. Vaidyanathan. 2001. Efficient Implementation of All-Digital Interpolation, *IEEE Trans. Image Proc.*, 10, 1639–1646.

18. P. Thevenaz, T. Blu and M. Unser. 2000. Interpolation Revisited, *IEEE Trans. Med. Imaging*, 19, 739–758.

19. T. Blu, P. Thevenaz and M. Unser. 2001. MOMS: Maximal-Order Interpolation of Minimal Support, *IEEE Trans. Image Proc.*, 10, 1069–1080.

20. M. Unser. 1999. Splines: A Perfect Fit For Signal and Image Processing, *IEEE Signal Proc.*

21. J. K. Han and H. M. Kim. 2001. Modified Cubic Convolution Scaler with Minimum Loss of Information, *Optical Eng.*, 40, 540–546.

22. A. Gotchev, K. Egiazarian, J. Vema et al. 2000. Edge-Preserving Image Resizing Using Modified B-Splines. In *Proceedings of ICASSP*.

23. A. Gotchev, J. Vesma, T. Saramaki et al. 2000. Digital Image Resampling by Modified B-Spline Functions. In *Proceedings of ICASSP*.

24. E. Meijering and M. Unser. 2003. A Note on Cubic Convolution Interpolation, *IEEE Trans. Image Proc.*, 12, 477–479.

25. T. Blu, P. Thevenaz and M. Unser. 2002. How a Simple Shift Can Significantly Improve the Performance of Linear Interpolation. In *Proceedings of International Conference on Image Processing*, III-377–III-380.

26. K. Ichige, T. Blu and M. Unser. 2003. Interpolation of Signals by Generalized Piecewise-Linear Multiple Generators. In *Proceedings of ICASSP*, VI-261–VI-264.

27. T. Blu, B. Thevenaz and M. Unser. 2004. Linear Interpolation Revitalized, *IEEE Trans. Image Proc.*, 13, 710–719.

28. G. Ramponi. 1999. Warped Distance for Space Variant Linear Image Interpolation, *IEEE Trans. Image Proc.*, 8, 629–639.

29. B. P. Lathi. 1998. *Modern Digital and Analog Communication Systems*. New York: Holt Rinehart and Winston.

30. S. E. El-Khamy, M. M. Hadhoud, M. I. Dessouky et al. 2003. Adaptive Image Interpolation Based on Local Activity Levels, URSI National Radio Science Conference, Cairo, March.

31. S. E. El-Khamy, M. M. Hadhoud, M. I. Dessouky et al. 2006. A Simple Adaptive Interpolation Approach Based on Varying Image Local Activity Levels, *Int. J. Information Acquisition*, 3, 1–8.

32. S. E. El-Khamy, M. M. Hadhoud, M. I. Dessouky et al. 2004. A New Edge Preserving Pixel-by-Pixel (PBP) Cubic Image Interpolation Approach, URSI National Radio Science Conference, Cairo, March.

33. S. E. El-Khamy, M. M. Hadhoud, M. I. Dessouky et al. 2005. An Adaptive Cubic Convolution Image Interpolation Approach, *Int. J. Machine Graphics Vision*, 14, 235–256.

34. S. E. El-Khamy, M. M. Hadhoud, M. I. Dessouky et al. 2006. A New Approach for Adaptive Polynomial-Based Image Interpolation, *Int. J. Information Acquisition*, 3, 139–159.

35. J. H. Shin, J. H. Jung, and J. K. Paik. 1998. Regularized Iterative Image Interpolation and Its Application to Spatially Scalable Coding, *IEEE Trans. Consumer Electronics*, 44, 1042–1047.

36. W. Y. V. Leung and P. J. Bones. 2001. Statistical Interpolation of Sampled Images, *Optical Eng.*, 40, 547–553.

37. Q. J. Zhang and K. C. Gupta. 2000. *Neural Networks for RF and Microwave Design*, Norwood, MA: Artech House.

38. C. Christodoulou and M. Georgiopoulos. 2001. *Applications of Neural Networks in Electromagnetics*, Norwood, MA: Artech House.
39. S. Hykin. 1998. *Neural Networks: A Comprehensive Foundation*, Upper Saddle River, NJ: Pearson Education.
40. P. Ojala, J. Saarinen, P. Elo et al. 1995. MSEE: Novel Technology-Independent Neural Network Approach on Device Modelling Interface, *IEEE Proc. Circuits Devices Syst.*, 142, 74–88.
41. Q. J. Zhang and M. S. Nakhla. 1994. Signal Integrity Analysis and Optimization of VLSI Interconnects Using Neural Network Models. In *Proceedings of IEEE International Symposium on Circuits and Systems, London*, 459–462.
42. M. Awad, S. E. El-Khamy and M. M. Abd Elnaby. 2009. Neural Modeling of Polynomial Image Interpolation. In *Proceedings of International Conference on Computer Engineering and Systems*, Cairo, 469–474.
43. B. E. Bayer. 1976. Color Imaging Array. U.S. Patent 3,971,065.
44. J. F. Hamilton and J. E. Adams. 1997. Adaptive Color Plane Interpolation in Single Sensor Color Electronic Camera. U.S. Patent 5,652,621.
45. R. K. Thakur, A. Tripathy and A. K. Ray. 2009. A Design Framework of Digital Camera Images Using Edge Adaptive and Directionally Weighted Color Interpolation Algorithm, *IEEE Trans. Image Proc.*, 1, 905–909.
46. W. Chaohong, S. Zhixin and V. Govindaraju. 2004. Fingerprint Image Enhancement Method Using Directional Median Filter, *Proc. SPIE*, 5404, 66–75.
47. S. Kasaei, M. Deriche and B. Boashash. 1997. Fingerprint Feature Enhancement Using Block Direction on Reconstructed Images. In *Proceedings of International Conference on Information, Communications and Signal Processing*, 721–725.
48. L. Hong, Y. Wan and A. Jain. 1998. Fingerprint Image Enhancement: Algorithm and Performance Evaluation, *IEEE Trans. Pattern Anal. Machine Intel.*, 20, 777–789.
49. H. Kasban, O. Zahran, S. M. Elaraby et al. 2010. A Comparative Study of Landmine Detection Techniques, *J. Sensing Imaging*, 11, 89–112.
50. H. Kasban, O. Zahran, M. El-Kordy et al. 2009. Efficient Detection of Landmines from Acoustic Images, *Progr. Electromagn. Res. C*, 6, 79–92.
51. H. Kasban, O. Zahran, S. M. Elaraby et al. 2010. False Alarm Rate Reduction in the Interpretation of Acoustic to Seismic Landmine Data using Mathematical Morphology and the Wavelet Transform, *J. Sensing Imaging*, 11, 113–130.
52. H. Kasban, O. Zahran, M. El-Kordy et al. 2008. Automatic Object Detection from Acoustic to Seismic Landmine Images. In *Proceedings of International Conference on Computer Engineering and Systems*, Cairo, 193–198.
53. H. Kasban, O. Zahran, M. El-Kordy et al. 2009. Optimizing Automatic Object Detection from Images in Laser Doppler Vibrometer-Based Acoustic to Seismic Landmine Detection System. In *Proceedings of URSI National Radio Science Conference*, Cairo, March.
54. E. A. El-shazly, O. Zahran, S. M. Elaraby et al. 2010. Automatic Detection of Buried Landmines Using Cepstral Approach. In *Proceedings of Electrical and Computer Systems Engineering Conference*.
55. E. A. El-shazly, S. M. Elaraby, O. Zahran et al. 2010. Cepstral Detection of Buried Landmines from Acoustic Images with a Spiral Scan. In *Proceedings of Sixth International Computer Engineering Conference*, Cairo, 97–102.
56. F. E. Abd El-Samie. 2009. Detection of Landmines from Acoustic Images Based on Cepstral Coefficients, *J. Sensing Imaging*, 10, 63–77.

57. U. S. Khan, W. Al-Nuaimy and F. E. Abd El-Samie. 2010. Detection of Landmines and Underground Utilities from Acoustic and GPR Images with a Cepstral Approach, *J. Visual Commun. Image Repres.*, 21, 731–740.

58. F. G. Hashad, T. M. Halim, S. M. Diab, B. M. Sallam and F. E. Abd El-Samie. 2010. Fingerprint Recognition Using Mel-Frequency Cepstral Coefficients, *Pattern Recog. Image Anal.*, 20, 360–369.

59. Z. H. Long. 2002. Image Fusion Using Wavelet Transform. In *Proceedings of Symposium on Geospatial Theory and Applications*.

60. V. S. Petrovic and C. S. Xydeas. 2004. Gradient-Based Multi-Resolution Image Fusion, *IEEE Trans. Image Proc.*, 13, 228–237.

61. I. Daubechies. 1996. Where Do Wavelets Come From? *Proc. IEEE*, 84, 510–513.

62. A. Cohen and J. Kovacevec. 1996. Wavelets: The Mathematical Background, *Proc. IEEE*, 84, 514–522.

63. N. H. Nielsen and M. V. Wickerhauser. 1996. Wavelets and Time-Frequency Analysis, *Proc. IEEE*, 84, 523–540.

64. K. Ramchandran, M. Vetterli and C. Herley. 1996. Wavelets, Sub-Band Coding, and Best Basis, *Proc. IEEE*, 84, 541–560.

65. P. Guillemain and R. K. Martinet. 1996. Characterization of Acoustic Signals through Continuous Linear Time Frequency Representations, *Proc. IEEE*, 84, 561–585.

66. G. W. Wornell. 1996. Emerging Applications of Multirate Signal Processing and Wavelets in Digital Communications, *Proc. IEEE*, 84, 586–603.

67. S. Mallat. 1996. Wavelets for Vision, *Proc. IEEE*, 84, 604–614.

68. P. Schroder. 1996. Wavelets in Computer Graphics, *Proc. IEEE*, 84, 615–625.

69. M. Unser and A. Aldroubi. 1996. A Review of Wavelets in Biomedical Applications, *Proc. IEEE*, 84, 626–638.

70. M. Farge, N. Kevlahan, V. Perrier et al. 1996. Wavelets and Turbulence, *Proc. IEEE*, 84, 639–669.

71. A. Bijaoui, E. Slezak, F. Rue et al. 1996. Wavelets and the Study of the Distant Universe, *Proc. IEEE*, 84, 670–679.

72. W. Sweldens, Wavelets: What Next? *Proc. IEEE*, 84, 680–685.

73. A. Prochazka, J. Uhlir, P. J. W. Rayner et al. 1998. *Signal Analysis and Prediction*, Basel: Birkhauser.

74. J. S. Walker. 1999. *A Primer on Wavelets and Their Scientific Applications*, Boca Raton: CRC Press.

75. P. Hill, N. Canagarajah and D. Bull. 2002. Image Fusion Using Complex Wavelets. In *Proceedings of British Machine Vision Conference*, 487–496.

76. M. Unser and A. Aldroubi. 1994. A General Sampling Theory for Nonideal Acquisition Devices, *IEEE Trans. Signal Proc.*, 42, 2915–2925.

77. H. C. Andrews and B. R. Hunt. 1977. *Digital Image Restoration*, Englewood Cliffs, NJ: Prentice Hall.

78. S. E. El-Khamy, M. M. Hadhoud, M. I. Dessouky et al. 2005. Adaptive Least Squares Acquisition of High Resolution Images, *Int. J. Information Acquisition*, 2, 45–53.

79. S. E. El-Khamy, M. M. Hadhoud, M. I. Dessouky et al. 2006. Efficient Solutions for Image Interpolation Treated as an Inverse Problem, *J. Information Sci. Eng.*, 22, 1569–1583.

80. H. I. Ashiba, K. H. Awadalla, S. M. El-Halfawy et al. 2011. Adaptive Least Squares Interpolation of Infrared Images, *J. Circuits Syst. Signal Proc.*, 30, 543–551.

81. S. E. El-Khamy, M. M. Hadhoud, M. I. Dessouky et al. 2004. Optimization of Image Interpolation as an Inverse Problem Using the LMMSE Algorithm. In *Proceedings of IEEE MELECON*, Croatia, 247–250.

82. S. E. El-Khamy, M. M. Hadhoud, M. I. Dessouky et al. 2005. Efficient Implementation of Image Interpolation as an Inverse Problem, *J. Digital Signal Proc.*, 15, 137–152.

83. S. E. El-Khamy, M. M. Hadhoud, M. I. Dessouky et al. 2003. Sectioned Implementation of Regularized Image Interpolation, *Proceedings of 46th IEEE MWSCAS*, Cairo, December.

84. S. W. Perry, H. S. Wong and L. Guan. 2002. *Adaptive Image Processing: A Computational Perspective*, Boca Raton: CRC Press.

85. N. B. Karayiannis and A. N. Venetsanopoulos. 1990. Regularization Theory in Image Restoration: The Stabilizing Functional Approach, *IEEE Trans. Acoustics Speech and Signal Proc.*, 38, 1155–1179.

86. N. P. Galatsanos and R. T. Chin. 1989. Digital Restoration of Multichannel Images, *IEEE Trans. Acoustics Speech and Signal Proc.*, 37, 415–421.

87. S. E. El-Khamy, M. M. Hadhoud, M. I. Dessouky et al. 2006. A New Approach for Regularized Image Interpolation, *J. Braz. Comp. Soc.*, 11, 65–79.

88. B. Zitova and J. Flusser. 2003. Image Registration Methods: A Survey, *Image Vision Comp.*, 21, 977–1000.

89. M. Xia and B. Liu. 2004. Image Registration by Super Curves, *IEEE Trans. Image Proc.*, 13, 720–732.

90. D. Robinson and P. Milanfar. 2004. Fundamental Performance Limits in Image Registration, 13, 1185–1199.

91. J. S. Lim. 1990. *Two-Dimensional Signal and Image Processing*, New York: Prentice Hall.

92. W. K. Pratt. 1991. *Digital Image Processing*, New York: John Wiley & Sons.

93. H. Moustafa and S. Rehan. 2006. Applying Image Fusion Techniques for the Detection of Hepatic Lesions and Acute Intracerebral Hemorrhage, *Proceedings of Fourth International Conference on Information and Communications Technology*, Cairo, December.

94. A. Wang, H. J. Sun and Y. Y. Guan. 2006. The Application of Wavelet Transform to Multi-Modality Medical Image Fusion. In *Proceedings of IEEE International Conference on Networking, Sensing and Control*, 270–274.

95. C. Pohl and J. L. van Genderen. 1998. Multisensor Image Fusion in Remote Sensing: Concepts, Methods, and Applications, *Int. J. Remote Sensing*, 19, 823–854.

96. A. A. Goshtasby, S. Nikolov and G. Editoria. 2007. Image Fusion: Advances in the State of the Art, *Information Fusion*, 8, 114–118.

97. M. Ouendeno. 2007. Image Fusion for Improved Perception, Ph.D. thesis, Florida Institute of Technology, Melbourne.

98. A. Jazaeri. 2007. Enhancing Hyberspectral Spatial Resolution Using Multispectral Image Fusion: A Wavelet Approach, Ph.D. thesis, George Mason University, Fairfax, VA.

99. V. S. Petrovic. 2001. Multisensor Pixel-Level Image Fusion, Ph.D. thesis, University of Manchester, U.K.

100. L. Hui. 1993. Multi-Sensor Image Registration and Fusion, Ph.D. thesis, University of California, Santa Barbara.

101. Y. Zhang and G. Hong. 2005. An IHS and Wavelet Integrated Approach to Improve Pan-Sharpening Visual Quality of Natural Colour Ikonos and Quick-Bird Images, *Information Fusion*, 6, 225–234.

102. Y. Shen, J. C. Ma and L. Y. Ma. 2006. An Adaptive Pixel Weighted Image Fusion Algorithm Based on Local Priority for CT and MRI Images. In *Proceedings of IEEE Instrumentation and Measurement Technology Conference*, 420–422.

103. A. Ben Hamza, Yun He, Hamid Krim et al. 2005. A Multiscale Approach to Pixel-Level Image Fusion. *Integr. Comp. Aided Eng.*, 12, 135–146.

104. F. E. Ali, I. M. El-Dokany, A. A. Saad et al. 2008. Curvelet Fusion of MR and CT Images, *Progr. Electromagn. Res. C*, 3, 215–224.

105. F. E. Ali, I. M. El-Dokany, A. A. Saad et al. 2008. Fusion of MR and CT Images Using the Curvelet Transform, *Proceedings of URSI National Radio Science Conference*, Tanta, Egypt, March.

106. F. E. Ali, I. M. El-Dokany, A. A. Saad et al. 2010. High Resolution Image Acquisition from MR and CT Scans Using the Curvelet Fusion Algorithm with Inverse Interpolation Techniques, *J. Appl. Optics*, 49, 114–125.

107. F. E. Ali, I. M. El-Dokany, A. A. Saad et al. 2010. A Curvelet Transform Approach for the Fusion of MR and CT Images, *J. Modern Optics*, 57, 273–286.

108. M. R. Metwalli, A. H. Nasr, S. El-Rabaie et al. 2011. Sharpening Misrsat-1 Data Using Super-Resolution and HPF Fusion Methods, *Proceedings of URSI National Radio Science Conference*, Cairo, March.

109. M. R. Metwalli, A. H. Nasr, O. S. Farag Allah et al. 2010. Satellite Image Fusion Based on Principal Component Analysis and High-Pass Filtering, *J. Optical Soc. Amer.*, 27.

110. S. M. Elkaffas, T. A. El-Tobely, A. M. Ragheb et al. 2006. An Integrated IHS and DWFT Fusion Technique for Improving the Spectral Quality of Remote Sensing Images, *Proceedings of International Computer Engineering Conference*, Cairo, IP-54–IP-61.

111. M. E. Nasr, M. Elkaffas, T. A. El-Tobely et al. 2007. An Integrated Image Fusion Technique for Boosting the Quality of Noisy Remote Sensing Images, *Proceedings of URSI National Radio Science Conference*, Cairo, March.

112. S. P. Kim and W. Y. Su. 1993. Recursive High Resolution Reconstruction of Blurred Multiframe Images, *IEEE Trans. Image Proc.*, 2, 534–539.

113. S. P. Kim, N. K. Bose and H. M. Valenzuela. 1990. Recursive Reconstruction of High Resolution Image from Noisy Undersampled Multiframes, *IEEE Trans. Acoustics Speech Signal Proc.*, 38, 1013–1027.

114. S. C. Park, M. K. Park and M. G. Kang. 2003. Super-Resolution Image Reconstruction: A Technical Overview, *IEEE Signal Proc.*, 20, 21–36.

115. P. E. Eren, M. I. Sezan and M. Tekalp. 1997. Robust, Object-Based High Resolution Image Reconstruction from Low Resolution Video, *IEEE Trans. Image Proc.*, 6, 1446–1451.

116. M. Elad and A. Feuer. 1997. Restoration of a Single Super-Resolution Image from Several Blurred, Noisy and Undersampled Measured Images, *IEEE Trans. Image Proc.*, 6, 1646–1658.

117. D. Capel and A. Zisserman. 2001. Computer Vision Applied to Super-Resolution, *IEEE Signal Proc.*, 20, 75–86.

118. N. Nguyen, P. Milanfar and G. Golub. 2001. A Computationally Efficient Super-Resolution Image Reconstruction Algorithm, *IEEE Trans. Image Proc.*, 10, 573–583.

119. D. Rajan, S. Chandhuri and M. V. Joshi. 2003. Multi-Objective Super-Resolution: Concepts and Examples, *IEEE Signal Proc.*, 20, 49-61.

120. C. A. Segall, R. Molina and A. K. Katsaggelos. 2003. High-Resolution Images from Low-Resolution Compressed Video, *IEEE Signal Proc.*, 20, 37–48.

121. M. Elad and A. Feuer. 1999. Super-Resolution Restoration of an Image Sequence: Adaptive Filtering Approach, *IEEE Trans. Image Proc.*, 8, 387–395.

122. M. K. Ng and N. K. Bose. 2003. Mathematical Analysis of Super-Resolution Methodology, *IEEE Signal Proc.*, 20, 62–74.

123. M. Vega, J. Mateos, R. Molina et al. 2003. Bayesian Parameter Estimation in Image Reconstruction from Subsampled Blurred Observations, *Proceedings of International Conference on Image Processing*.

124. S. E. El-Khamy, M. M. Hadhoud, M. I. Dessouky et al. 2005. Regularized Super-Resolution Reconstruction of Images Using Wavelet Fusion, *J. Optical Eng.*, 44.

125. S. E. El-Khamy, M. M. Hadhoud, M. I. Dessouky et al. 2006. Wavelet Fusion: A Tool to Break the Limits on LMMSE Image Super-Resolution, *Int. J. Wavelets Multiresolution Information Proc.*, 4, 105–118.

126. S. E. El-Khamy, M. M. Hadhoud, M. I. Dessouky. 2008. New Techniques to Conquer the Image Resolution Enhancement Problem, *Progr. Electromagn. Res. B*, 7, 13–51.

127. S. E. El-Khamy, M. M. Hadhoud, M. I. Dessouky et al. 2008. A Wavelet-Based Entropic Approach to High Resolution Image Reconstruction, *Int. J. Machine Graphics Vision*, 17, 235–256.

128. N. P. Galatsanos and R. T. Chin. 1989. Digital Restoration of Multichannel Images, *IEEE Trans. Acoustics Speech Signal Proc.*, 37, 415–421.

129. M. K. Ozkan, A. T. Erdem, M. I. Sezan et al. 1992. Efficient Multiframe Wiener Restoration of Blurred and Noisy Image Sequences, *IEEE Trans. Image Proc.*, 1, 453–476.

130. N. P. Galatsanos, A. K. Katasaggelos, R. T. Chin et al. 1991. Least Squares Restoration of Multichannel Images, *IEEE Trans. Signal Proc.*, 39, 2222–2236.

131. N. P. Galatsanos and A. K. Katsaggelos. 1992. Methods of Choosing the Regularization Parameter and Estimating the Noise Variance in Image Restoration and Their Relation, *IEEE Trans. Image Proc.*, 1, 322–336.

132. S. Baker and T. Kanade. 2001. Super-Resolution: Reconstruction or Recognition, *Proceedings of IEEE EURASIP Workshop on Nonlinear Signal and Image Processing*.

133. G. Harikumar and Y. Bresler. 1999. Perfect Blind Restoration of Images Blurred by Multiple Filters: Theory and Efficient Algorithms, *IEEE Trans. Image Proc.*, 8, 202–219.

134. H. T. Pai and A. C. Bovik. 1997. Exact Multi-Channel Blind Image Restoration, *IEEE Signal Proc.*, 4, 217–220.

135. R. Nakagaki and A. K. Katsaggelos. 2003. A VQ-Based Blind Image Restoration Algorithm, *IEEE Trans. Image Proc.*, 12, 1044–1053.

136. F. Sroubek and J. Flusser. 2003. Multichannel Blind Iterative Image Restoration, *IEEE Trans. Image Proc.*, 12, 1094–1106.

137. S. U. Pillai and B. Liang. 1999. Blind Image Deconvolution Using a Robust GCD Approach, *IEEE Trans. Image Proc.*, 8, 295–301.

138. S. E. El-Khamy, M. M. Hadhoud, M. I. Dessouky et al. 2001. Blind Deconvolution of Blurred Images from Multiple Observations Using the GCD Algorithm, *Proceedings of URSI National Radio Science Conference*, Manssoura, Egypt, March.

139. S. E. El-Khamy, M. M. Hadhoud, M. I. Dessouky et al. 2005. Blind Multi-Channel Reconstruction of High-Resolution Images Using Wavelet Fusion, *J. Appl. Optics*, 44, 7349–7356.

140. S. E. El-Khamy, M. M. Hadhoud, M. I. Dessouky et al. 2006. A Greatest Common Divisor Approach to Blind Super-Resolution Reconstruction of Images, *J. Modern Optics*, 53.

Index

Page numbers in *Italics* denote figures or tables